A Map of the Body, a Map of the Mind

Archaeopress Roman Archaeology 115

A Map of the Body, a Map of the Mind

Visualising Geographical Information in the Roman World

Iain Ferris

Archaeopress Archaeology

Archaeopress Publishing Ltd
Summertown Pavilion
18-24 Middle Way
Summertown
Oxford OX2 7LG
www.archaeopress.com

ISBN 978-1-80327-781-3
ISBN 978-1-80327-782-0 (e-Pdf)

© Iain Ferris and Archaeopress 2024

Cover: Artemis Ephesia, probably from Rome. Second century AD.
Detail of part of the Peutinger Map. Original early second century to early fourth century AD.

All rights reserved. No part of this book may be reproduced, or transmitted, in any form or by any means, electronic, mechanical, photocopying or otherwise, without the prior written permission of the copyright owners.

This book is available direct from Archaeopress or from our website www.archaeopress.com

For JMF.

Contents

Image Credits x

Acknowledgements xi

Preface xii
    Footfalls Echo In the Memory xix

Chapter One: Maps of the Mind 1
    Zone 1
    Interzone 6
    To the Heart of the World 8
    The Open Door 15

Chapter Two: Strangers in a Strange Land 21
    The Ceremony of Innocence 23
    Procession 27
    An Immersive Past 29
    Another Time 38

Chapter Three: Rome in Rome 43
    Seven Hills 43
    Mountains 48
    Water and Trees 52
    Streets and Buildings 54

Chapter Four: A River Without End 67
    Rivers of the Windfall Light 67
    Simultaneities 68
    Rising Waters 74
    Rivers of Deceit 81

Chapter Five: Staged Designs 87
    Short Distances and Definite Places 87
    The Inconstant Ones 97

Chapter Six: Landscape and Desire 108
    Possession 108
    (Dis)Possession 116

Chapter Seven: An Unseen Ruler 124
    Carving and Paring 125
    Crossing the Line 143

Chapter Eight: Maps of the Body 150
    The Widening Gyre 153
    Threading a Dream 156
    Bodies and Metaphor 166
    A Long Way from Here 171
    Craving for Oblivion 173
    Turning and Turning 177
    The Centre Cannot Hold 184

Chapter Nine: Moving Away from the Pulsebeat 189
    Odyshape 190
    Death's Echo 200
    Postcards from the Edge 210
    Postcards from the Future 213
    Petrified, the Landscape Grows 216

Chapter Ten: Slouching Towards Empire 225
    Shadowplay 225
    Unknown Pleasures 229
    Transmission 240
    Insight 245
    Opaque Manifesto 248
    Quiet Mapped Waters 252

Appendix 256
    I Remember 256

Bibliography 258

List of Figures and Tables

Figure 1. Relief depicting architrave supported by the figures of two caryatids, Pozzuoli. Early first century AD. *Museo Archeologico Nazionale*, Naples. (Photo: Author) xvi

Figure 2. Example of a Roman 'geography product'. Statue personification of the River Arno. Exact provenance uncertain, probably Rome. Hadrianic. *Musei Vaticani*, Rome. (Photo: Author) xviii

Figure 3. Example of a Roman 'geography product'. Nilotic scene on a terracotta Campana plaque. Mid-first century AD. British Museum, London. (Photo: Copyright Trustees of the British Museum) xx

Figure 4. Example of a Roman 'geography product'. Sarcophagus depicting the Indian triumph of Bacchus, Ostia/*Portus*. Second to third century AD. *Museo Archeologico Ostiense*. (Photo: Author) xxii

Figure 5. The Map of Bedolina at the rock art site of Bedolina, near Capo di Ponte, Valcamonica, Lombardy. Italian Bronze Age (c. 1500-1400 BC) and Iron Age (between 600-400 BC). (Photo: Angelo Fossati) 2

Figure 6. The Map of Bedolina at the rock art site of Bedolina, near Capo di Ponte, Valcamonica, Lombardy. Italian Bronze Age (c. 1500-1400 BC) and Iron Age (between 600-400 BC). (Photo: Angelo Fossati) 2

Figure 7. Survey drawing of the Map of Bedolina. (Photo: Angelo Fossati/*Footsteps of Man Cooperativo*) 3

Figure 8. The Liver of Piacenza. Etruscan, second century BC in *Museo Civico Palazzo Farnese*, Piacenza. Modern replica in *Museo Etrusco Guarnacci*, Volterra. (Photo: Copyright Jerónimo Roure Pérez) 7

Figure 9. *Roma/Tellus*, the *Ara Pacis Augustae*, Rome. 13-9 BC. (Photo: Author) 9

Figure 10. Vegetal decoration, the *Ara Pacis Augustae*, Rome. 13-9 BC. (Photo: Author) 10

Figure 11. Detail, the *Ara Pacis Augustae*, Rome. 13-9 BC. (Photo: Author) 10

Figure 12. The Prima Porta statue of Augustus. c. 20 BC. *Musei Vaticani*, Rome. (Photo: Author) 12

Figure 13. Detail of the decorated cuirass, the Prima Porta statue of Augustus. c. 20 BC. *Musei Vaticani*, Rome. (Photo: Author) 13

Figure 14. Portrait bust of Cicero. First century BC, Rome. *Palazzo dei Conservatori*, Rome. (Photo: Author) 24

Figure 15. Wall painting depicting treaty negotiations between Romans and Samnites. Esquiline Hill, Rome. 300-280 BC. *Musei Capitolini Centrale Montemartini*, Rome. (Photo: Author) 30

Figure 16. The bronze Capitoline Wolf, still widely considered to be of an Etruscan date, with Romulus and Remus added later. *Musei Capitolini*, Rome. (Photo: Author) 31

Figure 17. A Proto-Etruscan cinerary urn in the form of a model of a contemporary hut, perhaps like the *Casa Romuli*. Alban Hills, Lazio. 900-800 BC. British Museum, London. (Copyright Trustees of the British Museum) 32

Figure 18. Relief depicting Aeneas at the future site of Rome. The ship of Aeneas is docked on the right. Rome. AD 140-150. British Museum, London. (Photo: Copyright Trustees of the British Museum) 35

Figure 19. The *ficus ruminalis* depicted on one of the *Plutei* of Trajan or *Anaglypha Traiani*. Trajanic, possibly later and Hadrianic. *Curia Julia*, Roman Forum. (Photo: Author) 37

Figure 20. Bronze *sestertius* coin issue of Vespasian, Rome mint. AD 71. Reverse image of *Roma* seated on the seven hills. British Museum, London. (Photo: Copyright Trustees of the British Museum) 44

Figure 21. Face of altar to Mars and Venus, with depiction of Faustulus, Numitor, and Faustus, the She-Wolf nursing Romulus and Remus, and *Tiberinus*, Ostia. Trajanic or Hadrianic. *Palazzo Massimo Museo Nazionale Romano*, Rome. (Photo: Author)..... 45

Figure 22. Detail of face of altar to Mars and Venus, with depiction of the She-Wolf nursing Romulus and Remus and *Tiberinus* looking on, Ostia. Trajanic or Hadrianic. *Palazzo Massimo Museo Nazionale Romano*, Rome. (Photo: Author) 45

Figure 23. Apotheosis scene, with the youthful personification of the *Campus Martius* in attendance, on relief panel from the *Arco di Portogallo*, Rome. AD 136-138. *Palazzo dei Conservatori*, Rome. (Photo: Author).. 48

Figure 24. Wall painting of Bacchus in front of Mount Vesuvius. *Lararium* of the *Casa del Centenario* or House of the Centenary, Pompeii. AD 55-79. *Museo Archeologico Nazionale*, Naples. (Photo: Author) 51

Figure 25. Detail of buildings on reliefs from the Tomb of the Haterii, Rome. Late Flavian or early Trajanic. *Musei Vaticani*, Rome. (Photo: Author).................................... 55

Figure 26. Detail from the end panel of a marble sarcophagus, showing St Peter and his jailers in the city of Rome. Rome. Fourth century AD. *Musei Vaticani*, Rome. (Photo: Author) 55

Figure 27. Roman architecture on coins. Bronze *sestertius* of Titus, reverse the Colosseum from a bird's-eye view, Rome mint. AD 80-81. British Museum, London. (Photo: Copyright Trustees of the British Museum)... 57

Figure 28. Roman architecture on coins. Gold *aureus* of Nero, reverse the Temple of Vesta, Rome mint. AD 65-66. British Museum, London. (Photo: Copyright Trustees of the British Museum) 57

Figure 29. Roman architecture on coins. Gold *aureus* of Claudius, reverse a triumphal arch in Rome, Rome mint. AD 41-45. British Museum, London. (Photo: Copyright Trustees of the British Museum) 57

Figure 30. Relief from a sarcophagus, depicting a busy harbour scene at *Portus*. Rome. Mid-third century AD. *Musei Vaticani*, Rome. (Photo: Author) 61

Figure 31. Marble Christian sarcophagus relief from Rome depicting ship approaching the harbour at *Portus* and its lighthouse. Fourth century AD. *Musei Vaticani*, Rome. (Photo: Author) 61

Figure 32. Wall painting of *Roma*. Originally fourth century AD, San Giovanni in Laterano, Rome. Probably a heavily-overrestored Venus. *Palazzo Massimo Museo Nazionale Romano*, Rome. (Photo: Author).. 63

Figure 33. Face of painted altar bearing image of *Roma/Tellus*, Milan. Late first to early second century AD. *Museo Civico Archeologico*, Milan. (Photo: Author) 63

Figure 34. Detail of the Base of Tiberius, showing one group of the fourteen *Tychai* of Asian cities. Pozzuoli. AD 30-31. *Museo Archeologico Nazionale*, Naples. (Photo: Author)... 64

Figure 35. Head of Tyche, Classe. Second century AD. *Classis Ravenna Museo della Città e del Territorio*, Classe. (Photo: Author) 64

Figure 36. Tombstone of Lucius Aurelius Hermia, butcher on the Viminal Hill, Rome. First century BC. British Museum, London. (Photo: Copyright Trustees of the British Museum) 65

Figure 37. Massive statue of the personified Tiber (originally probably the Tigris), *Campidoglio*, Rome. Originally in the Baths of Constantine, Rome. Early fourth century AD. (Photo: Author) 71

Figure 38. Massive statue of the personified Nile, probably from the Temple of Isis in the *Campus Martius*, Rome. Very late first century AD. *Musei Vaticani*, Rome. (Photo: Author) 75

Figure 39. Massive statue of the personified Nile, *Campidoglio*, Rome. Originally in the Baths of Constantine, Rome. Early fourth century AD. (Photo: Author) 75

Figure 40. Gold *aureus* coin issue of Hadrian, Rome mint. AD 130-138. Reverse image of the personification of the Nile. British Museum, London. (Photo: Copyright Trustees of the British Museum) 76

Figure 41. Detail of relief panel showing submission of personification of Mesopotamia to Trajan, with personifications of the rivers Euphrates and Tigris in attendance. The Arch of Trajan at Benevento. AD 114-118. (Photo: Author) 76

Figure 42. The personified figure of the River Danube. Scene III, Trajan's Column, Rome. AD 113 (Photo: Author) 77

Figure 43. Bronze *sestertius* of Trajan, Rome mint, AD 104-111. Reverse of personification of *Danuvius* (River Danube) throttling and subduing the personified *Dacia*. British Museum, London. (Photo: Copyright Trustees of the British Museum) 78

Figure 44. Bronze *sestertius* of Trajan, Rome mint, AD 116-117. Reverse of emperor standing over seated personifications of the Rivers Euphrates and Tigris, with personified Armenia seated left. British Museum, London. (Photo: Copyright Trustees of the British Museum) 78

Figure 45. Stone head of Rhine god *Rhenus* from a mausoleum, Bonn. Second century AD. *Rheinisch Landesmuseum,* Bonn. (Photo: Carole Raddato) 79

Figure 46. Black and white mosaic depicting the Nile. *Piazzale delle Corporazioni*, Ostia. AD 150-170. (Photo: Author) 83

Figure 47. The rain god. Scene XVI, the Column of Marcus Aurelius, Rome. AD 180-192. (Photo: Author) 84

Figure 48. The Pesaro wind rose, *Via Appia*, Rome. End of second century AD. *Museo Oliveriano*, Pesaro. (Photo: Author) 84

Figure 49. Roman/Campanian landscape wall painting from Pompeii. Early first century AD to AD 79. *Museo Archeologico Nazionale*, Naples. (Photo: Author) 89

Figure 50. Roman/Campanian landscape wall painting from Pompeii. Early first century AD to AD 79. *Museo Archeologico Nazionale*, Naples. (Photo: Author) 89

Figure 51. Roman/Campanian landscape wall painting from Pompeii. Early first century AD to AD 79. *Museo Archeologico Nazionale*, Naples. (Photo: Author) 90

Figure 52. Relief depicting a sacro-idyllic landscape, Rome. First to second century AD. *Palazzo Massimo Museo Nazionale Romano*, Rome. (Photo: Author) 90

Figure 53. Wall painting depicting an idealised landscape, Pompeii. *Museo Archeologico Nazionale*, Naples. Early first century AD to AD 79. (Photo: Author) 91

Figure 54. Wall painting of rural estate and estate workers, *Palasgarten*, Trier. Second century AD. *Rheinisch Landesmuseum*, Trier. (Photo: Author) 92

Figure 55. Wall painting from the garden room of the Villa of Livia at Prima Porta. Second half of the first century BC. *Palazzo Massimo Museo Nazionale Romano*, Rome. (Photo: Author) 93

Figure 56. Christian sarcophagus with scene of Jesus preaching in Holy Land landscape defined by palm trees, Ravenna. Fourth century AD. *Museo Nazionale Romano*, Ravenna. (Photo: Author) 96

Figure 57. Scenes I-II, Trajan's Column, Rome. AD 113. (Photo: Author) 98

Figure 58. Scene XX, Trajan's Column, Rome. AD 113. (Photo: Author) 98

Figure 59. The Praeneste/Palestrina Nile Mosaic. First quarter of the second century AD. *Museo Archeologico Nazionale di Palestrina*, Palestrina. (Photo: Author) 111

Figure 60. Detail of the Praeneste/Palestrina Nile Mosaic. First quarter of the second century AD. *Museo Archeologico Nazionale di Palestrina*, Palestrina. (Photo: Author) 111

Figures 61-62. Detail of the Praeneste/Palestrina Nile Mosaic. First quarter of the second century AD. *Museo Archeologico Nazionale di Palestrina*, Palestrina. (Photo: Author) 112

Figures 63-65. Detail of the Praeneste/Palestrina Nile Mosaic. First quarter of the second century AD. *Museo Archeologico Nazionale di Palestrina*, Palestrina. (Photo: Author) 113

Figure 66. Nilotic relief with erotic scene, Rome. Date 30 BC-AD 100. British Museum, London. (Photo: Copyright Trustees of the British Museum) 119

Figure 67. Black and white mosaic, Nilotic scene with pygmies, Rome. First to second century AD. *Palazzo Massimo Museo Nazionale Romano*, Rome. (Photo: Author) 120

Figure 68. Marble statue of a black youth on a crocodile, Rome. First century BC to first century AD. British Museum, London. (Photo: Copyright Trustees of the British Museum) 120

Figure 69. Statue of the personified Nile carved in dark basanite, Rome. Flavian. *Musei Vaticani*, Rome. (Photo: Author) 121

Figure 70. Statue bust of Antinous, Rome. Hadrianic. *Musei Capitolini Centrale Montemartini*, Rome. (Photo:Author) 122

Figure 71. The *Canopus*. Hadrian's Villa at Tivoli. AD 133-138. (Photo: Author) 122

Figure 72. Marble ground plan of the tomb complex of Claudia Peloris and Tiberius Claudius Eutychus, Rome. Mid-first century AD. *Museo Archeologico Nazionale dell'Umbria*, Perugia. (Photo: Author) 127

Figure 73. The *Via Marsala* mosaic map. Late second to early third century AD. *Musei Capitolini Centrale Montemartini*, Rome. (Photo: Professor Lynne Lancaster, by permission of the *Musei Capitolini*) 128

Figure 74. Detail of the *Via Marsala* mosaic map. Late second to early third century AD. *Musei Capitolini Centrale Montemartini*, Rome. (Photo: Professor Lynne Lancaster, by permission of the *Musei Capitolini*) 128

Figure 75. The Mosaic of the Islands, Haidra, Tunisia. Third or fourth century AD. *Musée National du Bardo*, Tunis. (Photo: Author) 129

Figure 76. Portrait bust of Marcus Agrippa, Rome. 25-10 BC. British Museum, London. (Photo: Copyright Trustees of the British Museum) 132

Figure 77. Wall of the *Templum Pacis*/Temple of Peace, Rome. Now part of the Church of SS. Cosma e Damiano. (Photo: Author) 134

Figure 78. Wall of the *Templum Pacis*/Temple of Peace, Rome. Now part of the Church of SS. Cosma e Damiano. (Photo: Author) 135

Figure 79. The *Via Anicia* marble map fragment. Augustan or later. *Musei Capitolini*, Rome. (Photo: Author) 139

Figure 80. Detail of part of the Peutinger Map. Original early second century to early fourth century AD. (Photo of 1888: downloaded from cambridge.org/us/talbert/mapb.html TP1888seg1) 142

Figure 81. Detail of part of the Peutinger Map. Original early second century to early fourth century AD. (Photo of 1888: downloaded from cambridge.org/us/talbert/mapb.html TP1888seg2) 142

Figure 82. The Vicarello Itinerary Cups, Vicarello. First century AD. *Palazzo Massimo Museo Nazionale Romano*, Rome. (Photo: Copyright Ryan Baumann) 144

Figure 83. The Farnese Atlas, Rome. Second century AD. *Museo Archeologico Nazionale*, Naples. (Photo: Author) 146

Figure 84-85. The Dying Gaul. Roman copy of a Hellenistic original. *Musei Capitolini*, Rome. (Photo: Author) 151

Figure 86. Cybele/*Magna Mater*, Rome. AD 250-275. *Museo Archeologico Nazionale*, Naples. (Photo: Author) 158

Figure 87. *Artemis Ephesia*, probably from Rome. Second century AD. *Museo Archeologico Nazionale*, Naples. (Photo: Author) 159

Figure 88. Relief of Claudius and *Britannia*, from the *Sebasteion* at Aphrodisias. Julio-Claudian. (Photo: courtesy of New York University Excavations at Aphrodisias. Photographer G. Petruccioli) 160

Figure 89. Relief of Nero and *Armenia*, from the *Sebasteion* at Aphrodisias. Julio-Claudian. (Photo: courtesy of New York University Excavations at Aphrodisias. Photographer G. Petruccioli) 160

Figure 90. Relief depicting *Mauretania* from the *Hadrianeum* Rome. AD 145. *Palazzo dei Conservatori*, Rome. (Photo: Author) 168

Figure 91. Relief depicting *Hispania* from the *Hadrianeum* Rome. AD 145. *Museo Archeologico Nazionale*, Naples. (Photo: Author) 169

Figure 92. Relief depicting *Gallia* from the *Hadrianeum* Rome. AD 145. *Palazzo dei Conservatori*, Rome. (Photo: Author) 170

Figure 93. Ivory Indian figure from the *Casa dei Quattro Stili* or House of the Four Styles, Pompeii. First century AD, before AD 79. *Museo Archeologico Nazionale*, Naples. (Photo: Author) 172

Figure 94. Personification of India on a mosaic from Villa Casale, Piazza Armerina, Sicily. Fourth century AD. (Photo: slide collection of the former School of Continuing Studies, Birmingham University) 173

Figure 95. Leaf from an ivory diptych (the Barberini Ivory) depicting the emperor Justinian, Constantinople. First half of the sixth century AD. *Musée du Louvre* (Photo: Copyright *Musée du Louvre*) 174

Figure 96. Wall painting with depiction of Macedonia, Villa of P. Fannius Synistor, Boscoreale. First century AD, before AD 79. *Museo Archeologico Nazionale*, Naples. (Photo: Author) 178

Figure 97. Inner relief depicting Titus' Judaean triumph, Arch of Titus, Rome. After AD 81. (Photo: Author) 181

Figure 98. Marble sarcophagus depicting the Indian triumph of Bacchus, Rome. AD 260-270. Metropolitan Museum, New York. (Photo: Copyright Metropolitan Museum)..... 190

Figure 99. Relief depicting scenes of the Trojan War from the *Iliad* (one of the *Tabulae Iliacae*). First half of first century AD. Metropolitan Museum, New York. (Photo: Copyright Metropolitan Museum) 193

Figure 100. Gold *aureus* coin of Hadrian, Rome mint, AD 130-138. Reverse of personification of Africa. One of the travel series of Hadrianic coins. British Museum, London. (Photo: Copyright Trustees of the British Museum)... 197

Figure 101. Gold *aureus* coin of Hadrian, Rome mint, AD 130-138. Reverse of personification of *Aegyptus*/Egypt. One of the travel series of Hadrianic coins. British Museum, London. (Photo: Copyright Trustees of the British Museum) 197

Figure 102. Silver *denarius* of Hadrian, Rome mint, AD 117-138. Reverse of Hadrian raising the kneeling personification of *Gallia*/Gaul. One of the travel series of Hadrianic coins. British Museum, London. (Photo: Copyright Trustees of the British Museum) 197

Figure 103. Statue of the personified Nile. AD 133-138. *Museo Villa Adriana*, Tivoli. (Photo: Author)... 199

Figure 104. Statue of the personified Tiber. AD 133-138. *Museo Villa Adriana*, Tivoli. (Photo: Author)... 199

Figure 105. Statue of a Nile crocodile. AD 133-138. *Museo Villa Adriana*, Tivoli. (Photo: Author)... 200

Figure 106. The ship of Odysseus, massive sculpture, Villa and Grotto of Tiberius, Sperlonga. *Museo Archeologico Nazionale di Sperlonga*. (Photo: Author)... 201

Figures 107-108. The blinding of Polyphemus by Odysseus, Villa and Grotto of Tiberius, Sperlonga *Museo Archeologico Nazionale di Sperlonga*. (Photo: Author)..................... 202

Figures 109-110. Sarcophagus decorated with the abduction/rape of Proserpina/Persephone, Rome. Third century AD. *Musei Capitolini*, Rome. (Photo: Author) ... 203

Figure 111. End panel of a sarcophagus, with Cupid as Charon rowing across the River Styx in the Underworld, Milan. Third century AD. *Museo Civico Archeologico*, Milan. (Photo: Author)... 204

Figure 112. Detail from a marble sarcophagus, showing Hercules exiting the Underworld with the dog Cerberus, Rome. Third century AD. *Musei Capitolini Centrale Montemartini*. (Photo: Author) 204

Figure 113. Gold *aureus* of Caracalla, Rome mint, AD 214. Reverse of Serapis with the dog Cerberus seated at his feet to left. British Museum, London. (Photo: Copyright Trustees of the British Museum) 205

Figure 114. The *Bocca della Verita*, Santa Maria in Cosmedin, Rome. Date uncertain, possibly as early as first century AD. (Photo: Author)... 206

Figure 115. The Velletri Sarcophagus. AD 140-150. *Museo Civico Archeologico Oreste Nardini*, Velletri. (Photo: slide collection of the former School of Continuing Studies, Birmingham University) 207

Figure 116. The Rudge Cup, schematically depicting Hadrian's Wall and naming some forts along the frontier. Second century AD. Alnwick Castle, Northumberland. (Photo: Tullie House Museum Carlisle and Professor David Breeze) 211

Figure 117. The Pilkington Bottle, a souvenir from, and depicting, Puteoli. Third or fourth century AD. Pilkington World of Glass, St. Helens. (Photo: Pilkington Glass Collection. The World of Glass) 212

Figure 118. Sardonyx cameo of the Tyche of Constantinople crowning the emperor Constantine with a laurel wreath: known as the *Gemma Constantiniana*. Probably AD 315. *Rijksmuseum van Oudheden*, Leiden. (Photo: Author) 218

Figure 119. Example of a Roman 'geography product'. The silver Parabiago plate, bearing an image of Cybele and Attis in a cosmic setting. Mythological figures present include river deities and *Tellus*. Fourth to fifth century AD. *Museo Civico Archeologico*, Milan. (Photo: Author) 227

Figure 120. Example of a Roman 'geography product'. Small sarcophagus decorated with images of personified river deities, probably from Rome. Second or third century AD. (Photo: Duke's Auctions) 229

Figure 121. Example of a Roman 'geography product'. Small bronze figure of the Tyche of Antioch. First century AD. Metropolitan Museum, New York. (Photo: Copyright Metropolitan Museum) 231

Figure 122. Example of a Roman 'geography product'. Black and white mosaic of ships approaching the lighthouse at *Portus*, outside Tomb 43 *Necropoli di Porto, Isola Sacra*, Ostia. Second to third century AD. (Photo: Author) 234

Figure 123. Example of a Roman 'geography product'. Part of frieze depicting captured, bound barbarians, Trier. First century AD. *Rheinisch Landesmuseum*, Trier. (Photo: Author) 237

Figure 124. Example of a Roman 'geography product'. Relief from sarcophagus depicting a banquet scene and a wind god, possibly Rome. Third to fourth century AD. *Rijksmuseum van Oudheden*, Leiden. (Photo: Author) 239

Figure 125. Example of a Roman 'geography product'. Cupids carrying a crocodile, Oxyrhynchus. Fourth to fifth century AD. *Rijksmuseum van Oudheden*, Leiden. (Photo: Author) 241

Figure 126. Mosaic panel bearing a personification of the Tyche of Antioch merged with a portrait of a Hellenistic ruler (Arsinoe II perhaps), Thmouis, Egypt. As early as 200 BC. Graeco-Roman Museum, Alexandria. (Photo: Copyright *Centre d'Études Alexandrines*) 245

Figure 127. Example of a Roman 'geography product'. Nilotic-themed wall painting, Pompeii. First century AD, before AD 79. *Museo Archeologico Nazionale*, Naples. (Photo: Author) 247

Figure 128. Example of a Roman 'geography product'. Ceramic oil lamp decorated with image of eroticised woman (caricature of Cleopatra?) on the back of a crocodile. Provenance uncertain. AD 40-80. British Museum, London. (Photo: Copyright Trustees of the British Museum) 249

Figure 129. Example of a Roman 'geography product'. Mosaic depicting a Nilotic scene, Pompeii. First century AD, before AD 79. *Museo Archeologico Nazionale*, Naples. (Photo: Author) 251

Figure 130. Example of a Roman 'geography product'. Sarcophagus carrying image of cupids/*putti* operating ships in a busy harbour, with buildings behind forming an urban backdrop, Rome. Third century AD. *Terme di Diocleziano Museo Nazionale Romano*, Rome. (Photo: Author) 253

Image Credits

All images are by the author, with the exception of: 3, 17, 18, 20, 27, 28, 29, 36, 40, 43, 44, 66, 68, 76, 100, 101, 102, 113, and 128 Copyright Trustees of the British Museum; 98, 99, and 121 Copyright Metropolitan Museum; 95 Copyright *Musée du Louvre*; 5, 6, and 7 Dr Angelo Fossati and the *Footsteps of Man Cooperativo*; 88 and 89 courtesy of New York University Excavations at Aphrodisias through Professor R.R.R. Smith; 126 Copyright *Centre d'Études Alexandrines*; 94, and 115 slide collection of the former School of Continuing Studies, Birmingham University; 45 Carole Raddato; 116 Tullie House Museum through Professor David Breeze; 73 and 74 Professor Lynne Lancaster, with permission of *Musei Capitolini*; 80 and 81 downloaded through open access at cambridge.or/us/talbert; 82 Copyright Ryan Baumann; 8 Copyright Jerónimo Roure Pérez; 120 Copyright Duke's Auctions; and 117 Pilkington Glass Collection, the World of Glass.

Copyright: I have attempted in all cases to track down holders of copyright of images used in this book, and as far as I am aware no attributions of copyright are ambiguous or unresolved. I would be happy, though, for any copyright holders that I may have inadvertently not credited to get in touch with me through *Archaeopress*.

Acknowledgements

In writing this book I have received help from a number of individuals and organisations and I would like to take the opportunity to sincerely thank them all here.

For providing images for reproduction in the book I would like to thank: Carole Raddato (of the *Following Hadrian* website and blog); Dr Angelo Fossati of the *Footsteps of Man* Archaeological *Cooperativo* of Valcamonica and of the Catholic University of the Sacred Heart, Milan; Hannah Billinge of The Pilkington Glass Collection: The World of Glass; Professor Lynne Lancaster; and Professor David Breeze. Julian Parker once more gave help in locating a number of images for the book in an old slide collection and is thanked for his technical expertise in turning these old but invaluable colour slides into sharp digital images.

For answers to a number of enquiries about artworks in their care and information on viewing them I thank the staff of the *Pontificia Commissione di Archeologia Sacra* in Rome. In addition to visiting the many museums and sites of Rome over the last thirty years when I gathered information I was unaware I would one day use, new study visits have also been made to: Lyon and Paris in France; Leiden in the Netherlands; Trier in Germany; and to Classe, Ferrara, Milan, Naples, Palestrina, Perugia, Pesaro, Piacenza, Ravenna, Sperlonga, and Tivoli, in Italy.

Dr Penny Goodman of the University of Leeds very kindly answered two queries about evidence for the creation and advertising of local identities within Rome. Professor Richard Talbert answered an email query about his work on the Peutinger Map and kindly directed me to a relevant web resource. Professor Simon James helpfully replied to an email about the Dura Map. The late Professor Amanda Claridge very kindly alerted me to a number of vital sources of which I was then unaware.

As usual, the staff of the Institute of Classical Studies Library, London were unfailingly helpful in obtaining books and journals for my reference while researching this book. Sue Willetts at the ICSL is particularly thanked for patiently scanning and emailing articles to me when I could not visit the library in person during the recent lockdowns and closures. Many thanks to them all.

As always, my colleague and wife Dr Lynne Bevan read and commented on a draft of the book, much to the benefit of the finished work. The book is targeted at undergraduate students of Roman archaeology and informed visitors to Rome. At Archaeopress I would like to thank Dr David Davison for commissioning this book in the first place, Robin Orlić for typesetting, and Ben Heaney and Mike Schurer for editorial work and advice in seeing the book into print.

Note: As with any study there comes a point at which a cut-off time is reached, when bibliographic research and reading have to stop: for this book that point was March 2023. Therefore books and papers published after that time have not been consulted, with the exception of the important studies by Richard Talbert (Talbert 2023) and Andrew Fox (Fox 2023).

Preface

This study is about the relationship between geography, topography, and power in the ancient Roman world, and most particularly about the visualisation of ideas about geography, at the interface between art and environment, though not necessarily between nature and culture.

This is not going to be a work about centre and periphery but rather about ideas circulating at the centre itself or emanating from there. The centre, the Greek *agora* and the Roman *forum*, was probably the most essential space in the ancient world both because of its intrinsic sacrality and because of its very functionality, assuming different forms depending on what was taking place there or who was present there at any one time. As Rome broke its political bounds and headed towards empire the whole city became the centre and the Roman world-view changed with it.

The Roman state then needed to present to the Roman people an easily-digestible narrative about its imperial ambitions and its imperial possessions, in a way that went beyond the fact that servitude, enslavement, and misery for many underpinned this expansion. There needed to be a publicly-guided discourse centred around the smoothing out of difference, rather than its obliteration or elimination, and the presentation of very different lifeworlds in a familiar way. It marked a way of directing how change could be managed and a way of reimagining how the world might be and might work, at the intersection between selection, presentation, knowledge, and insight. Reflection and communication sought to create a communal sense of belonging.

There was a number of stages to this process, and indeed the first stage will have been the presentation of Rome as an entity in itself, and later as a regional power within Italy, but this is a programme that is difficult to see in the archaeological record even if it can be plotted through written historical accounts. For instance, Cato the Elder in his *Origines* of the 160s BC related the origins of a number of Italian peoples and discussed their relationships with Rome, clearly setting out what it then meant to be Roman.[1] The early Roman expansion into Italy might have been more of a process of elite negotiation and the promotion of elite family agendas, as quite recently suggested by a number of academics, rather than expansion by war and conquest.[2] A well-known wall painting from a fourth century BC tomb on the Esquiline Hill in Rome, now in the *Musei Capitolini Centrale Montemartini*, depicts what are generally considered to be treaty negotiations during the Samnite Wars and catches the progress of Rome's expansion *in stasis*. The so-called Social Wars have sometimes simplistically been seen as Rome's Italian allies turning on Rome in order to gain more rights rather than actions arising out of dissatisfaction at Roman rule and colonisation.

As Rome's power grew it acquired territory outside of Italy. The process of empire naturally led to the importation and adoption of goods and ideas from a wider network beyond the Italian peninsula itself, leading to what can be termed a deterritorialisation of Rome the city. By intensifying practices that stressed Roman individuality this process was mitigated. Rome after all was not the first power to establish wide Mediterranean networks beyond home

---
[1] Cato *Origines*.
[2] See, for example: Terrenato 2019; and Van Dommelen and Terrenato 2007.

territory: indeed far from it. Rome's external relationships were underpinned by strongly-articulated localised expressions which served to demarcate identity both internally and externally. Local and global practices interacting together played a significant and important role in Rome becoming what might awkwardly be called Mediterraneanised, as had previously and variously happened for the Greeks, Phoenicians, Etruscans, Assyrians, and Egyptians.

A point that needs to be made from the outset is that in the ancient world Rome's recourse to constructing a vision of the contemporary world using the kind of 'geography products' discussed in this book was not necessarily unique, though the motives, preparation, and presentation certainly were. The Hellenistic world was hyper-connected in a similar way to Rome's world. The Hellenistic Ptolemies in Egypt through their building of a great library at Alexandria, the collecting of books, the forging of a culture of intellectual classification, the assembling of scholars, poets, and artists at court, and the display of collected exotic animals in one particularly famed street parade in the 270s BC made grand statements about links to the world of Alexander the Great.

It is true to say that the turning point for Romans in terms of their engagement with new and original geographical knowledge came after the fall of Carthage to the Roman general P. Cornelius Scipio Aemilianus in 146 BC. Even though the Romans now had an empire, up until that time their world had to all intents and purposes been the world known to the Greeks and written about by them. Greek maritime and shipping itineraries, known as *periploi*, would have been available for use by the Romans criss-crossing the Mediterranean. After 146 BC the Romans began to break the bounds of their world and of their geographical knowledge.

Rome was an inland city, admittedly with a river connection to the coast, yet up until c. AD 300 it could still talk about *mare nostrum*-'our sea'-as if geographical location was a state of mind rather than a fact, a barrier to the exploitation of their environment as they pleased. Interestingly, the term was originally applied specifically to the Tyrrhenian Sea after the capture of Sicily, Corsica, and Sardinia from the Carthaginians in the Punic Wars, but by the mid-first century BC, around 30 BC, as Roman territorial ambitions grew, it came to be applied to the whole Mediterranean. Use of the phrase by Julius Caesar, Livy, Pomponius Mela, and Sallust suggests that it was a commonly-applied and widely-accepted term.[3] Thus a geographical epithet changed as a reflection of political ideology and military expansion. Yet curiously, despite this self-identification with the Mediterranean, the Romans never altogether identified themselves as a seafaring people in the way that the Greeks had done. Yet Rome became like the sea, its immensely deep sky and all its movement, all its houses, domes, and arches imitating the tumultuous waves of the ocean. The sound of the city awakening, of its bustle and street traffic must somehow have seemed to mimic the rhythmic crash of ocean tides. A true image merged with a metaphor.

The Roman attitude of 'owning' of the Mediterranean is especially interesting today in terms of comparisons with nineteenth and twentieth century European colonial discourses which helped create an imagined geography of the Mediterranean as somehow exclusively Greek and Roman in the past, and which then became part of the strategy for legitimising European colonial incursions and appropriations at that time.

---

[3] Julius Caesar *De Bello Gallico* 5.1; Livy *Ab Urbe Condita* 26.42; Pomponius Mela *De Situ Orbis* 1.5.1; and Sallust *Bellum Jugurthinum* 17.

A pivotal point in the history of the Roman empire, and highly germane to this present study, was the accession of Septimius Severus and the Severan dynasty in the late second century AD. Most probably because of the Severans' roots in the east their reigns saw concerted efforts to reach out to provinces and communities in the eastern empire, but at the same time to stress unity between east and west in a way that perhaps had not been done since the time of Augustus. The cultural syncretism that underpinned their policies reached a singular and significant peak in AD 212 with Caracalla's extension by law of Roman citizenship to virtually all of the empire's free people. The implications of this were extraordinary in terms of trying to foster a sense of empire-wide unity and Roman identity.

The creation of the Roman empire and its expansion almost inevitably led to a recalibration of spatial relationships between Rome and Italy and between Roman Italy and the rest of the known world as it was then. It is true to say that the detail is often in the small print of Roman culture, in its undercurrents. Geographical thinking shaped the forms of contemporary art then, even if only at the level of sub-genres, thus making this present study a geography *of* art but not a study of geography *in* art, a political geography of the Roman world told through images, a strange spatial ontology layered onto Rome's fixity in defined physical space. Extraterritoriality such as this can signify openness, freedom, imprisonment, or subjugation.

The history of Roman imperialism to some extent could be described as being a history of fragmentation and a history of exclusion. The city of Rome became a space where an attempt was made to create and present a kind of collective memory. If we can also then talk about moves towards inclusion our discussion has to be tempered with awareness of the constant undertow of cultural dislocation and alienation there. If environments can be said to inhabit us, then Rome became alive with peoples and products of the whole known world: it was not where you were but where you could be, through movement, transformation, becoming. However, a Roman when away from the city would have carried with him or her the air they once breathed in Rome, the waters of the Tiber, and the warmth of the Italian sun. The sun had not trapped all shadows there, without the city's refuge of silence its vision was fugitive, composed of silence, matter, and compact form. The potential level of synaesthetic power involved must have been considerable, with light, music, texts, images, and architecture probably headily combining.

It is becoming increasingly fashionable among ancient historians and archaeologists to write about geography and cartography, and indeed in the last few years a number of books have been published on conceptual ideas centred around ancient experiences of space and geography and the mapping and recording of space. However, this present book has to some extent been gestating for over twenty years, ever since I wrote my first book on images of barbarian peoples in Roman art-*Enemies of Rome. Barbarians Through Roman Eyes*.[4] Presenting such images to the Roman viewer was to all intents and purposes a way of coming to terms with the world in which Roman power held such a sway. In many contexts this was as much an exercise in self-representation as it was the deliberate and accurate dissemination of geographical information about distant lands and non-Roman peoples. Again, in a later book of mine-*Cave Canem. Animals and Roman Society*-attention was turned in part of the book to how the capture, collection, and display of exotic animals and birds, and often their killing in the

---

[4] Ferris 2000.

arena, became a defining characteristic of Roman culture's coming to terms with the wider world.[5]

As always, while researching and writing a book my mind has been sparked by various modern or contemporary artistic and cultural sources that provided both ideas and stimulation with regard to the subject at hand. One has only to read the works of Margaret Atwood, J.G. Ballard, W.G. Sebald, Amitav Ghosh, and Chang-Rae Lee, for example, to find parallel thinking about the relationships of peoples, places, and spaces to that in the writings of ancient authors such as Herodotus, Homer, Virgil, and Livy. Journeying, travelling, belonging, and being an outsider underpins much of the poetry and other writings of Blaise Cendrars. The interconnectedness of emigration, immigration, and identity can be gauged from works by writers as varied as Michael Ondaatje, Kristjana Gunnar, Noreen Masud, and Dan-el Padilla Peralta. The experimental, and often very funny, writings of Georges Perec together provide a fragmented but somehow coherent view of Paris in the 1960s and 1970s. Indeed, in a short Appendix to this present book I have produced my own attempt at replicating Perec's idea of utilising repetition (as in his book about Paris *Je ne souviens/I Remember*) to produce an imagistic vision of the ancient city of Rome and its reception. The consideration of films about space and place, most particularly those of Agnès Varda as discussed further below, but also others such as John Smith's *The Black Tower*, Patrick Keiller's *London*, Andrew Kötting's *Gallivant*, Sarah Maldoror's *Regards de Mémoire*, and almost any film by Andrei Tarkovsky, can add to the interpretative arsenal on the subject. Michelangelo Antonioni's way of representing places, of constructing narratives and telling stories by depicting squares and streets and the urban environment has been hugely influential for me. In his cinema such places became almost metaphysical arenas where knowledge was displayed and exchanged. Looking at published portfolios of photographs by Ansel Adams, Edward Weston, Sebastião Salgado, and Dorothea Lange has also opened my mind to thoughts about moving through, and being caught in, place and space. For me the post-punk music of Joy Division, The Pop Group, The Fall, and Wire from the late 1970s to early 1980s proposes many routes through interzones into clear space, as their music moves through time and maps a geography of revelation and resolution. Low-frequency sounds rumble up like suppressed memories of hauntological places. Less dark, but equally vivid in evoking place and time, past and present is the early 1980s music of The Go-Betweens, as best heard on their seminal, haunting *Cattle and Cane*.

Importantly, viewing films by the French director Agnès Varda has helped me to theorise around many past situations relating to ideas of place and home, and to throw up transferable possibilities of interpretation. For instance, in her short documentary film *Les Dites-Cariatides-The So-Called Caryatids-* of 1984 Varda drifted around Paris photographing numerous examples of architectural caryatids, cognisant of the descriptions of the Roman architectural writer Vitruvius who contextualised and sought to normalise the use of such images of captured, subservient women in Greek and Roman building practice (Figure 1). Accompanied by the poetry of Baudelaire and the music of Offenbach, Varda, as she so often did, here created a topography of the human experience mediated by the camera lens and her piercing feminist perspective. The rolling of the film gave motion where for the caryatids there was just paralysis. This was as much an exploration of ideas about place, a discourse on desires and the city form, as it was a study of architectural traditions. These images of women inhabited

---

[5] Ferris 2015.

Figure 1. Relief depicting architrave supported by the figures of two caryatids, Pozzuoli. Early first century AD. *Museo Archeologico Nazionale*, Naples. (Photo: Author).

these multifarious urban spaces, and Varda used them to reclaim the streets of Paris from the tyranny of the modern male *flaneur*. As Varda herself later said in her semi-autobiographical film *The Beaches of Agnès* of 2008 'if we opened people up, we'd find landscapes'.

Ideas relating to identity, alienation, assimilation, diaspora, and exile could be manifested and presented in the form of images in the Roman world without overt references to geography and origins, and yet inform viewers of just those very things through a mixture of lyricism and dialectics. These images were not simply part of a reflection of a separate or separated world of art: rather they were part of the passionate, rational, and dramatic aspects of everyday life at the time, sparking imaginations that were to be turned on the transformation of reality itself. Viewers were encouraged to discover within themselves desires for other, particular environments and places in order to make them seem real, to regenerate the nature of imagined experience under other skies. The tensions implicit in such strategies are obvious: the city became the total work of art, playing with the presentation of time, space, and place each in turn, then in tandem and combination. The solicitation of the city's architecture and monuments was seductive and informative to those who were susceptible or open to suggestion. Information gleaned in this way reflected the absence of more practical means to orientate oneself in a changing and expanding world. The study and correlation of accepted snippets of geographical information obtained by cultural osmosis or sought out in a targeted manner on the city's streets created new and what must sometimes have been very individual and idiosyncratic mental and emotional maps of both the existing cityscape and of distant imagined cities and places. These geographies framed Roman cultural practice. Representational and sometimes direct and sometimes almost abstract, intimate and monumental, systematised and impulsive, together these works did not break the rules of contemporary Roman art but they pushed the boundaries by signing up to all of them. If asked to say what they were about, I would say 'everything'.

New forms of communication, new messages, and deconditioning from misunderstood or jumbled images must also have gone hand in hand with all of this. Many individuals as viewers were given information, they and others must have also sometimes found it or stumbled across it. We should not imagine a merely one-way transmission of experience in a city such as Rome or in other Roman cities and towns: thinking of viewers as just an audience, non-creative, purely receptive, passive, and isolated, surely misdirects us away from understanding open forms of cultural communication that in fact must have been active as well as reactive. Although occasionally imprecise, these images conveying geographical information in the form of architectural expression were often highly charged with emotionally-evocative power and representing desires, control, events from the past, the present, and the future, rational extensions of religious experiences and myths. Imperial Rome ushered in a period of city planning seen as a means of knowledge exchange. Parts of the city could have corresponded to the feelings usually experienced by chance, but here managed or even manipulated. One could leave the realm of direct experience for that of representation and presentation. The passivity of the old, pre-imperial Rome needed to be reconstituted in some respects by a collective project explicitly concerned with confronting every aspect of the audience's lived experiences, by drawing attention to the contrast between what contemporary life was actually like and what it could be. Rome could only find its poetry in the present, if informed by the past.

While planning this study I read Katja Pilhuj's brilliant book of 2019 *Women and Geography On the Early Modern English Stage* and was particularly struck by her use there of the term 'geography product' to describe different kinds of objects and texts in the early modern period which each contained some element of spatial or geographical information.[6] There was an immediate realisation that the use of such an umbrella categorisation of disparate 'things' offered a number of potentially critical openings for looking at ancient Rome and its defining of place and space, that a new strand of critical conversation and dialogue could be started. Indeed, therefore I have adopted the phrase 'geography product' here with regard to the study of geography and ancient Rome, using it more broadly to include a slew of visual sources which acted as mnemonic triggers to spatial awareness and extending the definition even to encompass tastes, sounds, and smells which might have had the same effect. Written histories, geographic texts, ethnographic studies, poems, epigrams, plays, maps, drawn surveys, and inscriptions could also be geography products. Bodies and images of bodies could also on occasions be geography products or carry on them such information as to qualify in this respect: land, space, and place literally could be written on the body. In other words this book will set out to discuss the full range of geography products that would have appeared or circulated in the city of Rome and elsewhere in the Roman world from the time of early

---

[6] Pilhuj 2019.

Figure 2. Example of a Roman 'geography product'. Statue personification of the River Arno. Exact provenance uncertain, probably Rome. Hadrianic. *Musei Vaticani*, Rome. (Photo: Author).

Rome up to Late Antiquity. Adopting this strategy of definition has allowed narratives, plots, structures, and themes in the evidence to emerge.

Roman ideas and concepts about geographic space, about topography, about landscape, about foreign peoples, and about barbarians were developed, one might even say workshopped, in the theatrical sense, through a process of almost trial and error in terms of creating and presenting a coherent series of geography products which utilised words and images to telescope distance and space and to create maps of the body and maps of the mind. The famous and canonical statue of the emperor Augustus from Prima Porta, now in the *Musei Vaticani* in Rome, can very much be viewed and read as a kind of prophetic document and archive like the body of Ray Bradbury's fictional *Illustrated Man*. The decorated cuirass worn by the emperor contained in its roster of discrete images a narrative of sorts that helped set the agenda for the presentation of the geography of the Roman empire to its viewers in all its complexity.

Highly-relevant to the idea of traversing a city to read its buildings, streets, and monuments and to decode their messages is the concept of psychogeography as developed by the Situationist International and in more recent British culture best represented by the writings of Ian Sinclair, such as in his book *London Orbital* of 2002. Even though the psychogeographic concepts of the *flaneur* and of 'drifting' principally relate to modern urban or industrialised environments, nevertheless there are certain strategies of analysis that can satisfactorily be applied to the study of other types of cityscape, landscape, and topography in the deeper and more distant past.

Did people 'drift' (*dériver*) through the streets of Rome, wandering without purpose, intent, obligations, or destinations in mind? In the late nineteenth century the idea of the solitary, disassociated city *flaneur* took hold in avant garde circles, with walking the city being presented now as oriented towards some goal, some deep revelation. But, on the contrary, a city could become specific and non-specific, all at once, as Georg Simmel's early twentieth century city-dwelling 'blasé person' found, indifferent to their surroundings and often unresponsive to them.

As we travel around a modern city, wandering its streets and boulevards, exploring its arcades and alleyways, if not happy to be lost in urban space we are often glad to find a mounted street-map that displays our position in the city with a red arrow that tells us 'You are here'. It situates us at an exact location, a certain unique spot. Many of the urban and civic artworks discussed in this study played the same role, spatially and conceptually locating Romans who viewed them both in their city and in the wider world. Rome was marking out its own position in the world by inviting viewers to linger in the city and reaffirm its significance. Rome was an urban architectural map built from lived trajectories, histories, truths, and fictions. We can retrospectively search its spaces for the voices and stories buried within them, to create topographies of the human experience. Artworks conveying geographical information still managed to embed their locations with a sense of their individual human resonance. If women could not move as freely in Rome's public spaces as its male citizens and inhabitants then their active participation in the creation of the idea of the city as a nexus for a world beyond came to be largely represented by images of the female body in and imposed on masculinised spaces.

Footfalls Echo In the Memory

The Roman state, whether in the Republican era or during the years of imperial authority, was not an inventor and user of 'information technology' as we understand it today. Just as there can be seen to have been two strands to healthcare in the ancient world-a 'rational', quasi-scientific one relating to medical practice and an 'irrational', or perhaps emotional one linked to the recourse to the gods to seek help when sick- so space could be controlled and understood by measuring and mapping it, but equally could be confronted by its conceptualising in other ways by the use of images and, to a lesser extent, by metaphor.

The school of ancient historians who write about 'common-sense geography' or who use the even less-appealing term 'illiterate geography' often, unintentionally, present readers with an either/or dichotomy: that is between what might be termed scientific or descriptive geographical works, for the literate elite, and popular images presenting geographical information for the masses. Yet both types of geography co-existed in Roman society and each interacted with the other, informed the other, fed in to the other, enhanced the other, contradicted the other, and so on, depending on time, circumstance, and context. As in all of my previous books this work primarily will be about Roman visual culture which was central,

Figure 3. Example of a Roman 'geography product'. Nilotic scene on a terracotta Campana plaque. Mid-first century AD. British Museum, London. (Photo: Copyright Trustees of the British Museum).

perhaps the central, means by which the Romans forged their geographical identity and through which they ordered and transformed the world around them.

When you start out writing a book on a particular subject you know what the book is going to be about, even if only broadly. As work progresses you discover what the book is most certainly not going to be about, in other words which particular topics are either not going to be discussed at all or which topics are going to be raised in certain specific ways only, committing to omit other lines of enquiry or certain approaches to particular topics. Every study, like a flood, leaves a residue, a silt of unused case studies and undiscussed artworks. Not because they are irrelevant: simply that less is indeed often more, and these undiscussed cases remain otiose to the main arguments propounded, but not because they would necessarily be invalid as examples. This book is not a history of geography in the ancient world, of Greco-Roman geographers, or of maps and map-making at the time. A great deal of recent academic writing about Roman geography indeed has dwelt on the topic of maps, very specifically the absence of accurate scale-mapping at the time, and indeed has sometimes consisted of quite pedantic debates about the definition of what actually constitutes a map in the first place. Other studies appear to me to have become slightly bogged down in the mire around questions of accuracy and of measurement in Roman times. The Roman compilation and use of travel itineraries is not viewed here negatively, because they were not what we would understand as a map today. That some portable Roman sundials were not hugely accurate again is not an issue here. Nor is the fact that most Greco-Roman geographical writing is viewed by some as somehow being compromised and having had no 'utility on the ground' because of its structures, tropes, and themes. All of these things relate to context, to differences of scale and ambition, an unusual degree of lack of concern over precision and regularisation, and differing interests in the uses and currency of information at the time.

That the Hellenistic and Roman periods were the first great eras of mass mobility suggests that travelling without pinpoint-accurate scale maps was not an issue at the time. If the concern of the contemporary traveller then was to get from City A to City B along a single road route, stopping at five designated intermediate places, then an *Itinerarium* or itinerary was more than able to provide a sequential list of places making up this particular journey and to provide the distances between each stopping point, and thus provide a total mileage for the trip. Rather, the book is about what I will call geography products after Katja Pilhuj, principally in the form of images of peoples, places, and landscapes, but also in the form of written texts or inscriptions, public display maps, monuments, and objects of various kinds including surveying instruments and sundials: in fact any thing that could be said to be intended to convey or to have been conveying a piece or pieces of geographic information to a viewer or reader, even if that information was allusive rather than necessarily always factual or correct, sometimes impressionistic rather than always clear or detailed, and more often than not open to different interpretations or a number of interpretations all at the same time. The book will ask how the creation of these geography products came about, though that cannot always be explained or surmised, what their creation and dissemination were intended to convey in terms of knowledge transfer and outcomes, who the intended audience was for these geography products, and how the intended audience might have reacted to contact with the products, as single, stand-alone entities or as part of a series or sequence. In the main the action is set in the city of Rome itself but many other examples and case studies from elsewhere in Italy and from around the Roman empire will also be discussed. Inevitably the politics of geography will feature heavily here, while issues relating to mobility, travel, connectivity, interconnectedness, standardisation, sameness, and difference will also be addressed. People, ideas, and images travelled in the Roman world and around the Roman world, in a dance of dizzying complexity, and everywhere the ceremony of isolation and its attendant innocence was drowned.

By calling attention to the distortions, inflections, disidentifications, confrontations, dissatisfactions, and recombinations of a diverse range of codes in the preparation of geography products, this book attempts to contribute to the discussion of their innovative potential as vehicles of information about what it meant to be Roman.

In subject order consideration will be given to: pre-Roman Italian ideas about conceptualising geography, topography, and space (Chapter One); a discussion of an ancient rumination on belonging and an introduction to the broader subject of the study (Chapter Two); the contemporary significance of Rome's famed seven hills and of mountains in general to the Romans, and how Romans conceptualised Rome (Chapter Three); the significance of rivers and their personification in Roman art (Chapter Four); landscape art, especially painting, and

Figure 4. Example of a Roman 'geography product'. Sarcophagus depicting the Indian triumph of Bacchus, Ostia/*Portus*. Second to third century AD. *Museo Archeologico Ostiense*. (Photo: Author).

the idea of control and surveillance (Chapter Five); the particular Roman discourse about the city's relationship with Egypt and its colonial landscapes (Chapter Six); the practice and politics of mapping in the Roman world (Chapter Seven); the use of images of barbarian peoples in Roman imperial art as geographical markers for the definition of *Romanitas* (Chapter Eight); the process and concept of journeying in the Roman world as a way of establishing conceptual links and mnemonic chains between places (Chapter Nine); and more theoretical and comparative issues relating to the conceptualisation of space and place in the Roman world and in more recent societies and cultures (Chapter Ten).

Almost inevitably, images of barbarians will loom large in this study, as will writings about non-Roman peoples. Roman ethnographies were not all about barbarians and the other. It must be remembered that by often writing about non-Roman peoples using the terms *gens* and *populus* these writers were often doing more than simply telling tales about the barbarians. They were trying to define what it meant to belong to a defined territory with a particular character rather than simply being on some nondescript, interchangeable piece of land.

Who could have known that certain non-Roman peoples in some contexts would journey from image to subject, that they could pivot or move from being the object of the gaze to the subject that looked. The terms that modern academic archaeologists, historians, and classicists use to describe some of the subject peoples of the Roman empire-Romano-Britons, Gallo-Romans, and so on-were not used by the Romans themselves and therefore really can sometimes confuse the discourse around the relationships between the conquered and conquerors, between the subjects and the rulers. By using such terms of hybridity the origins of these peoples are sutured to Rome and masked by being described through fractions of other places. Designating in this way defines and contains them in a manner that will have been very different to ancient contemporary perceptions. No model of Roman belonging can function unproblematically based as it was on the constant recourse to violence inherent in founding any empire encompassing someone else's lands.

The idea that Latin literary and historical texts were artefacts-objects-of Roman culture is widely accepted among academics today, but here such texts when they include spatial information or ethnographies will also be treated as geography products, indeed as one such product among many in the Roman era. I certainly do not consider that the analytical blurring of boundaries here is a bad idea. The idea that we can speak of Ovid's Rome in the same way that we might about Dickens' London is attractive, but potentially misleading. Everybody had their own Rome, not just Ovid, even if his listing of hotspots around the city to meet women and find love or sex read like emotional maps for personal journeys of discovery. Ovid's Rome was indeed a mixture of interconnectivity and alienation, the latter intensified during his time in exile from the city at Tomis on the Black Sea.

On some occasions we can see indisputably that the presentation of geographical information was an imperial initiative but not necessarily a coherently thought-out and planned project with dedicated aims and objectives, with a beginning, a middle, and an end. Enquiring and answering how and why the Roman state and its military and bureaucratic agents shared and subdivided real and imaginary spaces goes some way towards understanding the making of the transnational cultures of the empire. Rather, information was placed in the public space and spaces in a piecemeal fashion, with the creation of a series and its maintenance not being always apparent. Pliny the Elder dedicated his monumental encyclopedia, the *Naturalis Historia*, to the emperor Titus and throughout the work can be found references to the benefits of the Roman peace for the whole world, while at the same time stressing the totality of all nature and not just that of those lands under Roman rule. It is also true to say that a Greek geographer and ethnographer such as Strabo, writing his *Geographia* during the reigns of Augustus and Tiberius, the pivotal era when the Roman conception of empire changed subtly but irredeemably, was not strictly 'writing the Roman world', and certainly not to order.

Roman culture needed to locate itself in both quotidian time and in a cosmic order. There needed to be a structure for aligning events in both time and space. In order to do so its social system defined itself by expressing boundaries which also involved the recording of social pathways through the pursuit of knowledge about ancestral relationships and lineage. To a member of the Roman male elite the idea that he was placed within a map of familial connections would have been certain and indisputable. Garrett Sullivan's idea of what he called 'affective spaces', that is locations imbued with some form of significance by those that occupied them or who were in some way connected to them is highly pertinent here. This combining of bodily and terrestrial space created two parallel geographies, of the wider world or known world and of others which were made up of affective or local places. This geographically-inflected thinking would have led men of this class to (over)identify with a family house or estate, or with a place linked to family origins. Its political and military system needed to centre Rome itself within its contemporary world by recording geographic pathways and their history. This meant that the Romans chose to visualise both qualitative data in their art relating to expressions of social hierarchy, their cultural and religious beliefs and values, and quantitative data relating to economic trends that were themselves defined by the management of overland travel and transportation, and trading by river and sea. Geographical knowledge in the Roman world was actually part of a vast network of cultural codes, rather than simply a specialised branch of general knowledge.

There were very varied models for conceptualising the world beyond Rome. Geographically they bled together, from text to text, image to image, region to region, race to race. The globe, an unbroken circle, could be seen to represent conquest, unbroken in time and space, the circular endlessness contrasting with a map demarcating finite, limited space: this was to all intents and purposes a contrasting of ambition with realism. These geographic products were in a language, both textual and visual as well as imaginative, and were not simply static representations or allusions to space: rather, they helped map out new ways for people to create conceptions of themselves and of the world.

The interplay between ideas of impermanence and permanence also must have been significant. A picture of a personified river carried in a triumphal procession-seen fleetingly if at all by the spectators lining the route of the triumph-was very different indeed to an image of a personified river in the form of a permanent statue, a static image. It will be argued that geographical information could be visualised in a number of ways in the Roman world and that the categories of representation were to some extent fluid depending on context.

Anyone familiar with the song *Roadrunner* by Jonathan Richman will be aware that the song's great success lies in the universality and relatability of its theme of carefree, youthful driving on American highways in general, startlingly contrasted with the geographically micro-specific experience of being 'out in Needham now, out on Route 128, by the power lines'. This astonishing switch of scale represents a good example of imagistic mapping that situates the local and perhaps unfamiliar in a broader context that feels inclusive and understandable.

In many ways ancient Rome became a kind of National Park, turning in on itself in order to display its character externally. Rome was a monolith, a monument to an idea, composed of thousands of individual monuments, each an accretion that nuanced a central theme and trope. Rome acted to frame art about the world around and its peoples. It was part of a more general and much broader programme to transform and inform everyday life in the city. It became a montage of attractions, with the real and imaginary, the physical and the mental, the objective and the subjective, description and narration, actual and virtual, reflecting each other, ultimately becoming confused, indiscernible one from the other. While each individual geography product tells us something about specific aspects of Roman geographical knowledge and the ordering of information, taken together they help provide a window onto broader issues relating to the coexistence of streams of thought relating to the pictorialisation of knowledge and the dissemination of ideas through objects and images. To understand the world was to see it and conceptualise it at the same time.

This would seem to have marked a critical conjunction between the poetics of representation and the poetics of ideological conceptualism that opened up a fluid, intertextual aesthetic and theoretical space for distinctive moments of engagement with the world beyond the city. The individual thus memorialised his or her own encounters with geography. Metaphors of storage and preservation, of legend and myth, of historical events, were reflected back at the population as monuments and as archival materials. In these geography products space became freed from objective context and became the subject in itself. A concern for tragedy, ecstasy, and doom in Roman battle monuments, marrying violence and anxiety, was contrasted here with flat grounds on which more subtle ideas were projected. Simultaneously dividing and uniting the composition through repetition they looked forward to a new and seemingly calmer future. The fragmented and compartmentalised nature of the then-contemporary spatial experience was not somehow a unique experience or some kind of exception, as history might be thought to be telling us: rather, utilising different levels of revelation of geographical information did not prevent the development of unifying narratives around the place of Rome in the world. Quite the opposite in fact it would appear.

Away from Rome, in first Italy and then in the conquered provinces, it will also be considered how the armature and infrastructure of Roman imperialism- its roads, forts and fortresses, towns and *coloniae*, ports, and harbours-served to create layered landscapes. Exploration of the stratigraphy of the empire's geography will reveal shadow landscapes under this superimposed, mesh-like grid. Beneath the rule(r) countries hid. What will be more difficult to extrapolate will be the different layers of lives that would have been lived contemporaneously in the same places: in other words how the same places and landscapes-the same topography and geography of a place-would have been experienced differently by soldiers and traders for example, or by colonists and indigenous peoples.

A number of twentieth and twenty-first century politicians have used remarkably similar phrases to each other to voice their concerns about modern internationalism. For instance, Hitler in 1933 railed against an international elite- '[the] clique…people who are at home both nowhere and everywhere', while British Prime Minister Theresa May in a speech in 2016 told pro-European British people that 'if you are a citizen of the world, you are a citizen of nowhere.' Both of these stances served to make very different points about citizenship, belonging, and the idea of home, and yet both adopted an either/or position, with there being no nuance, no compromise, and no recognition of the complexity of national identities. Time and time again throughout this book it will need to be stressed that Roman identity was astonishingly fluid as a concept, both geographically and chronologically: ever-changing and at the same time always staying the same. To paraphrase the political speech sound-bites quoted above Roman citizens came to think of themselves as at home both nowhere and everywhere, while simultaneously being citizens of the world and of nowhere.

Sexual exploitation, most usually of women, has been a marker of global empires throughout history, and the Roman empire was of course no exception, with travel, mixing, and integration a potent mixture that helped create a new blend of practices and ideas that constituted a new form of Roman sexuality. If sex, men, and imperial expansion were interlinked, so were sex, women, and imperial control. Surviving abuse, captivity, and exploitation was difficult but possible. The common Roman habit of masters marrying their female slaves and thus emancipating them in one way, but legally ensnaring them in another, evidently somehow became a marker of status, a recontextualisation of both sexual and geographic cultures.

In Rome in particular, but in many other Romanised locations and contexts, images of female foreigners were a kind of geographic product, displayed so that the Romans could write a world where they asserted their status of moral and cultural superiority. One of the most active demonstrations of this relates to the Egyptian queen Cleopatra. For Julius Caesar, Antony, and then Octavian (Augustus) she represented territory that could be claimed by Roman men through possession of her body. Octavian's Egyptian triumphal parade was itself ultimately an exploitation of the symbolic importance of Cleopatra's body, even if she had asserted her control over her own body, and thus her identity, by her suicide. In death she maintained a physical presence in Egypt in terms of it becoming a monument she had constructed. It removed her from any potential sexual and political uses-she did though come to be framed, it could be said contained, on posthumous Roman coin issues.

Augustus's great *Ara Pacis* monument in the *Campus Martius* in Rome with its much-repeated decoration of multiple trailing plant tendrils called up a discourse of planting that went hand in hand with the project of empire, an agro-sexual appropriation of fertile land. As these plants stabilised and bound the soil, so the Augustan peace bound the known world together, and brought different and disparate peoples together in the same spheres of influence and interaction. If Hellenistic mathematical geographers had developed the idea of the spherical earth, the orb became a very significant and different kind of symbol, for world conquest, in the art of the Augustan court. This was a kind of triumphal geography, with ceremony and art highlighting contemporary spatial awareness.

It is known through textual references to names and works that there were around 250 geographical books written by Greek and Roman authors, and yet only four of these today survive in whole or in part, that is works by Strabo of Amaesia, Pomponius Mela, Pliny the Elder, and Ptolemy of Alexandria.[7] Each book built on pre-existing knowledge, and it is through the acknowledgement by the authors of these four geographies of the studies and travels of their predecessors that the one-time existence of these lost works is known. The works of these ancient geographers were not intended as travel guides as we might understand and use the term today. They were works to be read at home or in the library, reference works for the sedentary traveller. Plutarch, writing probably around the end of the first century AD, introduced his *Life of Theseus* by comparing his task of writing accurate ancient history to that of the geographer:

'Just as geographers…crowd on to the outer edges of their maps the parts of the earth which elude their knowledge, with explanatory notes that 'What lies beyond is sandy desert without water and full of wild beasts' or 'blind marsh', or 'Scythian cold', or 'frozen sea'……I might well

---

[7] On Greek and Roman geography and geographers see, for example: Adams and Laurence 2001; Adams and Roy 2007; Batty 2000; Bekker-Nielsen 1988; Bianchetti *et al.* 2016; Bishop 2019; Blum 2019; Clarke 1999; Damer and Myers Forthcoming; Dueck 2000, 2012, and 2021; Dueck *et al.* 2005; Ellis and Kidner 2004; Evans 2005; Geus and Thiering 2014; Mayer 1986; Merrills 2005; Myers and Damer 2021; Nicolet 1991; Raaflaub and Talbert 2010; Riggsby 2017; Roller 2006, 2015, 2019, and 2022; Romer 1998; Romm 1992; Shahar 2004; Skempis and Ziogas 2014; Talbert 2010b; Van Der Vliet 2006; and Van Paasen 1957.

say of the earlier periods: 'What lies beyond is full of marvels and unreality, a land of poets and fabulists, of doubt and obscurity.'[8]

This passage is particularly relevant to this present study in a number of respects. If the world was experienced vicariously by those who stayed at home and were receptive to messages mediated through geography products, then it was differently experienced by those who travelled away from Rome or around the empire from other locations. A study such as this cannot altogether ignore the potent Greco-Roman myth of the wandering hero such as Aeneas, Bacchus/Dionysus, and Hercules, and the idea of journeying as a metaphor. The travels of the emperor Hadrian, who the writer Tertullian called *omnium curiositatum explorator*, that is 'an explorer of everything interesting', explored both foreign lands and foreign bodies.[9] Travel became under him virtually an imperial virtue. Imperial authority was wherever the emperor happened to be, a principle that first found its *raison d'etre* under Hadrian, to become much later a rationale for some emperors never to visit Rome or to seldom do so. Roman geography products helped their users and viewers to make journeys through ruins, and to comprehend the creation of a global sense of place through the working of theatres of memory. The fleet concatenation of images was experienced as a great flow which swept the viewer along, making him or her almost a passenger or traveller rather than an active agent in the creation of a visual narrative.

Itineraries were highly practical guides for travelling along predetermined routes but it is interesting to consider that they also formed a record of absent places, places off the route, topographic features with no name, areas of unpassable swamp, desert, or wilderness. They remind us that no matter how efficient the Romans might have been at gridding an infrastructure over their empire in order to exploit its natural resources, nevertheless many areas still remained off the grid, and that the wilderness was not pushed back everywhere. Rational expansion based on the argument of necessity met other forces headlong: movements of tectonic plates and the volcanic activity that saw the destruction of Pompeii, was more than just a mechanistic folding and faulting of the earth's crust, and the swells and surges of the Mediterranean Sea disputed with the land and coast through its slow, crumbling erosion.

Around AD 333 a native of the city of Bordeaux (Roman *Burdigala*) in the Roman province of *Gallia Aquitania* set out on a pilgrimage to the Holy Land. This pilgrim, who may well have been a woman, travelled there and back in AD 333-334 using a pre-prepared itinerary. Remarkably this document, known to us today as the *Itinerarium Burdigalense* or, less commonly, as the *Itinerarium Hierosolymitanum* or Jerusalem Itinerary, survived and was copied in whole or in part in four manuscripts between the eighth and tenth centuries. The route took the pilgrim through northern Italy and the Danube valley to Constantinople, then through the provinces of Asia and Syria, before arriving at Jerusalem. The route back was through Macedonia, Otranto, Rome, and Milan. The Itinerary is annotated with some notes on the journey and even some topographic descriptions of Jerusalem itself, providing the journey with personalised elements beyond being simply a practical list and guide to routes, stopping points, and distances.

---

[8] Plutarch *Theseus* 1.1.
[9] Tertullian *Apologeticum* 5.7.

This journey seems particular significant to us today but we must not view it as a necessarily exceptional undertaking for its time. The ancient world was criss-crossed daily by soldiers, administrators, merchants, and pilgrims, the latter group creating their own devotional geography. What makes the story of the Bordeaux Pilgrim stand out as an exemplar is the fact that in this document politics, salvation, and geography were here intertwined in a way that clearly demonstrated how interconnectivity and mobility made the Roman world negotiable and understandable. The story of the Bordeaux Pilgrim somehow acts as the perfect induction into the delicate connection-making of the time, between the very close and the unreachably distant, between the seismic and the minute. It is attractive to view pilgrimages such as this as individual Odysseys, almost mythic journeys controlled by four defining tropes: diffusion and mixture (travelling from home and encountering others); separation and distance; passage and detour; and return and split ending.

This book then will consider the many ways in which archaeologists and ancient historians have attempted to categorise different types or strategies of presentation of geographical knowledge in Roman times. The concepts of 'pilgrimage landscapes' as just mentioned, coexisted with 'chthonic landscapes', 'landscapes of war', 'landscapes of defeat', 'sacro-idyllic landscapes', 'colonial (or imperial) landscapes', and 'Christian landscapes', all of these terms being useful framing devices for certain pockets of information presentation.

Any study of ancient Roman society, if undertaken using modern moral standards as a yardstick, is likely to be condemnatory in some way, criticising Rome for its anthropocentric world view, its contribution to the degradation of Nature, its indifference to human and animal suffering, the materialistic corruption of love and religion, and the alienation of many social groups. If in Rome there were streets of blacksmiths forging chains for tomorrow's children then we must envisage the city as being like Federico García Lorca's Wall Street, with 'rivers of gold flow(ing) there from all over the earth, and death (coming) with it.' The social and sexual geographies of Rome would be reflected in the geographies of the other, and of the world outside Italy. The quite recent turn towards introspection in Roman studies means that it is now appropriate to bring contemporary concerns of social justice into a work such as this, while previously it might have seemed merely self-indulgent.

This research project, its methodologies and outcomes, are to some extent dialogues in subjectivities that are not fixed, but which subtly move in empathetic exchanges, open to the possibility of not just change itself but constant change. But this is not an easy situation to imagine or interpret. It is highly likely that many Roman monuments were, or became, so familiar that they became impregnated against attention in a complex historical practice. Mapping land and conceptualising land as other kinds of images were not contradictory ideas: they ran parallel to one another at the time. Ultimately this study will be an exploration of variations on themes of alienation, oppression, and trauma, as well as meaning, belonging, and hope. It is a fractured narrative to present, episodic and spasmodic, about identification and disidentification, the parsing of potential lineages, vehicles of transmission each with its own benefits and drawbacks, while still revealing the potential for new forms of exclusion and complication. But as a study it does not aspire to the uninflected detachment that this might suggest. It is not after all a novel.

In a Huxleyan sense the contemporary Roman viewers' doors of perception were opened to other worlds, but not necessarily through feeling oneness with the universe. Such subtle narcissism fuelled the desperate human need to transcend the given: the love and beauty and ironic fraternity to be found in a great city such as Rome led to a broader, more unstable and frightening outside world where human aspirations could be satisfied or thwarted by perversion, blasphemy, and death. If not actually doors, these images were at least windows on self-identity and on otherness, letting light in on a sombre struggle against accidie.

Iain Ferris, Pembrey, January 2021-January 2024.

Chapter One

Maps of the Mind

Before focusing this study directly on Rome I would like now to first pivot away from there (and then) to consider two prehistoric Italian 'maps' which in their presentation of spatial data, in their disparate sizes, and in the material from which they were made could not be more different: both though can be defined as geography products of different kinds. Yet I believe that the Map of Bedolina and the Liver of Piacenza both separately and together represented systematised maps whose study can throw light on the later Roman practices of visualising knowledge and disseminating ideas through the careful editing of images. Discussion will then turn to the Augustan codification of geographical information through the incorporation of certain types of images on the two most canonical artworks of the time, the *Ara Pacis Augustae* and the Prima Porta statue of Augustus, and to the broad range of themes with which this book will be concerned.

I do not want to discuss here the semantic question of what constitutes a map, and certainly do not see issues such as measured accuracy, temporal contemporaneity of information, and the capacity for duplication, replication, or copying as prerequisites for deploying the term in discussion in this book. Mapping is considered here to be a drawn, measured, diagrammatic, or schematic record of an event, space, or place, a visual representation of geographical and spatial information, or a visualisation of knowledge as an aid to understanding the world and in creating or enhancing identity. A map can be a plan or a picture.

Zone

A short but steep walk up the hillside from the village of Capo di Ponte in Valcamonica, a region in the northern part of Brescia province in Lombardy in northern Italy, lies the *Parco Nazionale delle Incisioni Rupestri di Naquane*-Naquane National Park-where can be found the most concentrated groupings of prehistoric rock art in the whole of Europe and probably in the world. Dating to the Neolithic, Bronze Age, and Iron Age the 30,000 individual carvings here represent a window onto the cultural and social practices of the people of the region-the *Camuni*. Further up the valley side still, as the slope steepens, lies the hamlet of Bedolina where another concentration of carved rock surfaces includes the remarkable and unique rock (Bedolina Rock 1) on which is carved the Map of Bedolina[1] (Figures 5-7). This flat rock surface of approximately 9 metres by 4.3 metres bears carved, incised, and pecked petroglyphs covering a surface area of approximately 4.3 metres by 2.4 metres, and consists of around 110 separate images of various types, including six houses, pathways and trackways, streams or irrigation channels, about thirty enclosed rectangular or square spaces which may represent cultivated fields, with some armed human figures and animals, a ladder form, cupmarks, and a regional motif known as a Camunian Rose. Like so many of the larger carved rock surfaces in Valcamonica the Bedolina Map stone represents a veritable palimpsest, with the earliest carved figures here belonging to the Bronze Age (to c. 1500-1400 BC), but with most of the carvings being done in the Iron Age, probably between 600-400 BC.

---

[1] On the Map of Bedolina see: Delano Smith 1982 and 1990; Marretta 2013; and Turconi 1997.

Figure 5. The Map of Bedolina at the rock art site of Bedolina, near Capo di Ponte, Valcamonica, Lombardy. Italian Bronze Age (c. 1500-1400 BC) and Iron Age (between 600-400 BC). (Photo: Angelo Fossati).

Figure 6. The Map of Bedolina at the rock art site of Bedolina, near Capo di Ponte, Valcamonica, Lombardy. Italian Bronze Age (c. 1500-1400 BC) and Iron Age (between 600-400 BC). (Photo: Angelo Fossati).

Figure 7. Survey drawing of the Map of Bedolina. (Photo: Angelo Fossati/*Footsteps of Man Cooperativo*).

Ironically there grows here too *Rhizocarpon geographicum*, more commonly called the map lichen. Its peculiar beauty colonises many of the rock surfaces at Bedolina, nature mapping the very rocks as they outcrop here at the summit, in the same way that the *Camuni* and their forebears sought to map the land or the heavens, this or another world.

Many interpretations of the carvings on this rock accept the artwork at face value: that it was simply a representation of the fields, houses, settlements, and tracks that could be seen down on the valley floor from this commanding and lofty spot. Ultimately unconvincing attempts have even been made to correlate specific images to actual recorded archaeological sites down in the valley, in order to support the 'literal depiction' stance. However, according to Cristina Turconi, one of the foremost authorities on the map stone, the cartographic framework of the map need not necessarily have been simply a record of the tamed landscape, some kind of cadastral map, but rather its aim might have been more abstract in terms of magico-religious symbolism, ritual, and it might have had some kind of apotropaic value. In the broader context of Camunian rock art these abstract functions would not be altogether out of place. As the map rock represents an accretion of images, a collection, a palimpsest, it may have meant different things at different times, to different viewers. The reworking of the overall image on the rock over time marks this out as a special locale in the region: a spot such as this may have provided further context for the images by their display here. This map was not just *of* the landscape, it was *in* the landscape, pecked and incised on a large flat rock surface.

Of course the 'human' figures in the map's landscapes need not necessarily have been images of mortals: some or all of them could have been figures of gods. The Roman idea of the sacro-idyllic landscape in which the gods could be imagined, either on their own or in interaction with mortals, and its common depiction in Roman art in many media, including wall paintings, mosaics, on precious metalwork, and on carved gemstone intaglios, probably had its roots in pre-Roman Italic religious practices.

It is certainly very easy to imagine this rock at Bedolina as having been a focus for commemoration and ceremony, and for the periodic restatement of geographic realities and perceptions through recarving over an extended period of time. However, it must be noted that a network analysis of the inscribed image constituting the Bedolina Map has recently been undertaken by Craig Alexander who concluded that the analysis confirmed the coherent totality of the map, with no sub-maps being detectable. The images are rooted in a social world not that far removed from that of early Roman identity creation and its logical expansion into an era of Roman colonialism and urbanisation. A subtle evocation of boundary and contour on the map stone marked an attempt to colonise land.

Much writing about maps and the motivations for mapping relate to the political and ideological aspects of cartography and cartographic practice, particularly at key historical periods such as the English Elizabethan era, the Dutch sixteenth century, and Renaissance Italy. This book is suggesting that the geography products of ancient Rome, including pictorial maps, plans, and surveys, also were imbued with the spirit of contemporary ideological discourses. This represented an active incitement to actually feel the world, rather than just think about it. This cultural strategy was informed by both a subtlety of approach and a more direct, didactic strain of artistic production which was more often than not tied in to the production of imperial imagery as panegyric. The Roman penchant for representing bird's-eye views and other types of views from above supplied implications of pattern, purpose, understanding, or explanation not available if the scene was being experienced on the ground. The metaphorical transposition of place or landscape was not simply a case of seeing somewhere as somewhere else, rather it was seeing somewhere in terms of a picture or image of somewhere else, a significant distinction.

It is remarkable how the images, views, and scales of representation oscillate on the Bedolina Map, with human and animal figures present as generic signifiers, passing witnesses of the unfolding of the landscape around them. The focus alternates between things close up and far away, investing the artwork with character, incident, and history. Replacing the act of depiction with reflection, the syncopation in the placement of these images is reminiscent of careful mapping: somehow the map's creators felt deeply the need to recreate certain aspects of the visible world that were common to all, to see it freshly, and to make viewers aware of it in all its unexpectedness, an intricate weave of incident that refused to coalesce into any kind of stable hierarchy. Upon first consideration viewers might have looked at the Map as a kind of presentation of alienation, closer reading probably suggested rather that what might have appeared to have been *about* nothing much during or after first viewing could well have expanded to be about everything on subsequent reflection.

Things that bore the residue of use were seemingly blighted by time, or fallen into desuetude; the composition of the Map became almost a strange, plangent lyric to wind and water, an elegiac mnemonic for destruction and death, a warning against the distortion that accompanied remembering and sometimes forgetting, a satisfying balance between lightness and density, transparency and opacity, enriched by the reciprocity of shapes and incised lines and grids. The word map is here a label, an umbrella term, capacious enough to cover a wide variety of practices of spatial representation.

However, the Map of Bedolina and other Valcamonican rock maps are seldom discussed in this way, as if the question of the area being contested terrain cannot be seriously countenanced. A reading of the Map of Bedolina as a colonial/colonising text may be overdue and instructive.

But the Map of Bedolina was not the only possible topographic rock art representation either at Bedolina itself or at other locations in Valcamonica, actually in quite a relatively small area. A second major map rock (Rock 7), resembling a complex geometric composition at first sight, was discovered at Bedolina as recently as 2005, the rock surface area being approximately 25 metres by 10 metres. Although there are numerous human figures in the composition the emphasis on this second map rock is on the presentation of a topographical layout of enclosed fields in a grid-like pattern or interlinked by paths or tracks, most of the enclosures being filled by pecked dots, there being evidence for a particularly large number of episodes of superimposition of images through recarving on this rock. The topographical phase of carving has been provisionally dated to the Late Bronze Age to the Middle Iron Age, that is 1000-500 BC.

A topographic or map stone, seemingly decorated with a dense grid of what might be interpreted as enclosed and interconnected fields, has been recorded at Giadeghe (also known as Pià d'Ort) and another at Paspardo, both in Valcamonica. There is no concensus among the local archaeologists as to the date of these two other topographic rocks, although most seem to favour a much earlier date than for Rock 1 at Bedolina, placing them in the Late Neolithic or Early Copper Age.

The idea mooted by Lynne Bevan that many of the rock surfaces at Naquane and certain important and particular images must have been actively 'curated', as she terms it, seems highly persuasive.[2] Anyone who has visited Naquane, Bedolina, and other rock art foci in Valcamonica will appreciate that the images on many of the rock surfaces here could quite easily and indeed quite quickly become obscured by vegetation and the processes of soil build-up. The evidence of over-carving, of the chronological layering of images on many rock surfaces, sometimes evidently on numerous occasions, testifies to the cultural necessity for the local community of returning to certain individual rocks or to broader locales. In order to do this the rock surfaces must either have been kept clear and clean or have been prepared regularly for events including recarving. It must be borne in mind that while some of the over-carved surfaces can be thought of as palimpsests, like collaged images created anew through over-carving of additional images and motifs, it is also likely that the process of over-carving on some occasions and in some contexts could have represented replacement, defacement, or erasure. There can be no doubt though that these rock art maps in Valcamonica were display maps in just the same way that the world map of Agrippa and the Severan marble plan of Rome were, as will be discussed more fully in Chapter Seven.

---

[2] Bevan 2006: 161-171.

More difficult to understand is how the cultural shift inherent in this region coming under the direct rule of Rome following Augustus' victories over the *Camuni* and other Alpine tribes in the very late first century BC manifested itself. The founding of an urban centre, the *Civitas Camunnorum* at Cividate Camuno, probably required a shift away from upland or mountainous sites as loci of ceremonial activity to the forum and temples of this new centre. A number of Latin inscriptions have been found inscribed on rocks in the Capo di Ponte and Naquane area of Valcamonica, 'colonising' these particular rocks and possibly symbolically the whole upland region, but at the same time also suggesting that certain rock art sites were still somehow curated and returned to in the Roman period here, even if only for the incision of inscriptions rather than of pictographs. Indeed, a significant number of the individual rocks in Valcamonica would be 'Christianised' in medieval times by the incision of inscriptions and contemporary symbols: however, in Roman times the only pictograph image in the rock art of Valcamonica made at this period occurs on Rock 24 at Pagherina and consists of two men in duel combat, each with a sword and shield. The combatant on the left has been knocked to the ground, and is perhaps now dead. The other man is named in a Latin inscription below his figure, along with the word Victor. Possibly these were gladiators, maybe the commemoration or memorialising of a specific event in the amphitheatre in Cividate Camuno. So here in Valcamonica exchanges of geographical information were mediated through two different culture-specific art forms.

It is interesting to note that so significant was the series of victories by Augustus over the Alpine tribes during the campaigns of 16-7 BC that a large victory monument-the *Tropaeum Alpium*-was built at La Turbie in the hills above Monaco in 6 BC, the dedicatory inscription being at pains to list the names of all forty five Alpine tribes defeated by the Roman forces. It might have been thought that many of these tribes were quite insignificant in terms of their histories and the extent of their geographical territories-indeed, a few are only known to us through their naming in this particular inscription. The listing of names, their recitation, is hypnotic, an exercise in coercive control through naming. Even looking at the list was meant to elicit a sense of awe from the viewer/reader.

Interzone

The famous Etruscan bronze Liver of Piacenza is both part of a map of a body and a map of the heavens.[3] It provides remarkable insights into the ideas that probably helped shape and form Roman divination and, at the same time, adds information about the practicalities of its practice. This life-size model of a sheep's liver, found near Gossolengo in Piacenza province in the nineteenth century and now in the *Museo Civico* in the *Palazzo Farnese* in Piacenza, probably dates to the late second to early first century BC. The liver is flat, with three protuberances that correspond to the gall bladder, the portal vein (posterior *vena cava*), and the caudate lobe, and is inscribed on its upper surface-the visceral side of the liver. A small amount of writing is also etched on the underside of the liver. The upper surface is divided by incised lines into sixteen areas or zones, each of which bears an Etruscan god's name or sometimes names, and each zone is further sub-divided and labelled, giving forty individually annotated units. The sixteen main zones probably corresponded to the sixteen zones of the heavens as defined by the Etruscans, each the home of particular deities (Figure 8). Thus observations of areas of

---

[3] On the Liver of Piacenza see principally Van der Meer 1987.

Figure 8. The Liver of Piacenza. Etruscan, second century BC in *Museo Civico Palazzo Farnese*, Piacenza. Modern replica in *Museo Etrusco Guarnacci*, Volterra. (Photo: Copyright Jerónimo Roure Pérez).

disease, discolouration, or enlargement taken on the warm, probably still throbbing, liver of a dead sheep would be viewed within this preordained grid system and then transposed onto a grid of the skies and interpreted in cosmic terms for their relevance to contemporary life and events.

It is possible that the Piacenza Liver could have been a teaching aid of some sort, used in the training of divinators or as an aide memoire for a trained operative. It is also possible that the object was actually used on the spot during an examination of a sacrificed beast's organ as a reference tool. In whichever of these scenarios it was used its very existence tells us that the system of divination, no matter how much hocus pocus we might think was attached to the practice, was formally codified and when readings were given on the spot these were somehow informed by a system of knowledge, however flawed or arcane. While the Liver of Piacenza is a unique object, a similar item is in fact represented held in the hand of a reclining man, probably himself a *haruspex*, adorning the lid of a second-century BC alabaster cinerary urn from the Etruscan necropolis at Volterra, and now in the *Museo Etrusco Guarnacci* there. This suggests that such models were perhaps commonly in use among the ritual groups performing divination in Etruscan society, and maybe subsequently Romanised versions of these models were produced and used. The images carried by the Piacenza Liver which arose out of journeys between life and death, earth-bound realities and the heavens, between delineated ritual practice and the reading of warm, still-pulsing organs, amounting to both elegy and celebration, a sense of light and atmosphere, of both amazement and unease which have long affected the reception of these images, encapsulating the ambivalence at the heart of visual representations of place and space. Light and dark, surface and depth, movement and stillness: these were the formal and metaphorical oppositions which gave the images on the liver and the liver as an image itself their tension and power. But all was not solemnity, and a new kind of formal clarity would have been achieved, perhaps suggesting through use and process intimations of the brilliance of life as a brief flourish in time. It was not as if the images were intended as the symbol of a beautiful impermanence. Rather, they appeared as if vatic, the oracular voice of the outside world, articulating its otherwise secret meaning, with a phantasmagoric intensity and a completeness of conception that took it beyond the limitations of the diurnal. Of course there was ambiguity in some of these images, but not prevarication in their reading for interpretation: they sang the body electric, detailing a journey for which no itinerary was available beyond myth and belief.

The users of this prophetic object were also viewers, needing thus to be close readers of the images, who probably would have enhanced their state of consciousness as their experiences of these readings multiplied, achieving not so much a new state of awareness as a continuously expanding awareness of the world around and of themselves and their perceptive vision. What might have seemed at first sight like transient images must have made an impression on the mind, though the user/viewer might not have been aware of this at the time. Information would have been digested and the subconscious mind would have developed the impression into a quasi-visualisation, then the conscience would have moved in and with insistent pressure have made the viewer feel troubled unless they returned to the source. Worthy of renewed attention: without a return they may have been gently haunted by a strange sense of loss. But they did not live in a shrine, they lived in the world.

Other depictions of *haruspices* are known. A relief from Rome, now in the *Musée du Louvre* in Paris, is almost unique in carrying a depiction of the immediate aftermath of the pole-axing of an oxen at a sacrificial ceremony. The beast lies on the ground, turned on its back while a man leans over it, presumably to open up its stomach with a knife to allow its entrails and organs to be inspected by a *haruspex* who stands by to one side. At Ostia a mid-first-century B.C. votive relief to the hero god Hercules was dedicated by the *haruspex* Caius Fulvius Salvis.

It is interesting to note the existence of a Roman portable sundial from the atrium of the *Villa dei Papiri* or the Villa of the Papyri in Herculaneum and now in the *Museo Archeologico Nazionale* in Naples in the form of what has been suggested to be a side of ham.

To the Heart of the World

The Roman and Italian countryside did not always feature in the Roman imagination as envisioned through laws and proscriptive edicts. In earlier times the pastoral and political had met in idyllic harmony. If Virgil's *Georgics* painted a rosy picture of peaceful, bucolic country life under the benign reign of his patron Augustus, so some of the artworks on the emperor's own great monument the *Ara Pacis Augustae* or Augustan Altar of Peace equally contributed towards the politicisation of pastoral imagery and of an animal paradise in a similar way.

The *Ara Pacis Augustae* is one of the most significant surviving monuments from ancient Rome.[4] Restored and reconstructed today it sits within a glistening white, custom-built modernist

---

[4] The literature on the *Ara Pacis Augustae* is enormous. Therefore for a general discussion see, for example: Kleiner 1993: 90-99. For a detailed discussion of the vegetal imagery and its wider context see: Castriota 1995. For its place more broadly in Augustan art see: Zanker 1988: 188-192.

museum compound near the banks of the River Tiber and close to Augustus' mausoleum, a short distance away from its original location facing the *Via Flaminia*. Completed and dedicated in 9 BC the monument comprises a rectangular white Luna marble enclosure wall surrounding an altar inside, with doors or openings at both ends. The high enclosure walls are profusely decorated with relief sculpture.

The upper parts of the outer faces of the long sides of the enclosure walls are carved with processional scenes involving the imperial family and participating religious officials. A procession of Vestal Virgins also appears in relief on the inner altar itself. Due to damage in places and perhaps over-zealous restoration of some portrait faces not all the members of the imperial entourage can be identified with certainty, and even then there is not always accord among academics as to the acceptance of certain of these identifications. It is generally accepted that Augustus, Marcus Agrippa, Livia, her sons Tiberius and Drusus, Drusus' wife Antonia the Younger, their son Germanicus, Antonia the Elder, her husband Lucius Domitius Ahenobarbus, and their children Gnaius and Domitia Lepida all appear in the frieze on the south side of the precinct wall of the *Ara Pacis*. On the north side can be seen another family group whose members might have included Augustus' daughter Julia and his sister Octavia. The number of women portrayed here was unusual, as was the presence of children in the processional scenes.

Figure 9. *Roma/Tellus*, the *Ara Pacis Augustae*, Rome. 13-9 BC. (Photo: Author).

Figure 10. Vegetal decoration, the *Ara Pacis Augustae*, Rome. 13-9 BC. (Photo: Author).

Figure 11. Detail, the *Ara Pacis Augustae*, Rome. 13-9 BC. (Photo: Author).

The artistic programme of the monument aimed to demonstrate to the Roman viewers the legitimacy of Augustus within the continuum of Roman history and historic myth through the transformation of reality into symbolic political capital.

At this time Livia was not only wife to the first emperor but she was also a mother, but importantly of a child from her first marriage and therefore not a child of the emperor's. While messages about motherhood and dynasty were part of the propaganda programme of the monument, this was conveyed through another female image on the precinct wall, that of *Roma* or *Tellus* and by the more abstract vegetal decoration on the friezes that alluded to concepts of the purity of nature, and of growth and fertility.

The south-east panel, in the upper register to one side of the entrance, is dominated by images of female fecundity and plenty. At its centre sits a nursing mother figure, a personification, usually identified as being either *Tellus*-the Earth, *Italia*, or *Roma*. Less convincingly she has been argued to be Venus, *Pax*, or Rhea Silvia, the mother of Romulus and Remus (Figure 9). In many ways it is not actually important which of these figures she is because of the overall message conveyed by her being first and foremost female and a mother. She holds two naked babies. To either side of her are semi-naked females with their mantles blowing out and billowing in such a manner as to suggest that these are personifications of the winds. Behind and below the three women are luxuriant plants and flowers, while below them in the foreground are a seated cow, a sheep, and a large bird caught in flight (Figures 10-11). This is a scene of peace and harmony, of growth and abundance, of fertile crops, fertile mothers, contented animals, and contented children. This is the Augustan peace in an image. It is a seemingly non-political image that is actually intensely political.

To some extent Roman imperial culture looked for powerful signs to represent its ideologies, and in a beloved woman, the demure, reticent, and exemplary Roman matron, found its perfect expression. The image of the idealised woman excused or legitimised male mutability. Woman as transcendental and desirable, as a local spirit of place, was celebrated. The idealised woman, born of a river and of history, suggested Rome's association with nature, to form an explanatory figure. She was both society and nature, without there being any contradiction expressed about the possible incompatibility of the two. She belonged to sidereal time while most people had to deal in the quotidian. She helped allow a space in which politics, transgressive or not, might be practised. In an age of increasing political disorder the female image was often unmistakably an imperative, beautiful and at one with nature. So very often the style, composition, and subject matter of certain artworks were melded together in order to allow the viewer to look to the future, this being largely achieved by looking to the past and stressing the woman/nature duality that the figure of *Roma* herself encapsulated.

If the *Ara Pacis* was about Rome's place in Italy and about Roman lineage, both familial and mythological, then the famous statue of Augustus from the Villa of Livia at Prima Porta just outside Rome, and now in the *Musei Vaticani*, dealt with other kinds of political, geographical, and spatial relationships.[5] It is not the portrait element of the statue that will be discussed here, important and significant though that is, but rather its place in a canon of Augustan geography products that sought to give birth to the presentation of different ways of

---
[5] On the Prima Porta Augustus see, for example: Kleiner 1993: 63-67.

Figure 12. The Prima Porta statue of Augustus. c. 20 BC. *Musei Vaticani*, Rome. (Photo: Author).

seeing the world and Rome's place in it: at the very centre. That the Prima Porta statue's pose deliberately recalled Greek athletic statue types of the fifth century BC placed the composition in a Greco-Roman cultural category which identified the work geographically and philosophically according to Roman culture's chosen stylistic route of Greekness, as will be discussed in a later chapter. That the emperor was portrayed here as a military commander makes it a very different portrayal of the emperor than on the *Ara Pacis* where he appears as a non-military figure: that the Prima Porta Augustus in statue form is depicted wearing a highly-decorated metal breastplate or cuirass allowed the artist to present a great deal of information through the careful selection of images adorning the cuirass, and through the strategy of luring the viewer into 'reading' the breastplate in a way remarkably similar to the way in which in the microcosmic and cosmological readings of the decorated Shield of Achilles in Homer's *Iliad* and the Shield of Aeneas in Virgil's *Aeneid* were prompted.

Dating to 20 BC, though the Vatican statue could be a slightly later copy of the Tiberian period, this statue of Augustus is full size, the statue's raised right arm and gesturing hand, perhaps pointing a finger up to the heavens, making it appear larger still (Figure 12). The emperor

Figure 13. Detail of the decorated cuirass, the Prima Porta statue of Augustus. c. 20 BC. *Musei Vaticani*, Rome. (Photo: Author).

holds a spear in his left hand. Like many statues of this size the work required an integral support of some kind, in this case not the most common form of plain strut linked to the statue's base, here in the form of Cupid or Eros riding on a dolphin. This symbolises Augustus' power over both land and sea, and obviously alludes to his most significant military victory in the great sea battle at Actium. The emperor's claims of divine ancestry, linked to Venus, can also be seen as being laid out here for the viewer to absorb.

But as one looks at the chest of the martial emperor the decoration on the front of the cuirass draws the viewer in, conflating time (then and now) and place (there and here), by displaying scenes of decoration in a number of zones which can be viewed in a number of ways, taking in the whole composition perhaps or viewing one scene at a time (Figure 13). In the centre of the cuirass, the principal focus of the decorative scheme, can be seen a meeting between two military figures who stand slightly apart from one another: one is quite clearly an eastern barbarian general, though holding what is clearly a Roman military standard or *signum*, and the other of whom is either Augustus himself, or Augustus in the guise of the Trojan hero Aeneas, or a senior Roman military commander, or a personified figure representing the

Roman army, with a dog or wolf at his feet. It is generally accepted that what the viewer was being presented with here was a depiction of the handing back in 20 BC of the Roman standard lost by Crassus to the Parthians in 53 BC, a devastating and humiliating set-back for Roman power and might.[6] The return of the standard followed intense and prolonged diplomatic negotiation between the Romans and Parthians, and the handover was being presented here almost as if a military victory was being celebrated. But the scene is extraordinary for the very fact that the Parthian general or envoy is himself portrayed here, for seldom were those outside the Roman empire portrayed in Roman art except as victims in battle, and portrayed in a manner that in no way denigrates his dignity: he seems to be depicted here at the centre of the cuirass as an equal to his Roman counterpart. However, on the right of the Parthian envoy, as the surface of the cuirass curves away, is depicted a seated, captive female barbarian or female personification, perhaps representing *Gallia*, and to the left of the Roman emperor or commander sits another female captive-cum-personification, probably of *Hispania*. These figures were perhaps erotic politicians of a kind. The voyeuristic male viewer could travel to these lands in their imagination and could possess them in their mind, travelling through art to Gaul and Spain, onwards to Egypt, Dacia, *Germania*, and later *Britannia*. They would have learned how to inhabit these locations without being present in them.

Below the scene of the handing over of the military standard appear the god Apollo, on a winged griffin, and the goddess Diana, seated on the back of a stag. Further below still sits *Tellus*, or possibly *Roma*, holding a *cornucopia* and cradling two small infants. In the heavens above the scene involving the Roman standard appear images of *Aurora* on the back of a female personification, the sun god *Sol* riding in his four horse chariot, and higher above the sky god *Caelus*. There is some decoration on the back of the cuirass, including a trophy.

Thus within one artwork there are references, some overt and some covert, to different aspects of the geography of the world and cosmology of the world; on one axis action moves from the heavens down to earthly political action, and down to the ocean. Conquered lands within the empire are alluded to, while the proud Parthian envoy references lands outside and beyond the Roman empire. West and east are here united. The return of the Parthian standard and the allusion to victory at sea at Actium neatly balance what might be called landscapes of victory with the landscape of defeat as represented by proud Parthia. The overall work alludes to the Roman appropriation of eastern, Greek style and cultural tropes. A geography of familial lineage is evoked by the Augustus/Aeneas figure at the centre of the world and *Tellus/Roma* like the She-Wolf nurturing Rome's children (possibly Romulus and Remus) below.

Conceptions of memorial space in the city, highlighted by Aeneas' visit to the future site of Rome in the *Aeneid*, foreshadowed the role of topography as an instrument of power employed first by Augustus, though not without precedent in Republican times. Displays of geography products were probably not intended to be some kind of closing: rather they would seem to have been creating an opening, gesturing at a shared world. Open texts to be filled with meaning. Many captured the lived moments of human, diurnal time, taking it out of human time and relocating it in geological time: mobility and materiality together shaped geographies of devotion. Reinscribing Rome's cultural memories onto colonised spaces in this way was a deeply political act. Living within Rome's urban grid people were implicit in and dependent

---

[6] On Rome and Parthia see, for example: Reitz-Joosse 2021; and Rose 2005.

on this spatial context and how life had to therefore be lived there: the axial grid of the city and the horizontal and vertical planes of its architecture made its inhabitants like supplicants.

In his science fiction collection of eighteen linked short stories *The Illustrated Man*, published in 1951, Ray Bradbury used the narrative tattoos on the body of the titular *Illustrated Man* to relate the tales depicted in each scene. The stories centred around a number of themes, constantly returning to the idea of the local versus the global, even if that is played out on an intergalactic stage, and ultimately of ideas relating to presence and absence.

The Open Door

So, a trigonometry of journeying on the *Ara Pacis* defined like deracinated snapshots Rome imprisoned in its own legends, and perhaps unable to come to terms with an age in which it could so easily have become a stranger to itself in a climax of lacerating cruelty. Yet the contemporary production of images of other lands and peoples on Augustan artworks elsewhere in the city and on the breastplate of the Prima Porta statue of Augustus alleviated matters and broke the tension. We should view this accreted archive of images not as a window onto the past but as fragments from it, wild atavistic refractions, representing a collective cathexis, leaving a kind of morphic resonance for future emperors to work with and build on.

In Augustan geographical art empire was presented not only as the conquest of land and peoples but also of the discontinuity between individuals, as the city of Rome reconstituted itself as an oneiric location. Rome both hymned the *domus* and the complex places of transit that helped ensure its wealth-the Tiberside docks, the markets, the ports and so on, and displayed a map of Italy in the Temple of *Tellus* on the Esquiline Hill in Rome, as revealed in Varro's writings.

Like a digital nomad, the true identity of Rome under the empire came to be in its absence: the simple early lifestyle of its inhabitants was now largely absent, but perhaps continued somewhere else. People were able to inhabit the same space, to be physically proximate, and yet to live in different worlds. The spatial uses characteristic of the late Republic and early Empire were determined by its potential not just for sociability but also for secrecy and security.

In terms of reimagining something as big as the Roman empire at its greatest expanse artworks would later need to both catch the attention of the viewer and at the same time to restrict the field of vision. The art of freedmen and freedwomen also learned to engage in a demotic way with contemporary geography. For the artisan class of Rome the geography of its streets was not static: indeed, it might have been thought that the naming of a city street after the predominant commercial activity there might have set that street and its related activity in aspic. But as I recently pointed out in a book on the identity of Roman workers this was not the case at all.

Augustan geographic images represented the permanent reality behind the passing incident, presented as a kind of discovery. They oscillated between recognition and indeterminancy, an ambiguity that paradoxically assisted in the viewer's arrival at comprehension and understanding. This alternate play on holding and delivery invited associations, both physical and psychological, but closed nothing down due to an increase in clarity and spaciousness.

The land was often a metaphor for the body and vice versa. The untrammelled imagination of the work's creator was reflected in a subtle touch which uncovered interest and vitality in the most desolate spaces, where entropy seemingly had the upper hand. The art of the past, like life, is an object of observation in the words of Natalia Goncharova. Assertive and overtly powerful these artworks played a cleverly balanced game of revelation and concealment. The extreme economy and stylisation of many of the compositions was both vital to their success, and resplendent, striking in their assurance, radical form and content, and power. They signalled that there was something of importance beyond the limits of sight, enmeshed in a network of images and symbols as extensive as any grid, resulting in a monumental stillness.

The land/body/self was a source of creativity, but also a site of struggle and dislocation, a dark, beautiful, austere moment caught in time. Such images seemed to play with and subvert the viewers' expectations, being all about the positioning, with the deliberate side-lining of the motif in order to create an interesting dialogue with the rest of the visual space enclosed there. It was a process of exploration and experiment, not a strategy of settled expectation. Dealing with notions of transparency and solidity, presiding over a meeting of shapes and shadows in spatial complexity, the works articulated presence and an ineffable sadness in the transience of things.

Defining a sense of place remained crucial to the Roman articulation of the relationship between the heartland and the edgelands of empire. Geography products in the form of geographical images shared little in terms of their contexts and material base, each powerfully representing the ambivalent emotional register of wondrous difference which marked the disparate efforts of Roman artists to visually evoke the encounter with others. This art emerged not from a single point in time or space but from a confluence of discursive practices and image-making techniques, some with Italic rather than Roman roots, that were constantly remodelled to fit with Roman art's institutional and ontological structures and forms. While we might not identify with this practice in the way that contemporary viewers would have, our gaze is certainly partially mediated through their presence in the frame, their libidinal appeal negotiating the paradoxical valences of difference. Contemporary debates over which representational forms were best suited for the dissemination of spatial knowledge would have reflected the displacement of the desire for certainty.

When employed in imperial art the viewer could not experience geographical images here outside the framing discourses of the institution itself-they were in an ideologically-loaded space, the visual tracks left by historical observers. The study of Roman geography products draws upon the sedimented layers of evidence attesting to their ubiquity and subtlety, making them part of a broader picture relating to the portrayal of distant locales and ethnic groups, a process codified under Augustus but with deeper, earlier roots.

Pliny related that Pompey the Great commissioned the artist Coponius to produce fourteen statues of female personifications of countries and peoples he had defeated for display in the Theatre of Pompey as it became known and the *porticus* there.[7] Interestingly, Suetonius presents us with an outrageous account of a bad dream suffered by the emperor Nero in which these self-same personifications came to life and proceeded to menace the emperor, much to his distress, in a portentous harbinger of the violent provincial unrest that would mark his reign in some parts of the empire.[8] The images in his text came to life, their geographical specificity pre-empting distant events: bodies and places as one.

A number of precedents can be found for Pompey's didactic mission, and it can be set in the broader context of the use of certain categories of buildings and spaces as what has been called by Elizabeth Macaulay-Lewis ' a type of public, politicised museum'.[9] Alongside temples these were principally porticos and portico-temples, at once both an enclosed space and a controlled environment for the presentation of art, messages conveyed through art and their viewing, contemplation of, and reception. The first such 'political museum' was probably the *Porticus Metelli* built by M. Caecilius Metullus between 141-143 BC with monies accrued from the spoils of his Macedonian campaign and victories. Though nothing is known of the layout of the complex it was recorded that displayed there were the looted *Turma Alexandri* or Granicus Monument by the famous Greek sculptor Lysippus and a statue of Cornelia, daughter of Scipio Africanus and mother of the Gracchi. The *Turma* consisted of twenty five equestrian figures representing the *heteri* or battle companions of Alexander the Great: this was a very suitable work to publicise and commemorated Metullus' own Macedonian victory and to locate it both geographically and temporally in present-day Rome. The precise reasoning behind the display of the statue of the seated Cornelia here, now lost though its base is in the *Musei Capitolini* in Rome, and its juxtaposition with the *Turma,* is not clear, but not strictly relevant here. The *Porticus Metelli* was demolished and replaced by the *Porticus Octaviae*, named by Augustus in honour of his sister Octavia and constructed 33-27 BC, funded by spoils from his Dalmatian military campaign of 33 BC. Both the *Turma* group and the statue of Cornelia had evidently been seized and retained to be displayed here alongside a large number of other works acquired or commissioned by Augustus. The redisplay would have acted to completely recontextualise these two works and resituate them in the discourses of Augustan political ideology with regard to distant times, distant peoples, and distant places. These two 'political museums' therefore can clearly be seen to be the forerunners of the Augustan *portico ad nationes* and the other linked monuments and artworks of his reign.

Like Antonin Artaud's concept of The Theatre of Cruelty the combined raw power of the examples of geographical works discussed in this chapter might have forced the viewers, or some viewers at least, to become immersed in a performative rite that allowed individuals to process information about the world and to question and locate their own place in that world: in other words to perform their own identity in geographic terms. This language of space devoid of dialogue or explanation must have stirred powerful emotions and expressions. The idea of the spectator as an active collaborator as I am suggesting here cannot have been universal in all situations and contexts: sometimes he or she would just have been a passive consumer of images, oblivious to the information they could potentially convey.

---

[7] Pliny *Naturalis Historia* 36.41-42.
[8] Suetonius *Nero* 46.1.
[9] Macaulay-Lewis 2009: 1.

We will have to think about how those parts of Italy outside Rome came to be part of Rome itself and eventually part and parcel of Rome's project of empire beyond the Italian peninsula, as promoting ideas of these Italian regions and peoples becoming part of a kind of greater-Rome perhaps paved the way for later, more complex projects of mental integration of more distant lands and peoples.[10] Certainly infrastructural projects helped in connecting and thus somehow uniting disparate points on a map, quite literally, and not necessarily just as sequential names in an itinerary. We can also think about aqueducts as having a unificatory role in a way: the *Aqua Appia*, Rome's first aqueduct, brought water from Campania and connected the system to the cities of *Magna Graecia*. This late fourth century BC project was conceived by Appius Claudius Caecus who also masterminded the surveying and building of the Appian Way. Heading out south-east from Rome the *Via Appia* passed across the Pontine Marshes, continuing through Capua and beyond the Caudine Forks, on from there to Tarentum, and then to Brundisium. So the peoples of Italy in some cases became early ethnotypes for presentation to the Roman people that was intended to turn them from foe to friend, or at least make them somehow understandable and accepted. This did not mean that major events of history were recast and rewritten; the Punishment of Tarpeia and the Rape of the Sabine Women were still deemed appropriate moral tales to set before the women of Augustan Rome in visual form at the *Basilica Aemilia* in the *Forum Romanum*. We can clearly see different manifestations of Roman masculinities here, recreating their own cultural heritage despite, and even within, the silences imposed by the violent ruptures of both mytho-hisorical and colonial or imperial history. Here women's bodies became globes of fear, countries the viewers would probably never know or visit. These quite dramatically described the tensions between forces of masculine innovation and constraint as they were revealed in these narratives, demonstrating the opposition between cultures of origin and destination, the disruptions of cross-cultural refraction which recognised both continuity and schism.

After a while though there was no need for any deployment of Italian, non-Roman ethnotypes, and in fact people seem to have been replaced by ethnoplaces or ethnospaces, as something like the Russian idea of 'the near abroad' took hold. Indeed, as has been pointed out by Diana Spencer, some of the Roman writers on farming and the countryside, such as Varro, came to present certain Italian regions in terms of the speciality agricultural produce grown or produced there.[11] Local identity inevitably became subsumed within a greater Italy that was to all intents and purposes a greater Rome. Once Falernum became more famous for its vines and Campania for its spelt it would have been these premium regional products transported to and on sale in Rome that evoked memories of place in Roman minds rather than historical encounters and the dead hand of the past.

But the main focus of this study is the Roman period itself, rather than the era which preceded it in Italy. It is difficult to see how things moved from there to here. How pictorial visualisation at focal public rock art sites like Bedolina might have led to the production of written geographies by elite Greek and Roman men such as Strabo.

If it is accepted that pre-Roman Italic cultures produced artworks such as the Map of Bedolina and the Liver of Piacenza to present ideas about spatial identity, then it can be argued that

---

[10] On early Roman-Italian relationships see, for example: Brock *et al.* 2021; Dench 1995; Terrenato 2019; and Van Dommelen and Terrenato 2007.
[11] Spencer 2010: 73.

the Augustan Prima Porta statue and the *Ara Pacis* dealt with similar themes, both earthbound and cosmic, in a more codified way. Augustan Rome was a new, rebuilt city. The sight of Rome renewing itself and its buildings all around at this time must have suggested to its citizens that paradise would soon be lost, in terms of the obliteration of the underlying, visible topography of the site. This was all part of the project for Leaving the First Century BC, to adapt the title of the influential collection of writings by members of the Situationist International, by the reimagining and redesigning of the world through spectacle.[12] To get here both 'rational' and popular systems for the dissemination of geographical knowledge had to be recognised as potent tools employed in helping to birth a new world.

While many such geographical portrayals might simply appear to have been considered as genre pieces, others were uniquely inspired and detailed, representing imaginative responses beyond simply the recognition and acknowledgement of difference. As a medium for communication these could have signified the specificity of an encounter-contact generating souvenir, but which became solidified and validated by cataloguing and museum-like display.

The passing of time would have changed contexts of viewing, a point made in many places in this study and something that cannot be stressed enough, when the categories of stranger or other could change from less to more benign, generally reflecting the shift in the Roman definition or vision of empire from ownership of human labour and enslavement to the ownership and exploitation of land and raw materials.

The geographic images were never anodyne, never without meaning: the juxtaposition of one image to another might have informed or educated the viewer, or perhaps confounded and confused them. Some of this visual information must have sparked recognition, if not sometimes actually a revelation. This was not a sea of disembodied, uncontested, and uncontextualised images produced to elicit mystery. Understanding what was seen was though almost a process of demystification, because seeing was here an active choice. Blooded by immersion in a culture of images itself, the viewer would have developed the capacity to see the grid of forces in which the images were enmeshed. At once both didactic and poetic, rigid and fluid, these images were locked in to ideologies of identity. Rather than problematising the dominant Roman culture they provided conceptual mechanisms for viewers to address a violent and traumatic history. Through linking knowledge attained via geography products, the history of the Roman empire, cultural and social memory, we can understand the crucial role that these played in forming the understanding of the inter-relationship between place, time, and identity. It was both a fragile and disruptive process of presenting ideas, in that those ideas would not always necessarily have been understood or interpreted correctly: on the surface, to us many artworks in the geography series are beautifully poetic, but they were intended at the time to be deeply political and ideologically informed. Difficult pathways to self-actualisation. Unlike other empires and colonial powers in history Rome never mythologised a peaceful settlement of the provinces of its empire and never denied the brutality of its conquests and frontier wars, though there might have been an epistemology of ignorance among some emperors and some of the Roman elite class.

---

[12] On the Situationist International *Leaving the Twentieth Century* see: Gray 1974.

If not a planned sequence then at least the geographic images discussed in this book constituted referents to a precise and compressed bundle of distinct but elided events. This poetics of space acted as a metaphor for the art itself as a potent reality, as constructive of meaning, as regenerative of myth. It presented a means for the reinstatement of a mythic dimension to quotidian actuality. It constantly evolved in dynamic recurrences and connections that were essentially anagogic rather than purely illustrative. Here in the marble, brick, and stone heart of the city of Rome it was the natural, the fluid element, that embodied the essential spirit of place: restive, evanescent and self-renewing, irrepressible nature asserting itself with variegated exuberance.

The concept of 'the geography product' as a bearer of meaning and as a unifying force when it became part of a series seems persuasive. The theory of heteroglossia, as propounded by Marxist linguist Mikhail Bakhtin, could surely equally be applied to visual works. The multiplicity of languages and verbal ideological belief systems in the novel or in the case here in Roman art are not just formal, standardised elements of cultural production reflecting or derived from a separate social world: rather, they are acting and shaping elements of it and can be created or manipulated to do just this. Signs such as these are not static though, and more often than not would have had contested, varied, and evolving meanings, though such meanings were not necessarily relative and potentially limitless. It was the interaction of numerous individual elements, each element being multi-accentual, that would have dictated their reception. The idea of *détournement* posited by the Situationists involved a diversion of pre-existing aesthetic elements in order to politically problematise and resignify them, as could have occurred here.

Throughout this book it will be argued that in the late Republic a sense of uncertainty and dislocation, both caused by and exacerbated by the political instability of the time, greatly affected the discourse around Roman identity. It can be seen as no surprise therefore that this period saw a proliferation of what might best be called antiquarian studies of Roman culture by writers such as Cicero (*De Re Republica*) and Varro (*De Lingua Latina* and *Antiquitates Rerum Divinarum*), studies that tried to anchor Roman identity in a present past, immured against drastic future change.

In the next two chapters attention will be turned to the significance of a sense of Rome as both place and home, and to the creation of a physical manifestation of Rome's mythological origins, as a prelude perhaps to the more codified use of geography products to explain and explore Rome's place in the wider world.

Chapter Two

Strangers in a Strange Land

It will be argued that the geography products which form the subject of this book were more than a simple sum of them all considered one by one. There would seem to have been the potential for something more total, more full of meaning, the creation of a world that viewers could inhabit. Both a metatext and a paratext were created over and alongside the images, the paratext being particularly interesting from the point of view of this present study in that it marked zones of transition and transaction where meaning was generated. The images' articulation of anomie was key to their constant reappraisal.

The Romans did not set out to challenge Greek ideas and understanding about cartography as an objective art and science but their stances showed the way in which it became entrenched in Roman colonialism. What could not be mapped in the conventional sense became something else, as subjectivity inevitably created lacunae. Documenting the environment left out the poetry, mystery, and the marginalised.

It should not be a surprise that Rome became a constantly changing point in a fixed world: viewers must have thought that this geographic art asserted itself as an instrument of knowledge, a microscope, a telescope, creating ideas that had grandeur and exalted spaciousness enough to foster and provide some kind of spiritual sustenance. It contained multitudes, dead-end obsessions with surfaces and optics, and as the flow of juxtapositions alternately deepened, widened, and extended the series, the viewers would have felt a duty to always react and re-act, to rethink everything, to reimagine everything always, to imagine everything *everywhere.*

The skill in the creation and deployment of geography products in the form of visual art lay in tempering anxiety about transformation by stressing factors such as origin, foundation, and coherence. They formed an archive and, as John Berger has said archives are a way of keeping 'the company of the past.' Exploring one you enter a place where the past is still in the present. Roman art was defined as much by its exclusionary bar as by its ideological aspirations. There was no place here for bell hooks' concept of 'the oppositional gaze', no opportunity to reclaim space, to create a context for transformation, a decolonialised perspective.

As afterimages these had a powerful life of their own, an afterimage being not the thing itself but rather the memory of it. These series of images and afterimages echoed others, their sense of containment within the city's bounds being altered by repetition, somehow making them vehicles for an imaginative and physical sense of vastness. Depth, distance, and proximity were embodied by sensation. Rome's natural topography of hills came to act as a metaphor for adventure, freedom, and divinity: a wholeness of vision was produced by uniting disparate geographical and imaginative parts. As Rome expanded, so did the complexity of geography products change. In order to recognise the allure of border crossings, of a new frontier pulling away from the familiar and journeying into and beyond the world of the other, ideological art citing geography came to caution against un-introspective voyeurism while at the same time embracing it, devouring both difference and the different without having to experientially engage with the other.

If Rome's economy was placed in a world system its political and ideological world was necessarily smaller. The flow of materials and energy from the empire's resources through the city's economy was a 'throughput' of a very particular kind, borrowing here a term coined by the ecological economist Herman Daly. While this throughput had a concomitant fallout in terms of damage to ecosystems and the depletion of natural materials it was the human fallout that Rome sought to address and explain. This was not sustainable development in any sense of the term: but it was, however, manageable through the presentation of visual geographical information. All natural resources are degraded when used in economic activity on this scale.

Staging critical interventions in this way through geography products opened up new possibilities for interrogating the intersection of collective and individual thought, fostering a negotiation between local, national, and imperial intellectual spheres and positions. The cross-pollination of many different modes of representation of the same thing was galvanising. Deep mapping, a multi-scalar study that combined distant with close readings blended interest in general trends and in details. The simultaneous deployment of multiple analytical scales contributed towards a geography of place, of peoples, of tension. The works' creators and commissioners did not know that some of their works would later be considered canonical, a corpus of place and the peripatetic body. Compositional strategies attempted to make sense of things in nature. The foreign or foreigner was not something that could then be tracked down and marked on a mental map: the process in fact fostered a sense of imaginative ownership over a foreign land, a meditation on the networks which linked mental and environmental spaces. Thus deep mapping the complex sociocultural and geographical interactions that were taking place throughout the Roman period and throughout the Roman world allowed a discourse to emerge about the relationship between the cityscape of Rome and what lay both beyond and within it. This raised a series of questions about sensation and knowledge, reality and representation, words and things: viewers were encouraged to place themselves in imaginative and embodied conversation with the source of the image and the cityscape which effected a deeper sense of belonging, as there morphed into here and the lexeme of Rome became less pronounced in its isolation and exclusivity.

Even though there were various routes for those who were not from Rome to become 'Roman' that did not necessarily mean that there was a unity among those deemed Roman, or unanimity in their thinking. Yes, there would seem to have been an over-arching architecture to being Roman, a binding thread that united rather than divided, but it must be borne in mind that there are many signs that within that apparent unity there were also underlying divisions. Ian Haynes has pointed out how the Latin language in the Roman military allowed a fighting force of people from Rome, Italy, and from regions around the empire to operate effectively and largely efficiently.[1] But at the same time he has flagged up the comment made by Tacitus in his *Historiae* regarding the inherent differences and thus seeming incompatabilities between the armies of the AD 69 emperors Otho and Vitellius based upon the snobby observation that

---

[1] Haynes 2013: 301-336.

Vitellius 'whose habits and speech were so distinct' was an outsider: not one of us. Along the same lines we can cite Cicero's dismissal of the delegates from Gaul in the Roman Senate.[2]

Art was the thread of this activity of defining place and flowed through the empire and through the centuries with the authority of a river. To walk through Rome was as much to walk through the past as the present, as the past then, as now, was rarely immune from decay.

The Ceremony of Innocence

In order to introduce the book's main theme of visualising belonging and alienation-of *here* and *there*-contemporary written ruminations on the matter will first need to be considered in this chapter. Attention will then be turned to the Romans' material creation of their own mytho-historical geography in the city of Rome and on the subsequent use of geography products to enable the understanding of lands and peoples beyond. In Rome history was the future once: these images were creatively cast into a future yet to come.

A study such as this should really adopt a variety of forms and methodological approaches, including theoretical speculation, detailed historical research, essayistic criticism, memoir, poetic contemplation, and interviews and conversation with texts. The deployment of geography products in the form of images contributed towards a number of projects of world-making in Rome, as a kind of cinema of foreshadowing, redundancy, character arcs, typage (in the cinematic casting sense), conflict, climax, recurring motifs, and ellipses. All this must have had a normalising effect, creating in Rome a distanced place of exile and difference: home indeed was something that Romans took with them wherever they went.

What was presented at the centre, that is in the city of Rome itself, was not issue-based art and generally not concerned with the experiences of those beyond: experiences of environmental damage, contestation of territory, and patriarchal imposition. These social realities relating to land use, contested territories, environmental pillage, the overuse of natural resources, and nationhood in places remote from Rome now became broken up by Rome's global ambitions. Women's fates were often foregrounded in the art, in quietly effective meditations on violence.

That cultural and geographic identity were symbiotically intertwined and in need of almost constant redefinition in ancient Rome is well illustrated by a dilemma faced by the late Republican Roman politician, lawyer, and writer Marcus Tullius Cicero (Figure 14). Ostensibly writing in praise of his contemporary and friend Marcus Terentius Varro, Cicero was here making a number of significant and telling points about Roman identity at the time, the 40s BC:

> 'For when we were passing through and wandering as if visitors in our own city, your books led us back home, so that we were able at last to recognise who and where we were.'.[3]

Passing through, wandering, visitors, led us, back, home, and *where.* This is at once a simply expressed sentiment, an encomium of praise, and at the same time a highly-complex reflection on the nature of place, space, and identity. Cicero firstly noted the existence of a kind of urban

---

[2] Tacitus *Historiae* 2.37.
[3] Cicero *Academica* 1.9.

Figure 14. Portrait bust of Cicero. First century BC, Rome. *Palazzo dei Conservatori*, Rome. (Photo: Author).

alienation, that one could feel like a visitor in one's own city, indeed feel as if one was not at home there. This suggests both an inability to cope with change and perhaps a fear of change, of estrangement. The political instability of the times may well have contributed further to this sense of loss, this existential displacement. There is no greater feeling of disorientation than waking from a dream in one's own bed and momentarily not knowing where you are. For Cicero Varro's writings provided an anchor, a sense of stability, and allowed the Roman reader of the works to understand his or her place in the scheme of things, in the city, in the wider world. It allowed the individual to restate and possibly rediscover their identity: this is who I am, this is where I call home, this is my city. I am a Roman: this is Rome.

This dreamworld, of a stranger wandering in a strange land, is one that others must have moved through and have gone through too: streets, buildings, monuments, and statues acting together as a visual bombardment perhaps not deliberately designed to conjure deep feelings of inadequacy and desire but which must have triggered such responses from some individuals in certain situations, at certain times. This long seduction could have led to anomie and ennui, to familiarity and, beyond that, overfamiliarity. Rome itself was an artwork, part of how contemporary reality was built, and indeed made anew each day, as Cicero suggested. The vision of someone walking the streets of Rome, adrift and troubled, the city depopulated as if on a Hopper or De Chirico canvas, is a seductive image.

This short passage can be mined further for other suggestive motives for Cicero's confusion and absolution. In mentioning home, or rather being back home, along with very active states like passing through, wandering, and visiting, Cicero would appear to be presenting this episode as the account of a journey made, with Varro as a kind of guide, indeed almost a spirit guide of sorts, unbodied and represented by his written words alone. The great significance of journeying in Greco-Roman mythology will be considered in detail later in this book, and the wanderings of Odysseus, Hercules, and Aeneas only made sense in terms of them returning home or finding/founding a new place to call home. Each of these myths, in their retelling, was intended in some way to anchor Rome's mytho-historical past to its historical present and, like a journey, to demonstrate how Rome had got from there to here, and to explain its position and presence in the wider world, not just in Italy. In the *Aeneid* Aeneas is shown how Rome will develop and grow, how the place at his journey's end will prosper and thrive. Yet here we have Cicero confessing to being adrift in Rome, indeed adrift *from* Rome, and in need of guidance and revelation, like Aeneas.

Indeed, some modern academics maintain and endorse Varro as having been one of the best guides to being Roman.[4] Though we do not know which of Varro's books in particular provided Cicero with such great solace and brought him 'back home' it is likely that their general mix of myth, history, topography, linguistic analysis, and philosophy together worked towards this effect, this re-centring. In other words they were 'a geography product', helping to define identity, to define Romanness, what it meant deep down to be Roman. That Cicero may have felt as if at times he was but 'a visitor' in Rome almost places him in that category of other, of outsider, of alien, of foreigner, of non-Roman barbarian. Though born in Arpinum to the south-east of Rome Cicero's family was well connected in the city through his father's membership of the equestrian order, and Cicero received his education in Rome where his talents were soon appreciated and celebrated.

But it was not only *who* he was that Cicero had lost sight of, but also the question of *where* Rome was, not in any sort of cosmic way but rather in terms of what kind of city it was and its place in the world at this time. He felt that spiritually he had journeyed away from it without ever actually leaving, though of course physically leave he did on many occasions. The articulation of this dual sense of confusion is fundamental in furthering our understanding of how and how rapidly the definition of *Romanitas* was changing at this crucial time in its history, with the Republic drawing to its end. The strength of Rome in Cicero's time was its view that *Romanitas* was a cultural rather than ethnic concept, a definition of plurality if not necessarily inclusiveness. Being caught up in a process, with no end-point in view, was almost inevitably bound to engender bewilderment and disorientation to a greater or lesser extent. The discourse around Rome might often have been about defining and refining *Romanitas*, but *Romanitas* was itself also a discourse of course.

---

[4] Spencer 2019.

If a cultured, aristocratic Roman such as Cicero, the very definition of an insider, could lose his way in this manner and be in need of reorientation, then it is apparent that similar feelings of dislocation could have been common in the city's population. Some there would never have viewed the city as home in the first place: incomers, outsiders, visitors, the enslaved. This is perhaps why there emerged strategies for the use of 'geography products' of various sorts to direct or redirect thoughts and notions about the city and to bolster the identity of the city and its inhabitants. This brings to mind David, the narrator character in James Baldwin's novel 'Giovanni's Room', who declares that: 'Perhaps home is not a place but simply an irrevocable condition.'

Feeling unmoored from one's own purpose, feeling like a foreigner in your own life was and is a timeless conundrum, with fractured and hallucinatory images slamming image against image, chopping up time, reordering and reversing it. Sex and death, primal images. Narrative exploded. It foregrounded the fact that the viewer was looking at art, foregrounding the brute power of the viewers' edits to contract or dilate time.

The crux of the matter was how it felt in the memory, its mass and metaphor. The art looped around, retrenched, made tiny adjustments that did not alter the basic language, alternatively seductive, mainstream, and antagonistic. The Romans became particularly adept at projecting different kinds of masculinity through art, from the almost parodically effete to the unforgivingly tough. There was continuity here but not continuation as such. Isolation and solitude had begun to assume a significant place in Roman culture, but it came to be replaced by a less hermetic, not shifting vision of worldliness and world-weariness. There was a mystic's keen sensitivity for the sublime which was like a secret river under Roman culture, an aesthetic of endurance and repetition, of meditation. The idea of beginnings and endings being contrivances. It was a narrative illustrating Rome's own evolution.

If Varro's oeuvre can be said in this way to be all about origins and directions, ultimately it is more about destinations. To travel is to hope to arrive. Unfortunately, the survival of Varro's once vast corpus of attested and referenced works is surprisingly slight and patchy, so the overall effect of its unifying themes can only be imagined by us today. If his endless quest to search for Roman origins, of words and the Latin language, of people, places and monuments eventually constituted an almost encyclopedic work about what it meant to be Roman then we must accept Cicero's verdict on the evident success of his endeavour. In discussing world origins as well Varro broke the false cosmology/geography boundary and united the earth and heavens.

Rome's origins he defined in natural terms and it is in fact Varro who first suggested that the city site was constituted by seven hills -*montes*. As Diana Spencer has noted: 'Varro's built Rome will….give monumental form to what nature had already laid out. Rome's walls will frame and define the hills…as signs of a unique environment and a distinctive place.'[5] In his *Antiquitates* Varro put flesh on the city site.

I do not want to discuss in any detail issues around the processes of banishment or exile from Rome and resulting feelings of dislocation as Cicero himself must have felt, forced into exile

---

[5] Spencer 2019: 135.

for a time by Clodius in 58 BC. Again, famously, the poet Ovid was banished to Tomis on the Black Sea in AD 8 by the emperor Augustus. He died there nine years later in AD 17, failing in all his attempts to lobby for a return to Rome. Many other notable Romans were forcibly exiled too. Julia Hillner has mapped the places from which women are recorded as having been banished, along with the places to which they were banished, and indeed in Late Antiquity the banishment of women was a common phenomenon. Between AD 390-641 a woman was banished in this way once every five years on average.

It was a quite terrible and tragic irony that following his murder by beheading on the road between Rome and Tusculum in 43 BC Cicero's head and his hands which had also been chopped off at the scene of his assassination were brought back to Rome and placed on the *Rostra* in the *Forum Romanum*. His body was thus both at Rome and away from it at the same time, both somewhere and nowhere.[6]

It is indeed again in the writing of Cicero that we can find another example of the complexity and sinuous subtlety of Roman identity at this time.[7] In his *De Legibus* he set down a conversation, real or imagined is uncertain, between himself and his great friend Atticus as they walked on Cicero's beloved estate grounds at Arpinum, seeking to make a point about what may best be called here ideological geographies. Cicero called the Arpinum estate his *patria*, presumably meaning both home and homeland here, eliciting from Atticus the question as to whether one should only really call Rome, their *communis patria,* by this name. Could one have twin *patriae* he asked or must one make a choice? Interestingly Atticus illustrated the point by asking if Cato the Elder's *patria* was Rome or Tusculum.

Cicero replied: 'I think that both Cato and all other municipal men have two *patriae*: one by birth (*naturae*), and one by citizenship (*civitatis*).....Therefore, we consider as our *patria* both the place where we were born, and that place by which we were adopted. But that *patria* must be pre-eminent in our affection, in which the name of the *res publica* signifies the common citizenship of us all. For her it is our duty to die; to her we ought to give our entire selves, and on her altar, we ought to place and to dedicate..all that we possess'.[8]

Procession

The creation of a world explained by a collection of geography products was not merely random: their creation set up processes and embraced constraints, almost as if employed just to see what would transpire. The kaleidoscopic perceptions engendered in Rome and at other locations in the Roman world often probably managed to be close to the fractured way in which we all experience reality. Thus there was no attempt here to tell a comprehensive, linear story: neither of lives complete in the sense of being geo-locatable, nor its most interesting fragments. Rather, it was an extension of the facade of Rome, the city's character, and a glimpse of the longing, loss, and desires that shaped it.

The majority of the geography products produced and displayed or used in Rome were in most respects official things, created quite specifically to say something about the Roman

---

[6] Hillner 2015.
[7] On Rome's Italian identity see principally Dench 1995 and 2005.
[8] Cicero *De Legibus* 2.5.

state and its place in the world. Yet quite a sizeable minority of geography products were created and consumed by private individuals or were used or set up in private spaces and places. There was an unerring eye for figures stranded in the landscape, no longer interested in itineraries, existentialism, and geographies of failure. Rome was a city perpetually and always on the cusp of change. If Virginia Woolf's insight that 'books continue each other, in spite of our habit of judging them separately' is true, and I think it undoubtedly is, then this is all the more so with a series of images each on the same theme: the after-image created itself had an afterlife. An urban thoroughfare could become a site of latent social exchange and potential transformation, turning script to audition, and testimony to profile, blurring boundaries, breaking down oppositions, and taking its means as its end.

Art sponsored by the Roman state fostered a relationship between noise and meaning on the same wavelength just as Republic was turning into Empire, and silence of a kind, not to say acquiescence, was setting in everywhere. This called attention to how the most intense connections could come launching unexpectedly out of the background, fostering a competing sense of anticipation, longing, unfilled possibility, and a sense of hesitation, or at least hesitancy. Lives could connect through the shape of a statue or the sound of a voice, coursing through everything that was somehow lacking: what was implied but unsaid, that was trying desperately to speak to someone else, somewhere else, made a connection. The Roman state offered not just constant triumphs but a suspicion of change and the paralysis of self-doubt, a potent mix, but in between the gaps the viewer had to fill out the polemic. In her 1967 study *The Aesthetics of Silence* Susan Sontag wrote that: 'With the passage of time and the intervention of newer, more difficult works, the artist's transgression becomes ingratiating, even legitimate', so that in the case of Roman art we might suggest that the retrospective effects of extremity shallowed out over time, eventually often becoming simply routine or anodyne. What was then public art became all too often a disguise for the monologue of power, the concept of art as a prophetic tool. Some artworks were not making a paradoxical point by performing authenticity. In terms of the potential criticism, the sameness of some artistic presentations flattened out or obscured their impacts and meanings. Robert Rauschenberg's White Paintings of 1951 are all so much the same but each is very different. In confronting the sameness the viewer can find ways of constantly seeing something different, that living is a way of filling in the gaps between us and everything else.

A corpus itself is something altogether much more than a simple sum of its parts. What the viewers could react to was something more total, more full of meaning: suffused by the images created by the artworks they became a world for the viewer to enter and inhabit. Sometimes they bypassed convention to speak directly from a wild, unfiltered vision. Therefore we must ask if it was there longer, was it more potent than an ephemeral image? This probably gave a greater potency. The image must have seemed particularly expressive of the body or people or place, of geography. It gave the viewer access to it without mediation or explanation and succeeded by manipulating space and time and breaking down the conventional structures of time and linear narrative, reassembling them to create new meanings, new places- *here* and *there*-collaging impressions and breaking temporal relationships. Like a cut-up, each viewed item could be recombined with others to disrupt the linearity of thought, each splice a fissure through which an imagined future could leak. Certainties were evidently dissolving and new worlds coming into view, even if only as ideas of places, notions of space, visions of other peoples and other worlds. It constituted taking something familiar and presenting it in such a way that it had the potential to change viewers' attitudes to the world they lived in. Everything was about not necessarily a refusal, but a reversal.

Unfortunately I did not see the Italian photographic exhibition *Roma Negata: Postcolonial Routes of the City* of 2014 and my discussion of it here has had to rely on a viewing of its published catalogue and a secondary account of the exhibition's content and impact. Curated by Rino Bianchi and Igiaba Scego the principal theme of the exhibition was the contrasting of a hoped-for multicultural Italy today with its fascist era state, achieved by the juxtaposition in photographs of Italians of African descent with well-known fascist monuments, buildings, and sites in the city of Rome. A linking text in the catalogue presented a meditation on Italian colonialism and its marks on Rome by the recording of a number of routes through the city chosen and walked by curator Igiaba Scego. This reclamation of the streets, these interventions (to use a Situationist term), these reconfigurations and discoveries, produced remarkable images and deep resonances of unsettling histories and disturbing geographies.

The relevance of the *Roma Negata* project to the present study is obvious but not necessarily clear in terms of comparative balance.[9] Its consideration raises numerous questions both about fascist and post-fascist Rome and by extension about ancient Rome and its sites and monuments. The human subjects in these striking photographs have been posed at the chosen locations to make a point about reclaiming the past in the present and about the constant recontextualisation of monuments according to changing ideological positions. If Rome's imperial histories became embedded within urban geographies, monuments, and institutions then the subversion of this dominant and dominating historiography was possible through a viewer's mental reconfiguration of each site that made up the series. The resulting unsettling geographies were as valid as the promulgation of an official political narrative. The kind of decolonising interventions staged by the *Roma Negata* project might have also occasionally occurred in earlier times. If a monument or monuments was/were intended to generate and somehow guide critical conversations about Rome's power and extent then their didactic purpose took precedence over their aesthetic value, and might also have encouraged the emergence of a counter discourse. Presence could sometimes signify absence and intensify feelings of loss. One wonders if the feeling of emptiness experienced by Igiaba Scago at *Porta Capena* in Rome when considering that the looted Ethiopian Obelisk of Axum had once stood there might have been similar to an Egyptian visitor in ancient Rome viewing the city's repurposed and thus recontextualised, Egyptian obelisks.

An Immersive Past

But if Cicero's two sets of comments discussed above represented a literary response to the discussion of Roman-ness on the other hand it can be seen that the city had been trying to come to terms with its mythological and real identities by materialising certain episodes of its mytho-historical past within the built geography of the city itself. If the reign of Augustus saw the systematising of presenting geography products in a didactic manner intended to inform certain ideological positions, then earlier attempts to place Rome spatially in the minds of its citizens appear less calculated and less programmatic. This was not an attempt by Augustus

---

[9] On the exhibition *Roma Negata* see: Greene 2023.

Figure 15. Wall painting depicting treaty negotiations between Romans and Samnites. Esquiline Hill, Rome. 300-280 BC. *Musei Capitolini Centrale Montemartini*, Rome. (Photo: Author).

to identify with non-Roman peoples but a technique to somehow mythologise them. Rome conquered Italy and Italy became Rome (Figure 15).

Though undoubtedly it was at the Theatre of Pompey that the first series of images of subject peoples or subject countries came to be displayed, thus creating a visual geography of Rome in the wider world, other earlier geographies existed. Here, however, attention now will be turned to the creation in the city of a mythological geography of Rome whose purpose would seem now to have been to locate Romans in a secure historico-mythological and geographic place (Figure 16).

Roman society itself took a spatial turn, with the need to present visualisations of the encountered world to the city's population, drawing attention towards certain places and peoples, all of whom now existed within an attention economy. This parade of images, this series of statements about place, became something to check out at regular intervals to

Figure 16. The bronze Capitoline Wolf, still widely considered to be of an Etruscan date, with Romulus and Remus added later. *Musei Capitolini*, Rome. (Photo: Author).

see how values associated with imperialism and masculinity stood. The more modest and unassuming these images seemed the larger their rhetorical gestures loomed. These potent, almost oneiric images balanced the archetypal hero's quest with suppression of the overt sexism of Roman imperialism.

There was an evident interest in the disruption of space, often using replication and personification to do that literally, creating a sense of the cyclical, sometimes like more of an ellipsis. The Trojan/Aenean legend was surely a kind of spiritual song-line across Roman time and space, creating an heroic atmosphere of a lost empire with an unwritten future. This risked an easy nostalgia for fictional times. As quiet and measured as the city of Rome was, it must also have had the feel of circulation, of a constant churn of people and goods coming and going: a city of influences operating beyond the confines of traditional maps and mapping of knowledge and power as we understand it today. Within such a diasporic world places were connected by memory, fantasy, or myth, shared history, and exchanged culture. There has been identified by art historian Tobias Wofford a specific kind of diasporic temporality marked by anamnesis, that is a simultaneous looking to the past and the future, to roots or origins, and to a visionary path yet to unfold.

Figure 17. A Proto-Etruscan cinerary urn in the form of a model of a contemporary hut, perhaps like the *Casa Romuli*. Alban Hills, Lazio. 900-800 BC. British Museum, London. (Copyright Trustees of the British Museum).

Astonishingly, one could visit Romulus' hut on the Palatine, walk down to the *Forum Romanum* and sit under the shade of the historic *Ficus Ruminalis*, before walking down to the banks of the Tiber to the shipshed that housed the Ship of Aeneas. Could you enter inside the Palatine hut? Could you go onboard the ship? This will be considered below. This all rather brings to mind Guillaume Apollinaire's wry observation that in the Rome of his day even the cars felt ancient.

One of the most curious buildings or monuments in ancient Rome has to my mind to have been the so-called *Casa Romuli* or House/Hut of Romulus, sometimes called the *Tugurium Romuli*, of which there may well have been two versions, one on the Capitoline Hill and another on the Palatine Hill[10] (Figure 17). This curious structure (or structures) was written about by a number of Rome's historians. All kinds of contradictions are inherent in the very concept

---

[10] On the *Casa Romuli* see, for example: Balland 1984; and Siwicki 2012. On legendary Rome more broadly see, for example: Cifani 2018; Crofton-Sleigh 2016; Fantham 2012; Kondratieff 2014; Levene 2019; Rea 2007; Rodriguez-Mayorgas 2010; Roller 2010; Sandberg 2018; Spencer 2007; Totten and Samuels 2012; Touati 2015; and Wiseman 1974. On legendary genealogies see, for example: Wiseman 1974.

of there being a real, solid hut structure in Rome associated with the city's mythological founder. The fact that there might have been two of these 'unique' huts also defies any sort of historical logic. Who built this hut and who promoted it as a heritage structure of some kind? Did Romans actually believe that this structure (or structures) was a genuine historical relic or were Romans happy to collude in the presence of this Ur-hut in their city. A hut around which legend had it that the city then grew? While this is of course principally a historical or mytho-historical issue it also has certain facets that allow us to discuss the *Casa Romuli* as one of the myriad of geographic products in Rome with which this study is concerned.

The late 1940s excavation on the south-western corner of the Palatine Hill found traces of a number of circular post-hole constructions cut into the tufa bedrock here, the archaeological remains of hut-like structures of a form known in the Italian Iron Age, particularly in the so-called Villanovan culture when cinerary urns in the form of ceramic or metal huts reflected the form of contemporary dwelling houses. Could the largest of these have been the *Casa Romuli*, thus making it an actual archaeological structure that was subsequently endowed with a spurious mythological history?

Dionysius of Halicarnassus[11] writing in the later first century BC wrote of the hut being 'preserved....by those who have charge of these matters......They add nothing to it to render it more stately, but if any part of it is injured, either by storms or by the lapse of time, they repair the damage and restore the hut as nearly as possible to its former condition.' Plutarch, writing probably towards the end of the first century AD, also made mention of the hut, noting that it was near the Steps of Cacus.[12] The structure was reported as being damaged by fire in 38 BC and again in 12 BC.[13] Remarkably the hut is recorded in the *Notitia* and the *Curiosum*, together known to us as the Regionary Catalogues or *Cataloghi Regionari*, as still existing in the fourth century AD. On the Palatine the *Casa Romuli* might have been grouped with the *Tugurium Faustuli* -the cottage of Faustulus-another primitive hut imbued with historico-mythological significance that is recorded here by Solinus[14] and again merits mention in the *Notitia*.

Great doubt hangs over the nature of the *Casa Romuli* said to have been located on the Capitoline Hill by Vitruvius[15] and others, though reference to '*Capitolio post casam Romuli*' on a military diploma[16] does tend to suggest that a second site for a Romulan hut here was actually accepted, even if its presence there seems to us problematic and contradictory.

There are certain hauntological overtones to the whole story of the *Casa Romuli*: a hut (or huts) whose geographic location is not fixed or certain, conceived as a one-time shelter for someone who probably did not exist even in the created persona of the mythological figure of Romulus. The whole existence of the geographic space around the hut, that is the city of Rome, was predicated on the one-time presence here of this shadowy figure from the deep past. It is as if the notion that Rome would not have existed as an urban place without the past presence of Romulus in this space willed into being the creation of this hut with its own materiality which then dragged an embodied Romulus into being alongside it. However, it is

---

[11] Dionysius of Halicarnassus *Antiquitates Romanae* 1.79.11.
[12] Plutarch *Romulus* 20.4.
[13] Dio *Historia Romana* 47.43.4 and 54.29.8.
[14] Solinus *De Mirabilibus Mundi* 1.18.
[15] Vitruvius *De Architectura* 2.1.5.
[16] *CIL* XVI 23.

not as if we cannot find numerous instances where places take on a new meaning, or have that meaning thrust upon them, by their association with a mythological or fictional character, the latter being the basis for much heritage tourism today.

If one could enter inside the Palatine hut or go onboard the ship what might this experience have been like? The possibility of enjoying such immersive experiences can be imagined. The Palatine hut must have been a work of imaginative reconstruction, but at the same time an artwork rather than a replica. It was an image that by reason of its form, one could enter.

When thinking about this dilemma, I recalled my own experiences viewing (and thus entering) the German artist Kurt Schwitters' artwork known as the *Merzbau* (untranslatable really, but literally *Work Barn*) in its present manifestation and location in the Hatton Gallery at Newcastle University. Schwitters had been constructing various versions of *Merzbau* at different locations since 1923: the Hatton Gallery *Merzbau* wall comes from inside a stone barn at Elterwater in the English Lake District, where Schwitters was working and living in 1947-1948. The wall incorporates local Cumbrian stone and rocks, wood, and flowers, along with found objects, a collage set in plaster and painted over: each *Merzbau* was constructed to reflect its environment and its time. Each was historically geo-located. Many commentators have called the *Merzbau* the first immersive artwork, the first piece of environmental art, and repeated visits to view it in the Hatton Gallery between 1977 and 2022 have not dimmed its extraordinary power for me. My wife, who was with me in 2022, the last time I viewed/entered the *Merzbau* and who had not seen it before, was equally moved and said that she felt she was inside a prehistoric chambered tomb. While Schwitters said that 'all it is, is form and colour, just form and colour', it seems in fact like a complete work, a room, for both the living and the dead, much as I can imagine the *Casa Romuli* to have been.

The so-called Ship of Aeneas (Figure 18) may have been a museum monument perhaps built by order of Augustus, and housed in a riverfront building by the Tiber. The sole attestation of the ship's existence is provided by the Byzantine historian Procopius of Caesarea writing in the mid-sixth century AD in his *De Belli* or *History of the Wars*.[17] He recounts seeing the ship in Rome and describes it at great length :

> '...the Romans love their city above all the men we know, and are eager to protect all their ancestral legacy and preserve it, so that nothing of the ancient glory of Rome may be obliterated....all the memorials of the race that were still left are preserved even to this day, among them the Ship of Aeneas, the founder of the city, an altogether incredible sight. For they build a ship-house in the middle of the city on the bank of the Tiber and, depositing it there, they have preserved it from that time.'[18]

While a single source testimony such as this might be thought of as being perhaps suspect in some way, the very fact that Procopius was a political insider, accompanying Count Belisarius in Justinian's wars, suggests that he certainly was shown a displayed vessel in a museum-like setting. Of course, what Procopius saw, if indeed of Augustan date, was itself ancient if not actually mythological and dating to the time of the Trojan wars. There is the possibility that this ship house was one of a number of shipsheds that formed the *navalia* complex that is

---
[17] On the Ship of Aeneas see, for example: Finn 2020.
[18] Procopius *De Bellis* 8.22.5-8.

Figure 18. Relief depicting Aeneas at the future site of Rome. The ship of Aeneas is docked on the right. Rome. AD 140-150. British Museum, London. (Photo: Copyright Trustees of the British Museum).

attested as having been sited down on the Tiber waterfront near the *Campus Flaminius*. First built to house and repair Roman warships in the third century BC it is recorded as having been damaged by lightning in 44 BC but is referenced later by both Livy and Pliny. One of the *navalia* shipsheds may be a structure that appears on the *Via Anicia* marble map fragment and on the Severan Marble Plan of the city of Rome, and it has been suggested, without any real evidence, that this is in fact the very shed that housed the Ship of Aeneas. The most plausible date for the construction of the Ship falls in the Augustan period, when official claims of Roman links to Trojan heritage were most vociferous and numerous, and the Ship may have been dedicated in 2 BC when the emperor dedicated the *Forum Augustus* and staged his famous *naumachia* or staged naval battle in a specially-constructed basin straddling the Tiber.

Such curious monuments were moving laments to a disappearing world and the destruction of communities and cultures elsewhere. The process of looking, finding, and refinding, rescuing viewers and visitors from the silt of past lives while caught in the turbulence of the present, unleashed the pallor of dead things, bleached bodies, remnants and last traces. An elegy to lost language and wilful destruction. They collapsed past and present, the human and natural worlds, the real and mythological (fictive). As conquest and war in the name of Rome raged in distant lands did visitors feel them shearing, with a threnody of destruction, loss, and the passing of the natural world? Dead trees, shedding bark, desiccated shrubs, burning fields and villages, and the human detritus of war like majestic and miserable shrouds swaying in a dry wind.

Along the same lines, the mythical landscapers of Augustan Rome included the hero god Hercules. Literary accounts of his violent encounter with the monstrous Cacus were given both by Virgil and Ovid who true to their Augustan times saw such combat or war as cleansing, necessary for the reshaping and transformation of landscapes.[19] Hercules' flattening of the Aventine hilltop during his battle with Cacus was said to have prepared the site for future building. By using a dislodged rock as an altar at his *Ara Maxima* Hercules was said to have set a precedent for building in stone at Rome.

Again, myth and legend fed great interest around some of the older trees in Rome.[20] Pliny described the area of the Forum around the *Lacus Curtius* and related the presence there of a planted and tended olive tree to offer shade, and a self-seeded fig tree and vine, the three together representing an unusual juxtaposition between Nature and Culture here at the very heart of the city of Rome, as well as mentioning an ancient oak on the Vatican Hill 'older than the city', discussed further below. He also noted three very old lotus trees in the city, one in the precinct of Juno Lucina dating to 375 BC, a lotus tree where Vestal Virgins brought offerings of their hair, and the third one in the area of the Vulcanal, the same age as the city itself. A Cypress tree is reported as stood nearby, of equal age, but which fell near the end of Nero's reign. An ancient vine was noted at the Portico of Livia.

But it was the city's oldest fig trees that received the most attention from its historians and chroniclers. There were variously reported to have been four ancient sacred fig trees in Rome, as can be determined by combining accounts of a number of Roman historical writers. The trees were located at: the Temple of Saturn in the Forum (Pliny says planted in the fifth century BC, later removed); in the Forum near the *Lacus Curtius,* along with an olive and a vine (Pliny again); in the *Comitium* (according to Verrius Flaccus, Pliny, and Tacitus); and high on the Palatine Hill (according to Livy, Ovid, Pliny, Plutarch, Servius, and Varro). It is not clear if these were simply seen by Romans as old trees, or if some or all of them were imbued with an element of sacrality.

The wild fig tree-the *Ficus Ruminalis* (Figure 19)-that grew in the *Forum Romanum*, was mentioned by Pliny who went on to say that its sacrality derived from the lightning-struck objects buried under it and 'because of its memorial power'. This tree is also mentioned by Livy (I.4), writing in the Augustan period, but his contemporary Ovid wrote only of the vestiges of the fig tree.[21] The *Ficus Ruminalis* was indeed such a great Roman landmark that it featured in at least two artworks, something very surprising and unusual for a natural feature. Another tree-the *Ficus Navia*-was located in the *Comitium* and was sometimes confused with the Forum tree.

On one of the so-called *Plutei* of Trajan, a monument known to archaeologists by a confusing number of often-contradictory names, there appears an image of the statue of the Hanging Marsyas next to a tree, or indeed a statue of a tree, which it has been suggested is the famous *Ficus Ruminalis*. The tree also is depicted alongside the Lupercal cave in which the She-Wolf

---
[19] Virgil *Aeneid* 8 and Ovid *Fasti* 1.
[20] On the *Ficus Ruminalis* and other old trees in Rome see, for example: Evans 1991; and Hunt 2012.
[21] Pliny *Naturalis Historia* 15.78; Livy *Ab Urbe Condita* 1.4; and Ovid *Fasti* 2.411.

Figure 19. The *ficus ruminalis* depicted on one of the *Plutei* of Trajan or *Anaglypha Traiani*. Trajanic, possibly later and Hadrianic. *Curia Julia*, Roman Forum. (Photo: Author).

sheltered to suckle Romulus and Remus on a second century AD Campana Plaque-a ceramic plaque- now in the Pergamon Museum in Berlin. The shepherd Faustulus approaches the cave, caught just at the moment before he discovers and rescues the twins.

The *Ficus Ruminalis*, as the fig tree in the Forum was most generally called, and the *Casa Romuli* both acted like messengers from the past, shadows of brilliant days, locating old Rome in contemporary Rome, allowing both to somehow coexist, but not in a way that survival of an ancient building would have done. As manufactured or created relics they served to alter the spaces around them and interrupt the historical moment. They represented in many ways what has been called in other contexts the non-ordinary reality of contemporary life. The fact that both the *Ficus* and *Casa* are given two locations each by various Roman writers is curious in that it suggests that the idea of authenticity with regard to location, if a word like authenticity can be applied to manufactured props or recreations associated with myths, was a fluid concept, and that sometimes geographic accuracy just did not matter. If the old idea from early cosmologies that a sacred mountain, pillar, or tree stood at the centre of the earth, then allowing a simulacrum sacred fig tree to grow here in the *Forum Romanum* perhaps suggested to some viewers of it that here indeed was the very centre of the known world.

But were there other trees in Rome that acted as geographic markers or were presented as holders of geographic specificity, in other words could be seen in the context of this study as what we have called geography products? Pliny,[22] as mentioned above, related that on the Vatican hill stood an ancient holm oak tree-'older than the city'-which had a bronze plaque attached to it bearing an inscription in Etruscan characters which he stated informs the reader that 'even in those days it was an object of religious veneration'. Whether Pliny himself saw the tree or he was citing some of his many earlier sources that provided a record of the tree's existence is not important here. The tree on its own is simply an old tree, but the inscription attempts to place it both historically (even if that is not actually true) and geographically, in a rural environment before the city of Rome grew up here. Despite this rather piecemeal evidence for significant trees at certain specific spots in Rome, and I consciously avoid using the term 'sacred tree' to describe them, it would seem that trees could be chosen to act as situational or geographic markers, as monuments, as memorials, as signifiers, as a kind of rebus it would also appear.

Another Time

Looking at part (of the world) in order to return to the whole was a psychological strategy that can be seen in terms of Rome's ideological positioning as Republic gave way to empire. As was noted above, art was the thread that bound together disparate elements into a single activity and which flowed through the city and around its empire with the sinuous authority of a river. To portray or observe geographical artworks from this perspective would have made the viewer aware of his or her earth-bound limitations. As a consequence it might have been assumed to have been sanctified with atavistic properties. If on the ground in far distant places experiences were different, and instead of finding richness in the soil the Romans there found only aridity, citizens at home learned how to inhabit these locations without being present in them.

The generative images, like the curatorial strategies that produced them, accumulated site specific importance, making a mark on the fabric of the city. The architectural sites in which works were placed impacted on them and the ways in which architectural containers negotiated their past functions, sometimes superimposing traces of past exhibitions that took place in the same spaces. Reflecting both a historical moment and a paradigm, these curatorial practices led to absences, not just lacunae, exclusions, blind spots, and silences. Each new monument, each reused foreign monument, such as Egyptian obelisks for example, was an instrument of encounter, reflecting the spectacularisation of power at Rome.

The displaying of these geographical images sought out an understanding but left a struggle: Rome's world-making in this way must have both confused and enlightened-not necessarily contradictory positions. Vietnamese theorist and filmmaker Trinh T. Minh Ha has used the term 'speaking nearby' in terms of creating images from a situated and embodied perspective. While creating spaces for engagement and difference there is no doubt expressed about the

---

[22] Pliny *Naturalis Historia* 16.237.

connections between place and gender for example, and this is not necessarily an abolitionist cry against structures of colonial violence and patriarchy. Land and bodies, society, and the Roman self were inscribed as intersecting sites of struggle, though certainly not resistance. It was a project for imagining alternatives, for critical imagination. This kind of world-making entailed a struggle for representational orders and ways of living that did not constrain or subordinate: looking elsewhere to discover the marginalised other surely risked reinforcing precisely the territorialism the works resisted. Perversely, and surely not intentionally, images of foreign women in particular took up space in Rome's very centre like figures in a desecrated landscape, as protagonists then triumphantly embodying the emancipatory possibilities of foreign women reclaiming space simply by their presence as images. These images were looked at, but also were looking back, disrupting expectations of passive objectification, forming a parallel to the abused women's bodies familiar from so many stories in classical mythology. Between desire and gratification lay an empty space, defining the anatomy of a fall, and reliant on the fabrication of zones of non-being and zones of sacrifice. The competing visions of the female barbarian body presented oppositions: between enumeration and narrativisation, between privatisation and public, between fiction, documentary, and a more scientific ethnology.

To paraphrase another art critic, Laura Mulvey, in another context, the spectacle here was vulnerable. Each a cut, an interface where one image met another, forming a relation. In this way, as montage, they produced knowledge of the world through the images' analytic and synthetic functions. They were thus direct and poetically oblique, joyous yet devastating, intimate yet somehow expansive. If poetry was integrated into these spaces of museumification it could have made a potential catalyst of rupture and schism. The body of images became about exchange, about marking the place where peoples were together, but also where they might have been together elsewhere and otherwise. The Haitian anthropologist Michel-Rolph Trouillot has argued that rather than speaking of the past, it is more productive to speak of pastness as a field of practice, as we can envisage here at Rome the attempt to visualise entangled and multiple temporalities.

These geography products in the form of sculptures, architectural embellishments, or whole monuments must have always commanded a difficult territory lying between figuration and almost abstraction. The roots of their subjects were firmly embedded in Rome's urban landscape: not the monumental or spectacular, but rather the commonplace and everyday. Objects and their traces, seen fleetingly or closely observed, were transformed into new and distinct entities.

They inhabited a separate and precious area which deflected and denied, and gave so much at the same time. The Roman drive to constantly mythologise allowed reflective repetition across and within different media and images. There was here a struggle between detail and generality. The various layers of meaning, increased by repetition and seriality acted as both barrier and statement at the same time. Each a new solution to the same problem, a proposal coming from within an apparently controlled repertoire. They carried that strange relation between being something, and being *of* something. The city of Rome was perhaps always looking for a ritual to join its fragments: Rome had both true neighbourhoods but distinctive cultures stretched out horizontally along specific streets-linear neighbourhoods, its criss-crossing geometry of local colours. In the era of its empire building Rome was in the midst of an unprecedented cultural conflict and crisis of identity.

The Romans seem to have realised that images were part of war, and to them were essential elements of victory and, by extension, defeat. The idea of shock and velocity in art and warfare, and both as new forms of industry, were new, a kind of ecstatic theology, that in much later times we can find explored in the essays on war and photography by Ernst Jünger.

Under the pavement of this bellicose modernity, dreams of soil, and of all these in-between moments, spaces, and epochs, almost as if Rome had missed out on its own first century AD. If Rome already had a history it also worked for history, using accumulated memory as raw material for constructing a new kind of historical action. As well as creating and clinging to its colonies and provinces it also advanced a strategy of colonising everyday life, with the need for perpetual renewal that this managed to create. It must have been tempting to create a conflict of precedence between several aspects of a problem, when the newness and importance of this problem resided principally in its indivisible characterlessness. To unify falsely-separated categories in this way was difficult, as they then moved towards their own reversal, because each one was wholly experienced along with its negation and permanent supersession in time. For Rome was not just simply going to distant lands to conquer them: people from those lands were being forcibly brought to Rome and Italy or were coming there for reasons other than forced enslavement. As Seneca wrote in *De tranquillitate animi* 2: 'Come, look at this crowd, which all the buildings of the vast city can scarcely house: the great majority of them have left their birthplaces behind them. From their towns and colonies, indeed from the whole world they have flooded together.'[23]

The idea of process in art is usually applied to modern art and its creation. Ancient art is meant to be simply linked to ideas about commission, connoisseurship, reference, replication, copying, and aesthetics. But repetition of certain tropes, themes, subjects, and topics, actually was an artistic process itself. Such works can also be viewed in part, as elements of what was intended to be a discrete visual object: the compositional whole could be broken down into parts with autonomous value. Viewers engaged in an ongoing negotiation of both the work and the frame, that is the city of Rome, obliging them to be aware of an unsettling of boundaries, and to view all facets of the geography products not as discrete items but as ingredients in a living sculpture that tried to define the relationships between place and space. Sometimes the lack of identifying detail was telling rather than lazy, in that the lack of specificity highlighted the historical continuity of conflict and war in empire building.

It is difficult to see coherence here, certainly over the longer term, but (re)emphasis might just have been part of a one-time strategy of tension, constantly reminding people in the city of Rome's power over others. Most writing on viewers and viewing suggests that this was a silent process, but if it is accepted that sound might have thrived around that suffocating space of encounter, contaminating the fiction and expanding the artworks' original intentions. The idea of an exhibition of exhibitions is interesting in that the Romans systematically exploited exhibitions (sometimes single artworks, sometimes grand monuments, sometimes a series) to establish narrative points or arcs, seducing the compliant elite, entertaining the masses, and

---

[23] Seneca *De Tranquillitate Animi* 2.

aestheticising politics. Exhibition strategies varied, but first became codified under Augustus and the succeeding Julio-Claudian emperors.

Roma, the personification of the city, came to represent different aspects of allegiance and home as the nature of Rome itself changed.[24] From a representative figure with which citizens could identify she became instead a focus of a broader allegiance. The earliest images of *Roma* on coins dating to the third century BC did not mean the same as the image of *Roma/Tellus* on the Augustan *Ara Pacis* for example, or the painted *Tellus* on one of the four faces of a mid-first to early second century AD altar from Via Circo, Milan and now in the *Museo Civico Archeologico* there.

In 20 BC Augustus very specifically and quite literally set down a marker for measuring the whole world in terms of journeys across it. The bronze clad marble *Milliarium Aureum*-the Golden Milestone-was built in the *Forum Romanum* to be the point of origin for measuring all roads leading to and from Rome.[25] Augustan art was fond of using images of vegetal plenty, images of the sea, of the marine environment, representing an urge to portray, to record, to document, to list, to catalogue, to explain. The Augustan vegetal imagery was also a way of mapping the unmapped world, those areas around the mapped points and between them, touching from a distance, further all the time. The garden room of the Villa of Livia, his wife and Rome's first empress, represented an idealised geography of the natural world, of the world that surrounded and infused every dry name on an itinerary: this was the world around, as well as the world between points, places off the routes.

By building aqueducts, bringing water to Rome from elsewhere, Augustus made *there* here, confusing space and place. Rome's emperors and their architects were unlikely though to have been conscious that they were also creating an underground or subterranean Rome, of culverts, drains, sewers, and later catacombs, a geography of what lay beneath, perhaps an intermediary messenger between one world and another. There would have been in this case a contrast between ascent and descent, an exploration of what lay beneath, of this hollow earth: the subterranean imagination became linked to doomsday bunkers of a kind, places of horror and magic, refuge and entombment, a portable underworld.

Two of the major contemporary written accounts of ancient geography that have come down to us also partially conflate geography and ideology. Pliny's *Naturalis Historia* or *Natural History* did so by its dedication to the emperor Titus, while it has been noted, quite perceptively I think, that Strabo's *Geographica* or *Geography* is about the world as it related to Rome mainly in the age of Augustus, and not about the world per se. Uninhabited or barren regions were not covered or were barely addressed. Ending with a list of Roman provinces, this structuring could not have been more implicit of ideological considerations. Rome acted almost magnetically to pull in people and things: envoys (mentioned four times), meat (once), textiles (once), papyrus (once), wine (once), water (once), stone (once), marble (twice), books, (once), and statues (mentioned twice).

But not all geography products and geographies being created and presented to the citizens of Rome had official imprimatur. From the writings of Horace, Ovid, and Juvenal we can to some

---

[24] On the goddess *Roma* see, for example: Joyce 2014/2015; Robinson 1974; and Vermeule 1974.
[25] Plutarch *Galba* 24.

extent recreate sexual geographies of the city of Rome, many spots becoming locations for desire and (sexual) conquest, in a mockery of masculinist imperialism. After listing the female sexual types to be found/encountered in the city Ovid's urban tour[26] begins at the Portico of the Theatre of Pompey, followed by the Portico of Octavia/Marcellus, the *Porticus Liviae*, and the portico of the Temple of Apollo on the Palatine Hill. Then there follows religious sites and temples, then the *Forum Iulium* and the law courts, the theatres, the circus, gladiatorial combat at the arena, and triumphal processions. This would seem to have been a classic example of the scopophilia-literally a sexual pleasure in looking-which has recently been explored in relation to cinema by Laura Mulvey and others. Although there can be two types of scopophilic look, it is that of male on female that is most significant here, but not exclusively so. While a statue is an object and there is no movement, these walking tours of sexual hotspots by some of Rome's more voyeuristic poets brought movement to an otherwise static geography of desire and contested consent.

In this chapter discussion has centred around a few ideas relating to the presentation of Rome as home, without having strayed too far into the process of redefinition of Rome's place in the wider world that would take place in imperial times using a veritable barrage of geography products to demonstrate and explain this. Rome had somehow started this process by building and monumentalising its own mythological/mythic past, nudging people to think about geography in a particular way. A number of such highly-structured ways of seeing the world were correlated to achieve this. In the next chapter I will further explore Roman identity as created by the dissemination of further ideas about Rome itself.

---

[26] Ovid *Ars Amatoria*.

Chapter Three

Rome in Rome

In this chapter an examination will be made of various aspects of geography products relating to Rome itself, displayed in the city or focused specifically on the city. In other words how Rome visualised itself as a geography product, through depictions of its natural topography and of its urban character through images of architecture. It will be argued that at certain times the Roman people needed to be reminded of who they were and how their identity was inextricably tied up with their home city specifically. Initially discussion will focus on the contemporary significance of Rome's famed seven hills and on the broader context of hills and mountains in general to the Romans. Discussion will also turn to a consideration of other natural features that made up Rome's topography and which became signifying images on some occasions, particularly significant types of trees. Coins and sculpture bearing images of certain iconic buildings in the city will be analysed in terms of how this reflective architecture was marshalled and why. Finally, some discussion about instances of self-identification with regions and sub-zones in Rome will be presented for analysis and interpretation. Rome's river, the Tiber, will be discussed in Chapter Four.

Seven Hills

As to the evolution of the idea of Rome as a city of seven hills, the origins and meanings of the attribution have been analysed and to some degree deconstructed by Caroline Vout in her recent book *The Hills of Rome. Signature of an Eternal City*.[1] As she noted there, it is difficult to come up with a definitive list of which of Rome's many hills actually constituted the magic seven which so occupied the minds of many Roman writers, particularly poets, from the time of Varro (Marcus Terentius Varro-116-27 BC) onwards, most noteworthy Virgil (Publius Virgilius Maro 70-19 BC), Statius (Publius Papinius Statius AD 45-96), and Claudian (Claudius Claudianus AD 370-404), and indeed it would appear that the writers could not agree upon this themselves, but in the context of this present study this is not strictly relevant. Before Varro the equation of Rome as a city of seven hills did not appear to have been made.[2]

What is of great interest here is the fact that while poetic Rome relied upon the use of this topographic descriptor, visual artists and their commissioning patrons would appear not to have been particularly interested in producing views, studies, or even imagistic allusions to the city's seven defining hills. Why might this have been so? The poet Ovid wrote of *Roma* 'surveying the whole world from the seven hills'.[3] However, as Caroline Vout has suggested, we might see the domestication of the seven hills and their incorporation within the city as 'an early model for imperial expansion'.[4] Thus the city was humanised and its underlying topography relegated to the silvery words of the city's competing poets.

---

[1] Vout 2012.
[2] On Rome's seven hills see principally: Vout 2012, but also: Langdon 1999.
[3] Ovid *Tristia* 1.5.67-70.
[4] Vout 2012: 57.

Figure 20. Bronze *sestertius* coin issue of Vespasian, Rome mint. AD 71. Reverse image of *Roma* seated on the seven hills. British Museum, London. (Photo: Copyright Trustees of the British Museum).

A bronze *sestertius* coin of the emperor Vespasian minted in Rome in AD 71 had a reverse on which was depicted on the right the goddess *Roma* reclining on seven rather lumpy and unimpressive mounds, presumably representing Rome's mythical seven hills, while the She-Wolf suckled the twins Romulus and Remus at the base of the hills and a small figure of *Tiberinus*, the personified Tiber, appeared on the left (Figure 20). This peaceful, almost bucolic scene did not seem urban in the character of its imagery at first sight but quite clearly was. It returned the hills to their natural state, unencumbered by houses, temples, and streets. It was a scene of peace and harmony, of origins, that was perhaps intended to celebrate the era of peace now being enjoyed by Rome following the chaotic power struggles of AD 69, the so-called year of the four emperors. Of course, bucolic scenes involving *Roma* and *Tellus* had their origins in the artworks on the *Ara Pacis Augustae*.[5]

On one of the faces of an altar dedicated to Mars and Venus in AD 124 from the portico of the *Piazzale delle Corporazioni* at Ostia but now in the collection of the *Palazzo Massimo alle Terme* site of the *Museo Nazionale Romano* in Rome the figures of the shepherds Faustulus and Faustinus, overlooked from on high by a third male figure who may represent a male personification of the Palatine Hill, seem to be walking across a hilly landscape high above the Tiber, indeed almost stepping from crag to crag, and it may be that we are expected to imagine that they are traversing some of the hill's at the site of Rome, before the city was built here, even though archaeological excavation has shown that the history of the city was one of evolution rather than sudden creation. They are about to stumble upon the Lupercal, the cave of the She-Wolf who is shown suckling the baby twins Romulus and Remus. At the bottom of this face of the altar reclines the benign figure of *Tiberinus*. In the *triclinium* or dining room in the House of M. Fabius Secundus at Pompeii was a wall painting on a mythological theme that by necessity of its setting featured the Palatine Hill in Rome. It acts as a backdrop to the much-told tale

---

[5] For references to *Roma* see Chapter 2 Note 24.

Figure 21. Face of altar to Mars and Venus, with depiction of Faustulus, Numitor, and Faustus, the She-Wolf nursing Romulus and Remus, and *Tiberinus*, Ostia. Trajanic or Hadrianic. *Palazzo Massimo Museo Nazionale Romano*, Rome. (Photo: Author).

Figure 22. Detail of face of altar to Mars and Venus, with depiction of the She-Wolf nursing Romulus and Remus and *Tiberinus* looking on, Ostia. Trajanic or Hadrianic. *Palazzo Massimo Museo Nazionale Romano*, Rome. (Photo: Author).

of Mars and Rhea Silvia and the births of Romulus and Remus. The hill is topped with two temples and a river flows at its foot (Figures 21 and 22).

Two other artworks depicting scenes on one or another of Rome's hills can be briefly mentioned here, the hill in each case simply acting as a backdrop to the central dramas depicted. Firstly, a relief found on the Caelian Hill, and now in the collections of the *Palazzo dei Conservatori* in Rome, carries a depiction of three deities, named in inscriptions below as Caelian Jupiter, Hercules Juilanus and the *Genius Caelimontis*. They are pictured in a landscape setting defined by trees and a small rocky outcrop on which the *Genius* sits. This relief presumably came from a monument here dedicated one year at the *Septimontium* or Feast of the Seven Hills. The second image appears on a fourth century AD *opus sectile* panel from Bovillae, now in the *Palazzo Colonna* in Rome, on which can be seen the goddess *Roma* seated on a rocky outcrop surveying a pre-urban scene of Rome before Rome, with a stark tree, a male figure, birds, the She-Wolf and twins, and *Tiberinus*.

A marble statue base, probably dating to the first half of the second century AD, found on the Lechaion Road at Corinth in Greece may have supported a bronze statue of *Dea Roma*, something suggested by the fact that the base takes the form of a rocky outcrop inscribed with the names of Rome's hills, five prefixed by *mons* and two by *collis*.[6] The city was here being evoked in a foreign land through the employment of the device of equating the personified deity of the city of Rome with its seven hills.

There was a certain degree of license used by different Roman writers as to whether these hills were to be called *montes* or *colles*, suggesting a difference in nomenclature based on height. An analysis of the eight inscriptions from Rome incorporating a name of one of the seven peaks suggests that the Capitoline/Tarpeium, the Esquiline, the Aventine, and the Caelian were all classed as *montes* (the Palatine is not named in any inscription so far recovered), while the Viminal was classed as *collis* (the Quirinal does not appear in an inscription), perhaps because they were smaller and technically actually spurs rather than hills as such.

It is likely that as an amenity together the hills of Rome allowed its citizens the opportunity to gain a raised, perspective view down on to parts of their home city. The view was from the hills onto the city and not from the city to the hills. That view would appear to have elicited very little interest indeed. Upper floors of tall buildings provided the same viewing experience, the opportunity to look down upon the city and to appreciate its character and size, something that could not be done from many locations at street level, but such a perspective was not replicated in art. But there must have been some kind of sensory or aesthetic experience to be gained from looking down over the city, as suggested by key locations whose plan and construction included just such a viewing amenity. One has only to think of the emperor Nero in the Tower of Maecenas overlooking Rome as fires burned out of control, according to the account given to us in the scurrilous histories of Suetonius. Again, Trajan's Column was built with an inner staircase and a viewing platform, access to which must have been severely restricted. Control of the view was quite evidently an exercise in power in itself. Was it not only the gods who could look down onto the world of men from atop Mount Olympus, making

---

[6] On the Corinth *Roma* see: Robinson 1974.

the very act of viewing from above something that was inextricably linked to their divine status?

In order to build Trajan's Column part of the rocky spur that connected the Quirinal Hill to the Capitoline Hill was removed: indeed, it is interesting to note that part of the column inscription suggests that the sections of the Quirinal Hill removed acted as some kind of height marker for the column. The relevant text of the inscription reads 'ad declarandum quantae altitudinis mons et locus tantis operibus sit egestus', translated as 'in order to show how lofty the mountain had been, and the site for such great works was nothing less, which had been cleared away'. Exactly the same point was made by the historian Cassius Dio[7] who noted that the column's height was to act as 'an indication of the labour involved in making the Forum. For the whole of that area was hilly, but Trajan cut down as deep as the height of the column and thus made his Forum level'. Such stressing of the labour involved in building the column echoed the often-cited emphasis on the column's frieze lauding the energy expended by the Roman army in clearing forests, building camps, constructing roads and bridges, to say nothing of their prowess at fighting, depicted in many scenes on the helical frieze. By comparing the natural (the quarried-away hill) with the built (the column and forum) the dedicators of the column (not Trajan himself as Dio almost implied but rather the Roman Senate) were here hymning another victory, this time over Nature and topography. While the famed seven hills had a potent emotional currency with Rome's poets its rulers retained the right to alter or landscape portions of these hills as part of the forward march of Roman progress.

A similar exhibition of the way in which Roman ingenuity and engineering could conquer even Nature was recorded at Pisco Montano above Terracina, on the coast to the south-east of Rome. Here a huge quantity of solid rock was cleared and the ground levelled to allow the easy rerouting of a stretch of the *Via Appia* to and from the new port facilities established there in the reign of Trajan. This remarkable feat was visually commemorated by the Roman surveyors and builders by a number of numerical inscriptions being carved into the cut rock face to show the depth of rock cleared by hand by the teams of labourers. The best surviving of the inscribed markers reads *CXX* or one hundred and twenty Roman feet, marking the base of the cut through the solid rock. It has been estimated that 13,000 cubic metres of rock was removed here: a fitting demonstration of the power of Rome over Nature in the form even of hills and mountains.

Another Roman topographical landmark which held a great deal of emotional currency within the Roman psyche was the Tarpeian Rock, the location of one of the city's founding historico-mythological events, the Punishment of Tarpeia. A steep cliff on the south side of the Capitoline Hill, the Rock though received little attention from artists. Only the punishment scene itself-the crushing of Tapeia by shields-ever appeared as images, on a relief from the *Basilica Aemilia* and now in *Palazzo Massimo alle Terme*, and on the reverse of a few coin issues.

The *Campus Martius*, once an open plain outside the city and officially 'brought in' to the city under Augustus, featured as a personification in two major imperial artworks. A relief panel depicting the apotheosis of Sabina, wife of Hadrian, now in the *Palazzo dei Conservatori* in Rome, includes a depiction of a semi-naked reclining youth at the feet of the seated Hadrian who

---

[7] Dio *Historia Romana* 68.16.3.

Figure 23. Apotheosis scene, with the youthful personification of the *Campus Martius* in attendance, on relief panel from the *Arco di Portogallo*, Rome. AD 136-138. *Palazzo dei Conservatori*, Rome. (Photo: Author).

probably represents the first appearance in Roman art of the personification of the *Campus Martius* (Figure 23). Together they watch Sabina ascend to the heavens, her spirit carried aloft by an eagle, as was a standard trope in apotheosis scenes. The base of the Column of Antoninus Pius, from the monument originally set up in the *Campus Martius*, is now in the one of the courtyards in the *Musei Vaticani* in Rome. The personification of the *Campus Martius* here again as a youthful figure is altogether in keeping with the fact that this area was not originally within the city or *pomerium* bounds and that in terms of Rome's deep history it would even by the time of the Antonines still perhaps have been viewed as a newish region of the city. He holds in his hand the obelisk from the *horologium* or urban sundial of Augustus.

Mountains

Attention will now be turned to Roman imagery associated with hills and mountains in general, and consideration will be given as to how mountains figured in the Greco-Roman cultural imagination,[8] a subject that has recently been presented in an academic monograph by Jason König.[9] When Petrarch wrote about his climbing of Mont Ventoux in 1350, he claimed that he was the first person since antiquity to climb a mountain simply for the view: certainly the Romans left many literary references to mountains, enough to suggest their own fascination.

Mountains featured a great deal in Strabo's *Geographica* or Geography and it would appear that their discussion was not always focused on topographical issues alone. The Greek idea that mountains were places outside of civilisation was to some extent followed by the Romans. Romans viewed mountains as barriers to people and not as geographical natural wonders with interest in their own right. Strabo himself was Greek and therefore it is hardly surprising that he regularly discussed mountain peoples as being somehow uncivilised, a strategy that dated back to the writings of Herodotus in fact, and how mountains had become tamed and thus brought within the civilised world as represented by Rome, its empire, and its culture. When we might talk today about mountains being tamed and conquered it is usually in the form of the achievements of climbers and mountaineers, though certainly the discourses around 'the conquering of Everest' in the 1950s represented a very similar form of English ideological and nationalist rhetoric and discourse. In a number of passages in the book Strabo wrote as if viewing from a mountain top, giving him almost a cartographic view of the land around which he then proceeded to describe as if he himself was looking down upon it. Such feats of imaginary cartography are not altogether different to those bird's-eye views imagined by many of the battle painters whose works so enhanced the viewers' understandings of Roman triumphs.

In one extraordinary passage in *Geographica* Strabo[10] recounted the tale of Selurus, a bandit or probably actually an escaped slave, from Mount Etna on Sicily, who was brought to Rome for execution in the arena. In order to geographically locate him in the audience's minds he was placed on a scaffold or platform disguised as a model of Etna. Prisoner and model were then dropped out through the false bottom of the platform into a cage of wild and ravening animals who tore the brigand to pieces. Here an inhabitant of rough mountainous territory, thus a product of his uncivilised environment, was tamed by the barbarism of Rome. This is very contradictory and indeed highly disturbing. Did the model Etna then go on display, drawing crowds like the model of Mount Everest made for the Royal Geographical Society in 1953 as a 10 inch to the mile replica, produced by Stephen Tallent's 'Cockade' company, and still on display in the society's map room today?

Strabo's account of the seven hills of Rome formed part of a much wider narrative to discuss these topographical features as colonised and civilised features, and indeed as civilising in their own right, as symbolic of the city of Rome. He wrote of 'the crowns of the hills beyond the river…presenting the appearance of a stage painting'.[11] He presented a number of other instances in Italy and Greece where hills or mountains had been 'brought within' a city to the benefit of all, not least from a strategic and military point of view, once more making these topographic features seem like human hostages or prisoners.

---

[8] On Greeks, Romans, and mountains see, for example: Buxton 1992; Duffy 2021; Foss 2022; Hollis and König 2021; Hyde 1915; König 2016, 2018, and 2022; and Taufer 2019.
[9] König 2022.
[10] Strabo *Geographica* 6.2.6.
[11] Strabo *Geographica* 5.3.8.

Other significant mountains that attracted the attention of Roman writers and artists included Mount Olympus in Greece, the home of the gods of both the Greeks and the Romans, Mount Helikon, home of the Muses, and the volcanic mountains Mount Vesuvius and Mount Etna. Very often discussion of mountains by ancient writers were used as a literary topos to contrast the mortal and divine.

Mount Olympus is mentioned numerous times in both Greek and Roman texts, most particularly those of Homer, most usually with regard to it being considered the home of the gods, though Homer's numerous descriptive adjectives applied to the mountain-many-summitted, snowy, gleaming, steep, shrouded in cloud-really do bring the site to life as a looming physical wonder of Nature. When seated gods were pictured in Greco-Roman art it would have been assumed by the viewers of such works that they were seated in their home on the mountain. Plutarch, writing in the late first to early second century AD and St Augustine writing in the fourth to fifth century both alluded to an inscription set up at Pythion, on the west slope of the mountain, which recorded that Xenagoras, son of Eumelus, measured the height of the mountain as 'ten full stadia, and in addition a plethron less four feet in size'. The significance of this is considerable, in that this suggests the mountain was regularly visited by tourists, as may have been other peaks elsewhere. These written descriptions of visits made to sites that would have been off the routes of a traditional Roman itinerary suggest that the use of itineraries for planning journeys did not necessarily stymie events off-grid such as these.

In the life of Hadrian in the *Scripta Historia Augusta* (*SHA*) we learn that Hadrian climbed Mount Etna, or at least proceeded some way up the mountain, to view the sunrise from there. Perhaps tellingly the account also noted a visit by the emperor to Mount Cassius in Syria, notable for his almost being struck by lightning there, and to a mountain in Pontus. The geographer Strabo seems to suggest that visitors regularly came to Etna and hired guides from the town of Centoripa at the foot of the mountain, and indeed other sources give details of such visits, by Empedocles and Lucilius. The Roman geographer Pomponius Mela described sunrises at Mount Ida on Crete.[12]

Interest in volcanoes is also apparent from Roman texts and images. While this is mostly a concern with Mount Etna on Sicily and Mount Vesuvius on the plain beyond Pompeii and Herculaneum sometimes even minor volcanic peaks were mentioned or portrayed. For instance, a silver drachm of Tiberius, minted at Caesarea in AD 14-37, has on its reverse the volcanic Mount Argaeus in Turkey surmounted by a radiate figure with a globe and sceptre.

The volcanic Mount Etna on Sicily received a good deal of attention as a subject from ancient writers, including Pindar (writing 470 BC), Aeschylus, Hesiod, Virgil, and Lucretius. A first century BC Latin poem *Aetna* remains unattributed. While the first four of these wove geography and myth together, it was (Titus) Lucretius (Carus) in *De Rerum Natura* who called for explanations firmly rooted in the workings of the natural world to explain the mountain's ferocious eruption of 121 BC.[13] Of course, with regard to the eruption of Mount Vesuvius in AD 79 we have the remarkable first-hand account of Pliny the Younger recounting in two letters

---

[12] *Scripta Historia Augusta Hadrianus* 13.3 and 13.2; Strabo *Geographica* 6.2.8; Diogenes Laertius *Vitae Philosophorum* 8.69; Seneca the Younger *Epistulae* 79.2-3; and Pomponius Mela *De Situ Orbis* 1.94.
[13] Lucretius *De Rerum Natura* 6.647-654 and 6.680-682.

Figure 24. Wall painting of Bacchus in front of Mount Vesuvius.
Lararium of the *Casa del Centenario* or House of the Centenary, Pompeii.
AD 55-79. *Museo Archeologico Nazionale*, Naples. (Photo: Author).

to his friend the historian Tacitus both the death of his uncle Pliny the Elder as a consequence of the eruption and his own experiences of viewing the aftermath events from Misenum.

Despite these searing accounts being classics of Roman journalistic writing which would have garnered great attention in Rome at the time, there is only one possible depiction of Vesuvius known in Roman art, a wall painting found in the *Casa del Centenario* or House of the Centenary in Pompeii and now in the *Museo Archeologico Nazionale* in Naples (Figure 24). The painting, probably dating to the remodelling of the house around AD 15, was sited in the *lararium* in the house's second atrium which housed the building's household shrine. In this striking painting we see the figure of Bacchus standing to the right, with the looming presence of Mount Vesuvius to his side. In the foreground a massive serpent or snake-Agathodaemon, the spirit of the vineyards and fields-writhes on the ground, signifying protection and security. It carries an egg towards an altar. A garland is strung across the scene and two birds are also depicted. The depiction of Bacchus is unusual in that the god's body is covered in grapes (much like the breast-covered body of the *Magna Mater*): he carries a *thyrsus*, and a panther sits at his feet. Vesuvius is here a single-peaked mountain, its lower and middle slopes embraced by vineyards, recalling the poet Martial's encomium to the locale: 'Here Vesuvius is shaded green with vines; here the noble grape had exuded its juices in vats: these are the ridges which Bacchus loved more than the hills of Nysa.'[14]

Of course, the mountain depicted in the Pompeian painting could rather be Mount Ida in Crete to which the infant Bacchus was said to have been taken for his protection. From the *Casa dell'Atrio a Mosaico* or House of the Mosaic Atrium at Herculaneum comes an early first century AD wall painting, also now in the *Museo Archeologico Nazionale* in Naples, depicting the Trojan prince Paris overseeing a flock of sheep and some cattle on Mount Ida. He is seated on a low rock outside the Temple of Diana, the peak of the mountain visible beyond. The form of the mountain and its peak here bear little resemblance to the Pompeii Vesuvius.

Water and Trees

Attention will now be turned to other topographical/natural features at Rome whose presentation as images would seem to have been of some significance: the waters at Rome-the Tiber, springs and marshes-and particularly significant and diagnostic trees.

A fascinating and highly original recent study by Dylan Rogers[15] has proposed that certain aspects of Rome's 'aquatic past', as Rogers terms it, came to be reflected and thus honoured through memory in certain features in the Roman *fora* of different eras. This was 'a metaphysical topography', represented by the Sanctuary of *Venus Cloacina*, the *Lacus Curtius*, *Lacus Iuturnae*, and the *Rostrum*, all in the *Forum Romanum*, which linked Rome's past, present, and future together in an enchainment of memory.

The Sanctuary or Shrine of *Venus Cloacina* in the Forum was dedicated to the protective deity of the great Roman sewer the *Cloaca Maxima*, making Venus here both a symbol of purity and of ordure or dirt. The small circular stone building, now just represented by its foundations, that can be seen in the Forum today probably dates from the second century BC and replaced an earlier structure here. An image of the later shrine appears on the reverse of a coin issue of 42 BC and shows that the shrine was unroofed and open to the air and surrounded by a circular balustrade. Statues of the presiding goddess stood inside.

The *Lacus Curtius* and the *Lacus Iuturnae* in the *Forum Romanum* in their latest manifestations were small enclosed pond-like structures whose retention as distinct features possibly harked back to the time when the area here was simply marshy ground around a small lake. The *Lacus Iuturnae* was fed by a spring and its waters were thought to have healing properties. In Dylan Rogers' formula their retention and presence would have acted as a mnemonic device to body forth long past times. The *Rostra/Rostrum*, originally a construction derived from captured rams of enemy warships, was there to evoke the seas and waters beyond Rome now controlled by Roman naval might.

---

[14] Martial *Epigrams* 4.44.1-3.
[15] Rogers 2020.

Some discussion about Rome's historic trees[16] has already been presented in Chapter Two. The wild fig tree known as the *Ficus Ruminalis*[17] that grew in the *Forum Romanum* was mentioned by Pliny who went on to describe how its sacrality derived from the lightning-struck objects buried under it and because of its memorial power. Other historic fig trees were also recorded by various writers, including the *Ficus Navia*, located in the *Comitium*, and a fig tree before the Temple of Saturn. Today in Rome it is not so much individual trees that act as signifiers of Rome's identity but rather one type of tree, the Umbrella Pine.

The very day after I had written about the deep significance of certain trees in ancient Rome an article appeared in the *Guardian* newspaper highlighting a serious problem for the modern city's famed and ubiquitous Umbrella Pines (*Pinus Pinea*), of which there are probably somewhere in the region of a million within the city boundaries. Over eighty percent of the city's trees are now infected by 'pine tortoise scale', a deadly parasite with origins in North America, which can kill a mature tree in two to three years, and in many cases cutting down infected trees remains the only solution to its inexorable spread. Nicola Zingaretti, the president of the Lazio region, was quoted in the article as saying that the Umbrella Pines are 'a natural and cultural part of the city's heritage that must be preserved', placing faith in the removal of infected trees and the financing of a bold new planting project of young trees-*Ossigeno* (Oxygen). Silvia Barbati, a resident of Rome's Trieste district, told the reporter that 'the pine trees are the identity of the neighbourhood, and cutting them down is like taking away a piece of history.'[18]

To the Arval Brethren who oversaw the care of the grove dedicated to *Dea Dia* just outside Rome the trees here were part of a religious geography that was as much about place as it was about belief and religious practice. Their meticulous records of attending to and interacting with the trees here, and on one occasion removing a rogue fig tree damaging the temple roof, attested in numerous formal inscriptions set up at the site, constituted a kind of geographic product so specific to this one location and to the brethren themselves as an enclosed sect and their favoured goddess that transmission of this information beyond was never intended or sought and therefore will not be further discussed here.

A recent monograph study by Annalisa Marzano[19] has examined both what she has called 'botanical imperialism' and a broader engagement with arboriculture on the part of the Roman elite and landowning classes. Roman elite interest in plants and planting, the creation of new floral variants and hybrids, became reflected in the importance of the great *horti* or gardens of Rome. The *Porticus Pompeiana* can be looked upon as the first public garden in Rome, and here as in the *horti* that were set up later and in imperial gardens geography products attesting to Roman power took the form of plants and trees. Pliny[20] noted that in his day the plane tree's range then extended 'as far as Belgium and actually occupies soil that pays tribute to Rome.' This phrasing was unusual and implied a number of things. Firstly it would seem that the plane tree was given the power of agency by Pliny writing in this manner, itself occupying

---

[16] On Romans and trees see principally: Fox 2023; and Hunt 2016. But see also: Fox 2019; Gowers 2011; Hallett 2021; Hunt 2012; Marzano 2022; Meiggs 1982; Stoiculescu 1985; and Thomas 1988.
[17] On the *Ficus Ruminalis* see Chapter Two Note 20.
[18] *Guardian* 15 April 2021.
[19] Marzano 2022.
[20] Pliny *Naturalis Historia* 12.6.

land like a human conqueror. Again, the very soil of this land paid tribute to Rome, not just its people this phrasing implied.

Streets and Buildings

A textual source which through repetition stressed the moral value of building things was the emperor Augustus' *Res Gestae*. The *Res Gestae* probably owed its form to the tradition of funeral *elogia*, in the form of orations and later more permanent memorial inscriptions detailing the dead man's virtues and achievements. If at times routine and monotonous in its listing of detail the *Res Gestae* text surely reflected the spoken origins of the form and the hypnotic power of repetition. The three principal sections listing the buildings in Rome Augustus either built, completed after having been started by others, or restored, demonstrated the Roman drive towards commemoration in posterity through architectural benefaction.

Urban topographical scenes in Roman painting or sculpture either at Rome, in Italy, or elsewhere in the provinces are quite rare. However, sometimes Rome's size and multitudinous character, its bustling streets, was captured in crowd scenes of various kinds. For example, a Roman street scene appears on the reliefs from the Tomb of the Haterii, now in the *Musei Vaticani* in Rome; a scene set in a provincial town in Moesia appears on the frieze on Trajan's Column: and a number of busy harbour scenes at Ostia and Portus are known. Famously, a street riot outside the amphitheatre at Pompeii was the subject of a wall painting in the *Casa di Anicetus* or House of Anicetus in the town.

Large portions of a remarkable funerary monument, the Tomb of the Haterii, that once stood on the *Via Labicana* in Rome, are on display today in the *Musei Vaticani* in Rome.[21] Quintus Haterius, a rich freedman of the later first century AD, would appear to have been a highly-successful building contractor in the city. His family tomb carried intricate carved reliefs depicting the lying in state of his dead wife, images of numerous buildings along the *Via Sacra* in Rome probably traversed by his funerary procession, and of a huge building crane powered by slaves on a wheel caught in the process of constructing a tomb, along with other sculptural decoration (Figure 25).

The street scene relief from the Tomb of the Haterii is thought to be a depiction of the *Via Sacra* in Rome. Five buildings are portrayed here, with portions of two others. But why this street and why these particular buildings? It may be that in portraying the *Via Sacra* the family was here commemorating the very route followed by the funerary procession. Given that we have already been presented with an unusual, intimate, and graphic portrayal of the lying-in-state of a body in another relief scene it might have been expected that there could be no qualms in picturing the funeral procession proceeding down the street. Some academic commentators have argued that this was simply a portrayal of famous contemporary Flavian monuments and buildings in the city whose portrayal would have allowed the viewer to situate the Haterii in both time and space. A locational mnemonic device in other words. However, scholarly opinion presently favours the explanation that we are being presented here with images of actual buildings constructed by the family building firm, a kind of *Res Gestae* for the Haterii family. This certainly seems to make a great deal of sense. However, the identification

---

[21] On the Tomb of the Haterii see, for example: Kleiner 1992: 196-199; and Trimble 2018.

Figure 25. Detail of buildings on reliefs from the Tomb of the Haterii, Rome. Late Flavian or early Trajanic. *Musei Vaticani*, Rome. (Photo: Author).

Figure 26. Detail from the end panel of a marble sarcophagus, showing St Peter and his jailers in the city of Rome. Rome. Fourth century AD. *Musei Vaticani*, Rome. (Photo: Author).

of these buildings is problematic. They are thought to be, from right to left: the Temple of *Jupiter Stator* in the *Forum Romanum*; the Arch of Titus (helpfully identified with an inscription that reads *Arcus in Sacra Via Summa*); a large arch in profile with a chariot group statue on top; the Colosseum; a partial view of a small arch or colonnade in profile, with statues of horses or a *quadriga* on top; and the Arch to Isis (again identified for us with an inscribed tag '*Arcus ad Isis*') that formed the ornamental entrance to the Temple of Isis and Serapis in the *Campus Martius*. Though at first sight the viewer might have thought that the image of the buildings portrayed here represented a continuous street scene, this was not in fact the case. Indeed, these individual structures were not a continuous or contiguous set, and in fact were in some cases many miles apart. Buildings in Rome also occasionally appeared on Christian artworks (Figure 26).

During excavations in the late 1990s in the great Baths of Trajan a c. 10 metres square wall painting was discovered in a *crypto-porticus* demolished to make way for the baths. This provides an almost bird's-eye view of a walled port city. Known as the *Città Dipinta* or 'painted city' this may have been one of two painted city scenes on either side of the main entrance to the building which is suggested to have had some sort of administrative function, perhaps as offices for the urban prefect.[22] Which city was depicted here remains uncertain.

Of considerable interest here is the phenomenon of the appearance of images of specific and sometimes generalised buildings, monuments, and other architectural structures in the city of Rome on the reverse of Roman coins[23] (Figures 27-29). This was a specifically Roman cultural practice that began in 135 BC, became increasingly common from the mid-first century BC onwards into the imperial era, becoming less common in the third and fourth centuries AD, and which ceased in the fifth century. The later periods saw more abstraction in the portrayal of buildings than was formerly apparent or intended. In the Greek east of the empire the practice was taken up, but as a copy of Roman cultural practice: when architecture had previously appeared in Greek contexts, on these rare occasions it was as part of a suite of markers of civic identifiers of individual cities. The trend towards architectural images on coins being symbolic rather than specific coincided with periods of increased coin production and the establishment of numerous branch mints in the period after the death of Severus Alexander in AD 235.

It is not intended here to consider many of the individual buildings portrayed on coin issues, though a few will be discussed in passing. Nor will I consider the veracity of these images and their potential for aiding architectural reconstruction or validating the appearance of some so-called 'lost monuments'. Rather, discussion will centre on the use of these images as part of the suite of topographical and geographical indicators used in Roman art to impart geographical knowledge and to consider in what ways the presentation of coin images and the expectations of their designers and viewers differed from other strategies of presentation. It is almost Magrittean to point out that a coin is not a map and is not a statue (his 1929 painting *The Treachery of Images* warns us that *Ceci n'est pas une pipe* when it is in fact the *image* of a pipe).

If those 'architectural coins' issued by Republican moneyers in many cases reflected a desire on the moneyers' parts to advertise the construction schemes related to their ancestors'

---

[22] On the *Città Dipinta* see principally: La Rocca 2000 and 2001.
[23] On buildings and architecture as images on Roman coins see, for example: Elkins 2015; Harl 1987; and Hill 1989.

Figure 27. Roman architecture on coins. Bronze *sestertius* of Titus, reverse the Colosseum from a bird's-eye view, Rome mint. AD 80-81. British Museum, London. (Photo: Copyright Trustees of the British Museum).

Figure 28. Roman architecture on coins. Gold *aureus* of Nero, reverse the Temple of Vesta, Rome mint. AD 65-66. British Museum, London. (Photo: Copyright Trustees of the British Museum).

Figure 29. Roman architecture on coins. Gold *aureus* of Claudius, reverse a triumphal arch in Rome, Rome mint. AD 41-45. British Museum, London. (Photo: Copyright Trustees of the British Museum).

benefaction, and thus to enhance their own prestige, as the culture of public building increased in Rome, so the emphasis can be seen to have changed to the use of images to recognise architectural schemes promoted by senatorial leaders, not necessarily related family, and towards the end of the Republic those promoted by military men to whom power was inexorably shifting.

A single image of a classical building on a coin could be taken to represent the building itself or even the whole built fabric of the city of Rome itself. It could represent a new building programme or a restoration project, or a rededication. Images or names of roads and bridges also appeared on coins and could have related to the general idea of the Roman triumph over Nature. Statius wrote about the personification of the River Volturnus rising up to address the emperor Domitian as 'the perpetual conqueror of the river bank', and Statius himself later described the emperor as 'better and more powerful than Nature herself'.[24]

The architecture depicted on Roman coins included images of altars, aqueducts, arches, architraves, *basilicae*, baths, bridges, the *Circus Maximus*, cities in general, the Colosseum, columns and columnar monuments, forts and fortifications, *fora*, gateways, harbours, market buildings, the *Meta Sudans* fountain, *navalia* or shipsheds, pediments, porticoes, *rostra* including the one in the *Forum Romanum*, shrines, stadiums, statuary in an architectural context, streets and roads, temples, theatres and amphitheatres, tombs, and walls.

The vast majority of coin issues carrying architectural designs by the reign of Nero and thereafter for some considerable time were large format bronze coins, most especially *sestertii*, rather than issues in silver and gold. As has been noted by Nathan Elkins 'the larger flans of the bronze coinage allowed for more intricate depictions of monuments'.[25] Why might this have been the case? It is almost certainly that the intended audience of viewers and consumers of these images on *sestertii* were the plebeian class in Rome, among whom these lower denomination coins were more likely to have been circulating. Indeed, between the reigns of Nero and Trajan a number of images of monuments constructed for public benefit rather than just necessarily for individual prestige appeared on widely-circulated *sestertii* issues.

It has been observed through die analysis that in many cases where architectural images were used on coins more detailed rendering of those images corresponded with the earlier positioning of coins in the die series. Later coins in the series often carried murky, blurred, or otherwise unclear images which were to some extent divorced from the purpose of depicting specific buildings in the first place.

There is a certain amount of evidence that some coin types with architectural reverses were very specifically intended for an audience in the city of Rome itself and not necessarily for consumption elsewhere: Nathan Elkins has suggested that this was the case for the *Aqua Traiana* (Trajanic aqueduct) type whose distribution in terms of find-spots from archaeological excavations is heavily loaded towards Rome, with many fewer from locations in the north-western provinces. But a note of caution must be struck here: not every viewer of every architectural coin type would have taken away the same message from their viewing

---
[24] Statius *Silvae* 4.3.81-84 and 128-129 and 134-135.
[25] Elkins 2015: 169.

of the images, if they gave that image anything more than a cursory glance in fact. Coins struck at provincial mints, intended in most cases for local or regional circulation, most certainly differed from coin issues from the mint at Rome. During the reign of Septimius Severus over one hundred provincial mints were producing coins, something that most certainly would have contributed towards differences in messaging in terms of production of readily-understandable images on the coins. As has been pointed out by Nathan Elkins,[26] when architectural images appeared on provincial coins these images fell in to four general categories: that is images of monuments in Rome, often in a simplified form, local religious centres, local public buildings, and schematic representations of individual cities, apart from Rome, or of city gates, the part here representing the whole. An example of each of these categories can be given here, each example being one of many in most cases it must be remembered. Images of monuments in Rome, in a simplified form, are well represented in the case of issues from the mint at Alexandria in Egypt, such as a bronze drachm of Trajan depicting a triple arch with pediments (of which Roman arch is uncertain), and indeed the number of issues from this particular mint depicting monuments in Rome seems unusually large, for some reason as yet unknown to scholars. An example of a local religious centre is represented by the image of the temple and cult centre to *Roma* and Augustus at Nicomedia in Bithynia on a *cistophorus* of Hadrian. Local public buildings depicted on provincial coinage included the Lechaeum Arch at Corinth in Greece on Hadrianic issues and, interestingly, the harbour of Cenchreae, one of two that served the city, appeared as an image on coin issues from the city mint under Antoninus Pius. Schematic representations of individual cities, apart from Rome, or of city gates, included a number of second century AD issues under Marcus Aurelius for *Augusta Traiana*, Thrace, Greece. A city gate with tower can be seen flanked by two side towers and on some examples a local river god reclines outside the city walls.

A Trajanic reverse used on *sestertii*, *dupondii,* and *asses* of AD 111-113 carried an image of a reclining female personification holding a reed and in an arched grotto resting beside an urn pouring water. The accompanying legend AQUA TRAIANA tells us that the personification is of this new aqueduct. Similarly, another reclining female personification holding a wheel on *aurei, denarii, sestertii*, and *dupondii* of AD 112-113 is revealed by the accompanying legend to be the personification of the VIA TRAIANA, the new road that extended the line of the *Via Appia* from Benevento to Brundisium.

Two particularly interesting architectural images on coins are the depictions of the Colosseum and the *Circus Maximus*, the principal public entertainment venues in Rome, and the harbours of Ostia and *Portus*. These four sites also occurred as images in a number of other media in Roman art which suggests that their ability as images to gain location recognition from viewers was very high indeed.

The depictions of the Colosseum and the *Circus Maximus* are interesting in that a *sestertius* type of AD 103-104 shows the *Circus Maximus* in three quarters aerial view, rather like the way the Colosseum was shown on Flavian *sestertii*. Such a perspective really would have given the viewer a true sense of the scale and layout of the complexes, and would have made them immediately identifiable to contemporary viewers without the need for any explanatory labelling or naming. The architectural detail of the *Circus* is astonishing, and, as Nathan Elkins

---

[26] Elkins 2015: 146-162.

has pointed out, appears so specific that it places the viewer of the image as if looking on to the monument from up on the Palatine Hill.[27] Under Hadrian *aurei* and *sestertii* appeared in AD 121 carrying an image of the personified *Circus Maximus*, in the form of a reclining figure. holding a wheel. An earlier Trajanic version also personified the *Circus Maximus*.

It has been suggested that the quite commonly used images of bridges decorated with triumphal structures, including arches, gates, and statuary, on Augustan and Trajanic coin issues related to a narrative concerned with the Roman triumph over Nature, a theme that is also explicitly apparent on the frieze on Trajan's Column. Some of these coin issues would appear to have used background architectural settings as a locus for the demonstration of imperial power, and Rome as the centre of that power, with imperial travels, sacrifice, *adlocutio*, and *liberalitas* linked to both events and setting.

Turning now to depictions of the harbours at Ostia and *Portus*. On a *sestertius* of Nero struck in AD 64-67 there appeared an image of the harbour at Ostia (POR OST), seen almost as an aerial view, the harbour itself filled with ships and surrounded by port buildings, and with a lighthouse at the harbour entrance with a statue of Neptune on top quite appropriately. The additional image of the reclining personification of the River Tiber symbolically and topographically linked the scene to Rome. Although it was Claudius who instigated the major infrastructural works at Ostia Nero's coin was probably linked to the completion or rededication of those works or of further complementary works to extend the facilities there. The hexagonal harbour basin at *Portus*-labelled PORTUM TRAIANI-packed with boats and ships, appeared almost as an aerial image on *sestertii* of AD 112-114. The lighthouse at *Portus* appeared on a black and white mosaic in the *Piazzale delle Corporazioni* at Ostia, and another mosaic image appears at Tomb 43 in the *Necropoli di Porto Isola Sacra*, and on the so-called Torlonia relief now in the *Musei Vaticani* (Figures 30-31). The lighthouse at Ostia indeed even is depicted on a sarcophagus from the *Necropoli di Porto* at *Isola Sacra*.

Away from Rome, at *Hippo Regius* in Algeria in Roman North Africa, there has been found a particularly notable mosaic, now in *Annaba, Musée des Ruines d'Hippone*, depicting buildings in an urban setting and structures along the city's shoreline-trellises, bridges, arcades, and, most impressively, a triumphal arch with a bronze statue of a *quadriga* on top. Dating to the second century AD, the mosaic would appear to have been from a private residence. In a private, domestic setting such as this-though of course the room with this mosaic could have been used as a reception room-the picturing of part of the city could be interpreted as voicing or enhancing a sense of belonging to the city, that is using the architectural/geographical images to prop up personal identity.

As is so often the case with Roman artworks it is sometimes interesting to consider the balance of presence and absence in certain images or sets of images. It is difficult to believe that those buildings or architectural features in the city of Rome itself which were featured on coins represented all of the most easily identifiable or significant landmark features in the city, though some most certainly were, and that the absence of images of other highly significant buildings and structures is not therefore worth exploring. Of Rome's major arches: the Arch of Titus, the Arch of Septimius Severus, and the Arch of Constantine were not represented. Of

---

[27] Elkins 2015: 86-87.

Figure 30. Relief from a sarcophagus, depicting a busy harbour scene at *Portus*. Rome. Mid-third century AD. *Musei Vaticani*, Rome. (Photo: Author).

Figure 31. Marble Christian sarcophagus relief from Rome depicting ship approaching the harbour at *Portus* and its lighthouse. Fourth century AD. *Musei Vaticani*, Rome. (Photo: Author).

the columnar monuments: Trajan's Column, the Column of Marcus Aurelius, and the Column of Antoninus Pius were not common. As to the fact that the *Templum Pacis* or Temple of Peace in Rome does not appear as a coin image is certainly a little detrimental to my argument presented at length elsewhere in the book that this particular building was hugely important as a centre for the dissemination of a number of types of geographical knowledge at different times and yet which would appear not itself to have been any kind of visual signifier for the city.

But the city of Rome could also be a woman, or rather an image of a woman-*Roma* (Figures 32-33). She appeared in sculptural form, in wall paintings, and on mosaics, and on numerous coin issues. Her martial aspect was usually stressed, particularly in Rome, Italy, and the west. In the Roman east she could take the form of a figure wearing a mural crown, representing city walls, and holding a *cornucopia*, like the *Tychai* protecting individual cities there, the most significant being the Tyche of Alexandria, the Tyche of Antioch, and the Tyche of Constantinople[28] (Figures 34-35). It was hardly surprising, given his affiliation with so many eastern Roman cities and his general Hellenism that the emperor Hadrian commissioned the building of a vast temple to *Roma Aeterna* in Rome itself. The idea of Rome may have needed stating and perhaps regular restating elsewhere around the empire and even at the empire's frontiers. The building of a small temple or shrine to *Roma Aeterna* at Corbridge on Rome's northern British frontier surely reflected this need.

Finally, an opportunity will be taken to consider what evidence we have from inscriptions that indicate some degree or identification with Rome, or indeed of specific districts within the city, and how this might suggest that geo-local identificatory tags represented another way for Romans to identify with their city and thus with its place in the world.[29]

If Rome as a single, monolithic, indivisible entity was difficult to manage, let alone conceptualise, then analysis of its various administrative divisions might shed some light on the management of the city. The *vicus* (plural *vici*) was a neighbourhood or street and can certainly be seen to have been an evolved and recognised entity in the census at the time of Julius Caesar. If Servius Tullius had divided up the Republican walled city into the four voting/tribe regions of *Suburana*, *Esquilina*, *Collina*, and *Palatina*, then the *vici* were subdivisions. Then in 7 BC fourteen regions were defined and created by Augustus, separate from the voting regions. These territorial divisions probably both reflected and created a knowledge of the city that went far beyond administrative functionality. The *vici* were not though simply conceptual areas: rather they were to some extent marked out on the ground by an altar at the main crossroad or *compitum* in each *vicus*, hence them being known as compital altars, all set up in or after 7 BC. All were small, decorated, square or rectangular, and made of Luna marble. Each was unique to its *vicus*. The street altars dedicated to the *Lares Augusti* were set up to mark the boundaries of each *vicus* and a number of these have come down to us. Five definitely Augustan and eight possibly Augustan altars survive, the most famous of which is the *Vicus Sandaliarus* altar, now in the *Galleria degli Uffizi* in Florence.

---

[28] On *Tychai* see, for example: Belayche 2003; Cameron 2015; Matheson 1994; and Vermeule 1986.
[29] On the geo-locating of Romans in Rome see, for example: Goodman 2018 and 2020; Lott 2004; Malmberg 2009; Panciera 1970; Ross Taylor 1954; and Wallace-Hadrill 2003.

Figure 32. Wall painting of *Roma*. Originally fourth century AD, San Giovanni in Laterano, Rome. Probably a heavily-overrestored Venus. *Palazzo Massimo Museo Nazionale Romano*, Rome. (Photo: Author).

Figure 33. Face of painted altar bearing image of *Roma/Tellus*, Milan. Late first to early second century AD. *Museo Civico Archeologico*, Milan. (Photo: Author).

Figure 34. Detail of the Base of Tiberius, showing one group of the fourteen *Tychai* of Asian cities. Pozzuoli. AD 30-31. *Museo Archeologico Nazionale*, Naples. (Photo: Author).

Figure 35. Head of Tyche, Classe. Second century AD. *Classis Ravenna Museo della Città e del Territorio*, Classe. (Photo: Author).

Figure 36. Tombstone of Lucius Aurelius Hermia, butcher on the Viminal Hill, Rome. First century BC. British Museum, London. (Photo: Copyright Trustees of the British Museum).

The question is whether the *vici* of Rome were a purely Augustan invention for administrative convenience or whether his administrative reorganisation simply reflected a *de facto* situation with regard to neighbourhood identities in Rome. In other words was this a top down reform or a bottom upwards one? The number of private individuals, mainly workers and artisans of one sort or another, who stated their *vicus* on their tombstones is most interesting in this respect. Inscriptions naming twenty four individuals from fourteen *vici* are known. Yet the inscription commemorating the *pigmentarius* Q. Fabius Theogonus omitted a *vicus* name but included much more specific geo-locational detail: that he worked on the Esquiline, near the statue of Plancus. The funerary memorial to the butcher Aurelius Hermia identified him as being 'of the Viminal Hill' (Figure 36). The funerary monument of the shoemaker Caius Iulius Helius, dating to the early second century AD, found on *Via Leone* in Rome and now in the *Centrale Montemartini* outpost of the *Musei Capitolini* in Rome, located him and his business very specifically *at the Porta Fontinalis*. The funerary stele of Sextius Vetulenus Lavicanus, an employee at the *Circus Maximus*, declares that 'the seventh and sixth region loved (him) equally', again another interesting set of geo-locaters. Only three *vici* names appeared on the Severan Marble Plan, suggesting that by then these were not Roman sub-divisions that counted in terms of moulding broad identification and recognition.

In conclusion, there were many ways for Romans to identify with their city and appreciate it. For the elite this was through literary narratives and poetry, especially those relating to the equation that the city was inextricably linked to its underlying topography of seven hills.

The art of freedmen and freedwomen also learned to engage with contemporary geography. For the artisan class of Rome the geography of its streets was not static: indeed, it might have been thought that the naming of a city street after the predominant commercial activity there might have set that street and its related activity in aspic. But as I recently pointed out in a book on the identity of Roman workers this was not the case at all. Identification for the elite could also be expressed through a geography of family lineage, while administrative city divisions such as voting tribes, *pagi* (for those in rural areas), *vici*, and *regiones* were important tools to focus the minds of Rome's other classes.

In the next chapter attention will be turned to the personification of the River Tiber and of other rivers.

Chapter Four

A River Without End

Rivers of the Windfall Light

In this chapter discussion will centre on the significance of topographical geographical personifications, particularly of rivers, in Roman ideological art.[1] As Roman studies now seems to be taking the first tentative steps in what might best be called an ecological turn there looks to be the start of a trend for archaeologists, classicists, and ancient historians writing about rivers, as well as trees, as will be discussed elsewhere in this book, best exemplified in the recent book-length studies by Brian Campbell and Prudence Jones.[2]

When many Greek and Roman geographers wrote about rivers, Strabo being a particularly good example of this trend, they would consider the issue of navigability and distance, how the river marked out lands, how it came to define a particular region, even to define a particular locality. Rivers here were described almost as active agents in the formation of identities. But rivers were important to the Romans because they helped to define boundaries between what was in (the empire) and what was out(side), though there were highly-significant spaces which lay outside the borders which still served to enhance Roman state power. Rivers marked boundaries, edges of territories or worlds, places of transition, and thus were of great strategic and ideological significance.

It must be stressed that the linking of space, environment, and territory to the political ideology of power was not a uniquely Roman phenomenon and we can find numerous similar situations arising in Pharaonic Egypt, in Mesopotamia, in the Seleucid empire, and in the Hellenistic world.

But rivers were also a resource to the Romans, something to be exploited, a general view of the passivity of the natural world that indeed defined Romanness. Writing during the reign of Augustus, Dionysius of Halicarnassus in his *Roman Antiquities* stated that 'the extraordinary greatness of the Roman empire manifests itself above all in three things: aqueducts, paved roads, and the construction of drains'.[3] The *Aqua Appia*, the first aqueduct built to service Rome, was constructed in 312 BC and four more were constructed before the imperial era. But it was Marcus Agrippa who devised and managed the Augustan programme for both bringing water into the city by further aqueducts and for dealing with its disposal once there and dirtied or polluted by drains or sewers. Between 33-19 BC he oversaw the completion of two new aqueducts, an Augustan waterpark in the *Campus Martius*, comprising a pool, an artificial canal, a basilica dedicated to Neptune, a portico decorated with scenes from the *Argonautica*,

---

[1] On personification and personifications in the Greco-Roman world see, for example: Cameron 2015; Houghtalin 1996; Hughes 2009; Huskinson 2005; Juhász 2016; Lee 2006; Meyer 2021; Ostrowski 1990a, 1990b, 1991, and 1996; Shapiro 1993; Stafford 2000; Stafford and Herrin 2005; and Webster 1954. On Italian Renaissance river personifications see principally: Lazzaro 2011.
[2] Campbell 2012 and Jones 2005. On various aspects of rivers in the Roman world see, for example: Aldrete 2006; Braund 1996; Campbell 2009, 2010, 2012, and 2015; Coombe 2022; Franconi 2015; Gais 1978; Gall 1953; Goodfellow 1981; Hermon 2010; Huskinson 2005; Jones 2005; Lazzaro 2011; Lee 2006; Meyers 2009; Östenberg 1999; Ostrowski 1990b and 1991; Purcell 2012 and 2015; Rankov 2005; Taylor 2000 and 2009; and Tekin 2001.
[3] Dionysius of Halicarnassus *Antiquitates Romanae* 3.67.5.

and the Baths of Agrippa, along with the renovation and cleaning of the *Cloaca Maxima*, Rome's major sewer, in 33 BC. Aqueducts moved water from *there* to *here* and thus literally brought the assets of the countryside in the form of its clean waters into the city where the Tiber was too polluted to provide safe, clean drinking water, and where the number and capacity of the city's wells could no longer help sustain Rome's growing population and the ideological requirements of its rulers.

Cicero[4] proudly noted that 'We keep rivers at bay, channel them, and divert them. In the domain of Nature we are trying to bring into being a second Nature with our own hands'. Josephus[5] wrote almost in awe of the Romans' power over water. In Juvenal's third Satire he wrote about the Syrian Orontes flowing into the Tiber, bringing with it abundant human detritus in the form of foreign and, as he saw it, malign foreign influences.[6] This remarkable telescoping of distance and manipulation of geographical space and territory is hugely telling, as is the interconnecting of these two great rivers as part of one great riverine system, the Orontes now a mere tributary or feeder into the Tiber in Rome.

It will be seen in a later chapter on landscape art that water could be presented as a landscape element, with rivers, streams, lakes, and the sea regularly appearing in Roman and Campanian wall paintings without recourse to the use of personifications, not to mention the many very specific appearances in Roman art of the Nile and the landscapes that it created and nourished. But rivers and the seas were not always portrayed realistically or semi-realistically, and were often represented visually by anthropomorphised deities-river gods, Neptune, water nymphs and so on. Here, as so regularly in Roman art, geography was being written on and by bodies.

Simultaneities

The concept of the personification of a natural physical feature in human form had its origins in Greek art, but Roman personification was to some extent more complex. Images of rivers, including personified images, were common, among a plethora of visual motifs, used for illustrative purposes and as a trope in Roman imperial triumphs for example. In such contexts they came to represent the conquered lands in which they flowed and, by extension, the conquered peoples of those lands. Caesar paraded at his triumph in 46 BC 'the Rhine, the Rhone and an Ocean of gold'.[7]

Lucius Cornelius Balbus while governor in Africa in c. 20 BC made an expedition deep into Libya: at a triumph to celebrate this pictures of mountains and rivers were carried. As an illustration of this complexity one can consider the Roman poet Ovid in his *Tristia* of AD 10-11 where he describes a fictional triumphal procession to celebrate and memorialise victory over the Germans at which he ventriloquises the reactions of a bystander on the processional route who declares that a personified statue representing the River Rhine was discoloured by staining from its own shed blood, as if the river could somehow be wounded like a man or woman and bleed.[8] Water was thought by some in the Roman world to represent some kind

---

[4] Cicero *De Natura Deorum* 2.152.
[5] Josephus *Bellum Judaicum* 5.410.
[6] Juvenal *Saturae* 3.62.
[7] Florus *Epitome* 2.13.88.
[8] Ovid *Tristia* 4.2.

of freedom and self-dissolution, indeed as a way of making contact with the past and with what were thought to have been earlier, simpler stages of life. The flow of water in rivers was sometimes conflated with ideas and thoughts about time, memory, and intuition. The human life-course was like the topography of a river along its length, from spring to stream to mature river to the merciful release at the estuary out into the sea.

Trying to understand how the Romans viewed rivers in terms of their usefulness as barriers or frontiers can also provide further insights into the relationships between geography and power in the Roman world. But, of course, rivers were equally seen as vital for communication and trade: they were not barriers per se but rather routes for lateral communication. That the Rhine and the Danube became frontier boundaries need not necessarily be viewed as a deliberate actioning of a well thought-through policy, but rather as a pragmatic response to historical events and circumstances. These were not 'natural barriers'. The connections between rivers, springs, and water in general and certain Roman religious practices are generally accepted as part of a wider phenomenon linked to the idea of *locus* or place and to the spirits of these places-the *genii loci*. If there is no direct Roman equivalent to the writings of Homer on rivers-where rivers go to visit the gods at Mount Olympus, receive sacrifices, procreate, speak and fight-at least they were often perceived in Roman writings sometimes as paternal-*pater Tiberinus*, *pater Padus*, and so on. Bearded, elderly river deities were venerated for their age, their life-giving capacity, and their wisdom.

Some of the most interesting questions about the motives and strategies behind the creation and presentation of personification of rivers to ideological ends have been asked by Claudia Lazzaro in her 2011 study of river personifications in sixteenth century Italy. As she noted: 'artists explored how the human body could represent the natural world, how anthropomorphisms, rivers given human form, could function as personifications, figures who speak and act, and how the human body could embody the meaning imprinted on it. At the same time patrons discovered the potential of river gods as vehicles for political messages about rule over territory'[9] That Renaissance rulers, in dynamic cultural centres throughout Italy, at Florence, Mantova, and Ferrara for instance, utilised geography products such as these to express political power and ideology, is of huge importance to this present study centred on much earlier times.

Relating to a much later period still I have been given pause to thought at ideas about gendered personification as a process and ideological statement as expressed by Klaus Theweleit in his study of the images of women and women as metaphors in the collective unconscious of the German Freikorps of the period immediately after World War One. Woman, he writes: 'a river without end, enormous and wide, flows through the world's literatures: the women-in-the-water; woman as water; as a stormy, cavorting, cooling ocean'.[10]

Of course the Romans were not alone as a culture in using personifications of rivers as images to convey geographical information, in their case to a very large receptive audience keen to understand the world and Rome's place in it. For example, a series of fifth century BC coins minted for the town of Gela in Sicily carry depictions of the River Gelas as a potent, aroused bull with a bearded man's head, a most curious image indeed. The River Amenanos is depicted

---

[9] Lazzaro 2011.
[10] Theweleit 1987.

in a similar way on a c. 460 BC coin issue of Katane/Catania, again in Sicily. Yet the River Hipparis appears as a youthful, slightly-androgynous male, portrayed just by his head and shoulders as a bust, on two coin issues of Kamarina, once more on Sicily, dating to c. 410 BC, one image type attributed to Exakestidas and the other to Euainetos. The same strategy of youthful personification was applied to the River Gelas on two Sicilian coin issues of the 420s BC, one a coin of Gela and the other issued by Piakos. On the coins of Olbia on Sardinia, dating later to 280-270BC, the River Boristhenus is represented by the bust of a straggly-bearded, elderly man with flowing locks of hair. It is possible that a youthful, muscly reclining male figure on the West Pediment of the Parthenon in Athens, dating to 447-432 BC, represented Ilissos, a stream in Athens.

Greek river gods often took the imagistic form of the man-headed bull, possibly derived from Near Eastern prototypes, sometimes just a figure with short horns. One of the most famous rivers of antiquity was the River Acheloös or Achelous of north-western Greece. Rising in central Epirus and flowing out into the Ionian Sea, the river is known today as the Aspropotamo. The figure of Acheloös battling with Hercules, shape-shifting between a bull and a serpent, would seem to have been reflecting the ever-changing nature of a river itself. He was a very common figure in Etruscan art, on metalwork and even as terracotta antefixes of his horned head.

A wall painting from the *Sede degli Augustales* or Hall of the Augustales at Herculaneum depicts Hercules and Acheloös engaged in battle over Deianeira who stands in the background looking on. This particular panel on the west wall of the building's *sacellum* or strongroom faces a pendant panel on the east wall depicting Hercules again, this time being welcomed into Olympus, the home of the gods, by Jupiter in the form of a rainbow, Minerva, and Juno. This decorative scheme, and other elements including small panels carrying depictions of winged victories driving triumphal chariots, needs to be interpreted in terms of its relationship to ideas centred on the promotion of the imperial cult of the emperor Augustus. Of course, Hercules, the mythical founder of Herculaneum is an altogether appropriate figure to find here, pictured in life and in the heavens after death and divine transformation. Ovid in his *Metamorphoses* relates the tale of both the battle with Acheloös and the death/transformation.[11] That the battle between Hercules and Acheloös is also dramatically depicted on one of the third century AD black and white mosaic floors from the Villa of Nero at Anzio and now in the *Museo Nazionale Romano Terme di Diocleziano* in Rome shows the longevity of this popular story and trope.

It is also worth considering here the iconography of the Egyptian god Hapi, not a personification as such in the Greek or Roman sense of the process and not a river god *per se*, but rather a manifestation of the huge benefits of the annual flooding of the Nile. Not to be confused with the identically-named funerary deity Hapi who was one of the four sons of Horus, the god Hapi in question here was depicted in bodily form but in a manner which questioned gender identity in its presentation. Generally pictured as a long-haired, androgynous young man with blue or green skin wearing only a loincloth and with a ceremonial false beard Hapi also had large drooping, pendulous breasts and a large, distended belly, the latter two physical traits suggesting pregnancy and a generalised fertility. Yet epithets attached to him also called him

---

[11] Ovid *Metamorphoses* 9.

Figure 37. Massive statue of the personified Tiber (originally probably the Tigris), *Campidoglio*, Rome. Originally in the Baths of Constantine, Rome. Early fourth century AD. (Photo: Author).

the father of the gods, thus further blurring the boundaries. He was not the god of the Nile but rather of its annual inundation. In Lower Egypt his most common attributes were papyrus leaves and river frogs, while in Upper Egypt more common were lotus plants and crocodiles. He was usually depicted bringing offerings or pouring water out of an amphora, both acts symbolic of the blessings and benefits of the flood and the regenerative properties of the fertile alluvial silts left behind each year.

Though said to reside in a cave near Aswan, at the source of the Nile, the cult around him was probably centred on the area of the First Cataract of the river at Elephantine. The priests of the cult would also seemed to have played a particularly practical role in managing his yearly arrival by managing and controlling the flow of water with reference to an architectural measuring device of some kind known as the Nilometer. Future predictions based on accurate readings gave mystique and kudos to the priests overseeing what was in reality a broadly scientific approach to water management disguised as religious observance.

The Romans therefore had a considerable artistic heritage to draw upon in depicting and personifying their own home river, the Tiber[12] (Figure 37). That *Tiberinus*, always depicted as a reclining, bare-chested old, bearded man with long shaggy hair, sometimes crowned with

---

[12] On *Tiberinus* and the Tiber see, for example: Le Gall 1953; Meyers 2009; Warde Fowler 1916; and Warner 1917.

vegetation, most usually appears holding a cornucopia, a *canna palustre* or bundle of reeds, an oar, or a boat's prow. While Aeneas reveres *Tiberinus* as 'horned river god' horns never appear in the numerous images. The important thing to consider when discussing the Tiber is to not simply define the river as being of the city, in the same way that the seven hills came to topographically define Rome. The river ran through the city but connected it with a broader world both upstream and downstream. The Tiber formed one of the outer boundaries of Rome during much of the Republican period and then became integrated into the city during the empire. It connected Rome to a dispersed hinterland, comprising the Tiber valley leading to the northern ports and the Mediterranean Sea itself through the southern ports.

Quite tellingly, the figure of *Tiberinus* does not appear in Roman literature and art until almost the end of the first century BC: in other words contemporary times and the political ideologies of those times required his creation. Book 8 of Virgil's *Aeneid* starts with action at the mouth of the Tiber where the river god *Tiberinus* is encountered by the Trojans. *Tiberinus* appears to Aeneas[13] and he is variously 'an anthropomorphic male water deity, an abstracted spirit (genius) of the waters and a..zoomorphic creature'. In the extended description of the decoration on the Shield of Aeneas Virgil tells us how rivers encircle the world, with the Mediterranean Sea at its centre. Rome's world geography was being mapped out here.

Writers and historians lauded Rome's river as if a person. The poet, writer, and historian Ennius, probably writing in the 190s BC, called the Tiber 'the river most eminent of all'. Virgil, virtually the court poet of Augustus, called it 'the ruler of waters'. An unknown orator, possibly Fronto, called it 'the lord and ruler of circumfluent waters'.[14]

An account given by Tacitus of the debate in the Roman senate about flood-prevention measures proposed for the Tiber suggests that some considered this almost a form of emasculation or castration of the mighty river that helped birth the city of Rome itself: 'Tiber himself would be altogether unwilling to be deprived of his neighbour streams and to flow with less glory.'[15] The river Tiber had taken on almost a religious aspect. Indeed, the *Pons Sublicius*, the first bridge across the Tiber became a locus for periodic but regular rituals involving the Vestals and *pontifices*.

The Tiber as a personified figure appeared on reverses of Neronian, Flavian, Trajanic, Hadrianic and Antonine coins. A large second century AD statue of *Tiberinus* now in the *Musée du Louvre* in Paris was probably a pair with the figure of the Nile now in the *Musei Vaticani* in Rome: originally they would have stood in the vicinity of the Temple of Isis in the *Campus Martius*. *Tiberinus*' attributes include a *cornucopia* and the *canna palustre*, an oar or prow. Early on his image appeared in foundation scenes relating to Rome's deep, mythic past. For example a painted frieze from a mid-late first century BC *columbarium* on the Esquiline Hill features Aeneas, Romulus and Remus and others, along with *Tiberinus* who appears twice, observing Mars abducting Rhea Silvia and later reclining on the riverbank and gesturing welcome to the basket containing Romulus and Remus. It has been suggested that rather than *Tiberinus* this rather could be the god of the *Numicus* or *Anio* rivers. Other appearances include: with Mars and Rhea Silvia in the *Casa delle Origini di Roma* in Pompeii; on the *Ara Casali*; on a fresco from

---

[13] Virgil *Aeneid* 8.31-78.
[14] Ennius *Annales* 1.67 and Virgil *Aeneid* 8.77.
[15] Tacitus *Annales* 1.79.

Hadrian's Villa at Tivoli; on an altar from Ostia dated to the first century AD; and on a third century AD sarcophagi with Mars and Rhea Silvia.

A rare depiction of the Tiber as a flowing river, a natural entity, along with the personified body of the river god, appears on the Arch of Constantine in Rome. The easternmost section of the south face frieze depicts Constantine's victory at the Battle of the Milvian Bridge, illustrated in quite graphic detail, with the figure of Constantine, no longer extant, placed firmly in the action to the right of the scene, standing on a barge on the river with Victory and *Dea Roma* and viewing the carnage around as Maxentius's troops are slaughtered or drowned in the Tiber. The river god himself rears up in the churning water in front of the imperial barge. On the far left of the scene two musicians, possibly the same two from the depiction of the exit from Milan, strike up a victory fanfare. The events of 28th October AD 312 are shown in two registers. In the background are the emperor and gods and goddesses, the musicians, and the rest of his army, infantrymen, cavalry, and bow-firing Moorish auxiliaries, and in the foreground is the fast-flowing River Tiber with its waters enveloping Maxentius' heavily-armoured troops who have fallen or been driven into the river. The body of one soldier is shown tumbling off the riverbank, killed by one of Constantine's men. The writhing, interlocked bodies of the soldiers in the river assume the character of a netted shoal of fishes, agitating the waters, their scale armour flashing and glinting like fishscales. On the bank the depiction of a trio of Moorish archers, all posed in exactly the same way, suggests the regimented, methodical nature of the slaughter. On the far left a Maxentian soldier kneels begging for mercy. The Tiber here literally is fighting for the emperor.[16]

In the case of Rome the River Tiber linked the countryside to the city, and vice versa, both literally, physically, and metaphorically. Pliny the Elder[17] wrote that the Tiber 'almost on its own to a greater extent than all the other streams of all other countries put together, is the focus of the panoramas of country estates and the object of their cultivated attention'. To paraphrase Guillaume Apollinaire's 'Ocean of Earth', it seemed as if Rome had built itself in the middle of the ocean, rivers flowing from its eyes, an ocean it knew so well but which was never still. A trickle of images became a river, and then a flood. Like the regenerating Nile, that colonial landscape that so exercised the Roman mind, their fertile soil gave birth to ideas about the city of Rome and its place in the order of things, at the centre of the world, at the world's ends. It must have been both comforting and yet at the same time discomfiting to be both here and there, everywhere and nowhere, to understand that Rome was new and older at the same time. This simultaneity, this suite of effects, served to render disparate spaces, worlds, and peoples as the susurrus around connectivity, as hauntological moments and events, spectres of empire, spectral structures on the seven hills.

The great Severan marble map of the city of Rome, the *Forma Urbis Romae*, included no topographical features on it, meaning that the seven hills that Roman poets believed gave Rome its unique character, did not feature. Likewise the River Tiber was both present and absent, its position and course marked as a negative, that is an area marked principally by the absence of streets and buildings, a hauntological negative presence.

---

[16] Ferris 2013: 76-78.
[17] Pliny *Naturalis Historia* 3.54-55.

There was no Roman equivalent of the mighty British native oak trees whose very being and qualities came to be evoked as enabling and emblematic of British sea power in the mid-seventeenth to mid-nineteenth centuries. In this case part of the British landscape came to be seen as symbolic of the power of the nation itself. For Rome the Tiber and the Seven Hills played this role.

Very few other rivers in Italy were discussed by Roman historians and writers, with the notable exception of the *Rubico*(n) and the *Padus* or River Po of northern Italy, the latter's name usually being employed in a geolocational context. For example, Cispadana was the term used to refer to the area next to the Apennines and Liguria, and Transpadana the area to the north. Cornelius Nepos was described as *Padi accola*, that is resident along the Po.

In his *Bellum Civile* Lucan relates the folk tale of how, as Julius Caesar made to cross the River Rubicon and thus to all intents and purposes declare a civil war by doing so, the river swelled up-*tumidumque per amnem*- to try and hinder his crossing,[18] all eventually to no avail. This act of defiance by a natural feature lacking human agency and intent suggested that even the Roman landscape opposed Caesar's presumption and might. Rivers acting against humans in a landscape of war-a term that will be returned to later in the book-has a much earlier precedent as a literary trope through its appearance in the *Iliad*, where Homer tells the story of the River Scamander outside Troy and the Greek hero Achilles. The personification of Scamander angrily reacts to the pollution of his waters by the mass dumped bodies of the Trojan dead, hurls them out, and floods his banks with an angry torrent aimed to sweep away Achilles who has cause to call on Zeus for aid.

Rising Waters

As common in Roman art, if not more so, was the River Nile (Figures 38-40). Personifications of the Nile appeared in many media, a dense topographical portrayal of the river's various stretches from source to mouth most famously appearing on the staggering early second century AD Nile Mosaic from Palestrina near Rome, discussed in Chapter Five. The Nile often appeared twinned with images of the Tiber, in a complex ideological pairing that constituted the domination of a colonial landscape, again as discussed in detail in Chapter Five. Again, to give a third example, the bearded River Nile appears as one of many figures on the *Tazza Farnese* in the *Museo Archeologico Nazionale* in Naples. At Hadrian's Villa at Tivoli, just outside Rome, the famous *Canopus* portico surrounds a pool, a feature meant to be symbolic of the River Nile, while a statue of Hermes stood watch here, as might have been expected from the god who acted as the protector of travellers.

Personifications of rivers were employed on a number of imperial monuments, including the Arch of Titus, the Parthian Monument of Lucius Verus, the Arch of Septimius Severus, and the Arch of Constantine, and on numerous coin issues. The personified Euphrates and Tigris appear on the Arch of Trajan at Benevento (Figure 41).

The image of the personification of the River Danube on the lower register of the decorated helical frieze around Trajan's Column in Rome (Figure 42) echoed a fondness for the portrayal

---

[18] Lucan De *Bello Civili* 1.204.

Figure 38. Massive statue of the personified Nile, probably from the Temple of Isis in the *Campus Martius*, Rome. Very late first century AD. *Musei Vaticani*, Rome. (Photo: Author).

Figure 39. Massive statue of the personified Nile, *Campidoglio*, Rome. Originally in the Baths of Constantine, Rome. Early fourth century AD. (Photo: Author).

Figure 40. Gold *aureus* coin issue of Hadrian, Rome mint. AD 130-138. Reverse image of the personification of the Nile. British Museum, London. (Photo: Copyright Trustees of the British Museum).

Figure 41. Detail of relief panel showing submission of personification of Mesopotamia to Trajan, with personifications of the rivers Euphrates and Tigris in attendance. The Arch of Trajan at Benevento. AD 114-118. (Photo: Author).

Figure 42. The personified figure of the River Danube. Scene III, Trajan's Column, Rome. AD 113 (Photo: Author).

of similar personifications of this great river on a number of coin issues under Trajan (Figures 43-44). Landscape and geography played a significant role in the decorative frieze scenes on Trajan's Column. Indeed, it could be argued that the Roman forces at war in Dacia were as much in conflict with the natural environment here as they were with their flesh and blood Dacian opponents in Trajan's wars. Pliny the Younger in his Panegyric to Trajan wrote: 'Though he (Decebalus) is defended by the seas between, the mighty rivers or sheer mountains, he will surely find that all these barriers yield and fall away before your prowess, and will find that the mountains have subsided, the rivers dried up and the sea drained away, while his country falls a victim not only to our fleets but to the natural forces of the earth'.[19]

In this way the artists of the frieze situated the action quite specifically in an alien natural terrain: by fording and crossing rivers, by chopping down forests, and by storming a mountain stronghold the Roman army was not being presented here as a fighting force but rather emphasis was also being placed on the dignity of their labour and how this impacted on their efforts to confront and then tackle physical and natural barriers to progress and victory. It has been noted how (Plates CIX, CXIII) evident contrast is made between a woodland god contemplating a lake and forest rich with game animals and a scene of a single, desolate tree, stripped of its branches and standing forlorn and possibly irretrievably damaged. Together,

[19] Pliny *Panegyricus* 68.6.13-16.

Figure 43. Bronze *sestertius* of Trajan, Rome mint, AD 104-111. Reverse of personification of *Danuvius* (River Danube) throttling and subduing the personified *Dacia*. British Museum, London. (Photo: Copyright Trustees of the British Museum).

Figure 44. Bronze *sestertius* of Trajan, Rome mint, AD 116-117. Reverse of emperor standing over seated personifications of the Rivers Euphrates and Tigris, with personified Armenia seated left. British Museum, London. (Photo: Copyright Trustees of the British Museum).

Trajan's Column and the Column of Marcus Aurelius made a precise but perhaps tangential reference to the homogenisation of empire: the columns represented phenomenologies of verticality that were central to Rome's imaginative and physical experiences of war, conquest, colonisation, and exploitation.

Another example of personifications of both rivers and cities being used together on an imperial monument to provide geographical information to viewers is the so-called Parthian

Figure 45. Stone head of Rhine god *Rhenus* from a mausoleum, Bonn. Second century AD. *Rheinisch Landesmuseum,* Bonn. (Photo: Carole Raddato).

Monument of Lucius Verus (sometimes also known as the Great Antonine Altar) built at Ephesos in Asia Minor, though most of the sculpture from the monument is now in the Ephesos Museum of the *Kunsthistorisches Museum* in Vienna in Austria. Probably built soon after Verus' death in AD 169, of most significance in the context of this present study is one panel, known as Slab T, which depicts two standing female figures, one holding an ornamental object in the form of a ship's prow in the crook of her left arm but the attributes of the other female are unfortunately now missing. Behind them is a container full of ears of corn. Reclining at their feet is a bearded and straggly haired river god, his upper torso bare and a garment draped over his arms and covering his lower body. He has scales on his body and holds a fish. It has been suggested that we are seeing here personifications of the cities of Ctesiphon and Seleuceia, along with the River Tigris.

It is not altogether certain how the recovered pieces of the frieze fitted together originally, and indeed how many separate lengths of frieze are represented here in part. The reconstruction of a whole length of frieze dedicated to the portrayal of eleven pairs of standing personifications of cities, each accompanied by a reclining river god appropriate to the cities' locations by W. Oberleitner is interesting but ultimately unconvincing. However, the presentation of a gallery of personifications of cities and rivers such as this would certainly have been a way to present to the viewer an interpretative locus for placing the military manoeuvres and battle scenes on other frieze fragments in their geographical context, in the same way that the set of panels individually portraying personified pacified provinces-the *Provinciae Fideles*- on the decorative frieze on the *Hadrianeum* or Temple to the Deified Hadrian in Rome likewise took viewers on a whistlestop tour of the empire in the same way that the emperor Hadrian had famously journeyed all around it.

On a coin of Trajan the River Danube seizes Dacia by the throat, in the service of Rome. Trajan was said 'to conquer the Danube and Euphrates with bridges'.[20] Pliny both lambasted Domitian for failure in Dacia: 'Rivers witnessed this shameful travesty; the Danube and the Rhine were delighted...to play their part in your disgrace'[21] and praised Trajan in a similar way: 'You were scarcely more than a boy when...the Rhine and Euphrates were united in admiration of you.'.[22] The Nile, the Tiber, the Rhine, and the Euphrates all also appeared as named personifications on certain coin issues and in other media between the reigns of Hadrian and Constantine (Figure 45). It is possible, though I think unlikely, that one or both of the unnamed river gods (or marine deities) on a Hadrianic coin issue and one of Septimius Severus could also be personifications of the Tyne, given the particular links of both emperors to military activity in northern Britain.

Well away from Rome, on the north-west frontier in northern Britain, it is worth noting the finding here of a sculpture of a large bearded male deity, recovered by excavation from the bath suite of the commandant's house at Chesters Roman fort on Hadrian's Wall. The long-haired reclining figure has a bare upper torso, with the rest of his body covered by a draped garment of some kind. A mask of a similarly-bearded, long-haired male sits by his left elbow. There is nothing in the way of attributes to suggest that this is an image of the god Neptune, though such an image would have been entirely appropriate to a bath house setting, and the general consensus among academics is that the figure is a river god, probably in this context the personification of the North Tyne that flows nearby. It has been suggested that this particular statue could have been one of a series of such images that adorned the nearby massive Roman stone bridge over the river and that it was subsequently taken off that structure for some reason and repurposed in the bath house. This is an attractive idea, but one for which there is presently no supporting evidence. Dated to the second century AD, though it could in fact be as late as the early third century, the Tyne river god is quite literally an image created to evoke the northern environment, more allusively portrayed on many other sculptural works from the region. Given that northern Britain was a frontier region it is highly likely that the importance placed by the Romans in rivers was not just related to their status as landscape features, but also to their prescriptive role as natural boundaries, and thus as potentially-important defensive features, their existence and courses being integral to the planning and construction of frontier works.

In a purely Christian context the use of personifications of rivers can be seen to have continued as a practice in the case of a sixth century AD carved elephant ivory panel, possibly inlay from a box or casket, made in either Egypt or Syria and now in the collections of the British

---

[20] Ammianus Marcellinus *Res Gestae* 24.3.9.
[21] Pliny *Panegyricus* 82.4.
[22] Pliny *Panegyricus* 82.4 and 13.4.

Museum in London. Though damaged and fragmentary, the carved relief scene on the panel can quite clearly be seen to be the baptism of the young Jesus in the River Jordan by John the Baptist. An angel looks on from above and the hand of god can be seen in the sky beside her, a ray of light emanating from it, and a dove flies down from the heavens holding a bowl in its beak. Christ can clearly be seen up to his waist in the water, and in the foreground of the scene reclines the small, bearded figure of the personification of the River Jordan itself, crowned by reeds and raising his right hand up in salutation or in blessing. The contrast between the figure of the Jordan here taking part in what is presented as an epochal moment and the Jordan on the Arch of Titus, discussed above, could not be greater.

Just as Rome had been influenced in its use of geographical personifications by other cultures, such as those of the Etruscans and Greeks, so we can find an example of Greco-Roman influence on the process in one instance far beyond the limits of Rome's empire. Gandhāra was a region in Peshawar in north-west Pakistan and north-east Afghanistan, centred on the confluence of the Swat and Kabul rivers that for a time became a melting pot of local artistic traditions with Persian, Greco-Roman, and Buddhist influences and borrowings, reflecting its strategic position along the Silk Road.[23] In the present context of greatest interest is a first to second century AD statue of a reclining river god in the National Museum of Pakistan in Karachi. In terms of both the subject matter of the piece and its composition a direct Roman model is apparent. The naked, bearded god rests one hand on his leg and in the other holds a cornucopia or horn of plenty. A small lion sits at one side. In Indian art personifications of rivers were usually female, so this example is unusual not just because of its style but also because of its potential to have disrupted gender expectations.

There were a number of ways in which citizens of Rome could learn about other lands if they did not or could not access such information through the work of Rome's historians and chroniclers. They could attend a Roman triumph and, as will be explained in more detail elsewhere in this book, they would have geographical information thrown at them in a number of media, including captured barbarian enemies, sometimes exotic animals paraded as booty, and by displayed paintings with explanatory captions. They could see exotic animals from overseas-so-called charismatic megafauna-at the amphitheatre games. They could see images of foreign peoples on monuments all over the city.

Rivers of Deceit

In addition to rivers some mention at least needs to be made of other watery places-springs and lakes for instance-and to other landscape features created by water movement, percolation, or erosion, such as caves. The association between shrines and healing waters in ancient Italy is well known,[24] particularly in central Italy and Etruria, but these healing cults could also be associated with sacred groves, while fertility cults could also be centred on sacred caves or cave sanctuaries. The role of Hercules in healing is again of great interest. Many of these

---
[23] On Gandhāran connections with Greco-Roman source material and ideas see principally: Rienjang and Stewart 2020.
[24] On Rome and the ideology of water and the politics of water management in general see, for example: Bassani 2019; Bassani *et al.* 2019; Evans 1997; Henig and Lundock 2022; Koehler 2013; Kosso and Scott 2009; Purcell 1996; Rinne 2010; Rogers 2020; and Taylor 2000 and 2009.

markers in the Italian landscape linked topography with what we have called folk belief, as opposed to more scientific or rational modes of healing.

While there has been a certain amount of academic discussion of the Romans' views of certain rivers such as the Rhine, Danube, Tigris, and Euphrates as natural frontiers, as boundaries in other words and concomitantly as defensible positions in strategy and warfare, religious aspects also need flagging up. The environmental psychology of the time allowed for rivers not just to be seen as demarcation lines but also to be valued as natural highways, facilitating and linking areas, economically, socially, and culturally significant in this respect. Roman religion quite literally muddied the waters here in terms of introducing elements of awe and wonder into the mix, with the appeasement of river gods a fixation that belied more practical and pragmatic responses.

In James Joyce's *Finnegans Wake* Anna Livia Plurabelle is a woman transformed into a river flowing to the sea, replenishing the cycle of life, as man, also a river or the sea, is fed by rivers of deceit. The images were almost theatrical, a performed topographical atlas of metaphor and gender, like much of the Roman geographical artworks discussed in this chapter.

There is something almost hypnotic about the evocation of the inexorable and never-ending movement of water in a number of visual contexts in Roman art, the churning Tiber depicted on the Arch of Constantine in Rome being one that has already been discussed. In the unit or office designated *Statio 27* in the *Piazzale delle Corporazioni* in Ostia is a black and white mosaic decorated with almost abstract ripples and zigzag chevrons in three strands, probably depicting the three branches of the Nile Delta, calling up the onward flow of the Nile, so distant from Rome and yet always so near in the Roman imagination (Figure 46). In the church of Haghia Sophia in Constantinople, completed in AD 538 under emperor Justinian, the initially-perceived dull grey of the majority Proconnesian marble slab flooring is brought to life as its veining glitters, these flagstones broken in places by four green stripes of Thessalian coloured marble, evoking waters and rivers. The hypnotic effect of these stripes brings about a feeling in the viewer 'of walking on water', exacerbating a feeling of returning to watery origins or deeply-buried baptismal memories. Equally the stripes could each have been intended to represent the four rivers of Paradise: the Pishon, Gihon, Chidekel (the Tigris), and Phirat (the Euphrates).[25]

Personification of places, including built places such as cities, rivers and oceans, winds and weather phenomena, seasons and months, was really to some extent an extension of the pagan religious system that equated Jupiter's power with the thunderbolt, and had Neptune and *Oceanus* oversee the oceans. To this list can be added *Fulgora* (lightning), *Iris* (the rainbow), and *Tempestas* and the *Tempestates*, a temple to whom was dedicated by L. Cornelius Scipio in 259 BC, near the *Porta Capena* on the *Via Appia*, in thanks for her intervention in saving his fleet from a storm off the coast of Corsica.[26]

---

[25] On so-called cosmic floors see: Barry 2007.
[26] On wind and weather in the Roman imagination see, for example: Kienast 2014; Taub 2003; and Webb 2017.

Figure 46. Black and white mosaic depicting the Nile. *Piazzale delle Corporazioni*, Ostia. AD 150-170. (Photo: Author).

To some extent the Roman geographical view encompassed a totality formed of phenomena that included topography, geomorphology, geology, fluviology, oceanography, and climatology, marrying both the rational and irrational sides of these practices of knowledge. The very specific day on which the German landscapes of war depicted on the Column of Marcus Aurelius were altered by the visitation of the brooding Rain God in Scene XVI of the column frieze well illustrates this point (Figure 47). The Roman fascination with recording, surveying, and mapping of land and topography, of routes, distances, and itineraries was mirrored by a need to also categorise and sometimes name weather phenomena, including the winds. Yet the rational side of Roman interest in the winds was perhaps inspired by elite knowledge of the first century BC Tower of the Winds in Athens, with its wind vane and sundial, and its eight reliefs of various wind gods. The so-called Rose of the Winds was another public monument inscribed on paving in the Forum at Dougga in Tunisia, with inscriptions naming the winds *Auster*, *Argestes*, and *Africus*. The stone, inscribed Pesaro Wind Rose, actually found in Rome near *Porta Capena* on the *Via Appia* but now in the *Museo Oliveriano* in Pesaro in north-eastern Italy, may well have been inspired by the Vitruvian twenty four wind compass rose to map the direction of the major winds (Figure 48). Its relatively small size, compared to the Dougga wind rose, suggests that it was made for a private elite person. Its date is uncertain. Often the deep fragile mirror of the Roman viewer's eye seems to have been drawn to the skies as if the

Figure 47. The rain god. Scene XVI, the Column of Marcus Aurelius, Rome. AD 180-192. (Photo: Author).

Figure 48. The Pesaro wind rose, *Via Appia*, Rome. End of second century AD. *Museo Oliveriano*, Pesaro. (Photo: Author).

sole aim of the world was to glory in itself unceasingly by pushing along shining clouds above human heads: no matter what happened the wind of eternity simply continued to rush on.

A similar public and private interest among the Roman elite in certain types of geographico-scientific knowledge centred around the sundial. Indeed, over five hundred Roman sundials, most made of stone with bronze gnomen, are so far recorded through archaeological work, thirty six in Pompeii alone, which suggests great popularity and ubiquity. Richard Talbert has discussed this phenomenon in the context of the commissioning and manufacture of portable sundials.[27] Made of bronze, around a dozen geographical examples are known from the Roman world: a small number of other surviving ones lack a geographic element. Most are inscribed with place-names, sometimes a city, sometimes a province name. Each was inscribed with a sheet of coordinates for latitude to facilitate the use of the device in specific locations. That these objects were primarily intended for practical use is without doubt, even if the practicability and accuracy of some of them leaves something to be desired. However, this does not necessarily mean that they were used for the purpose for which they were intended: to calculate the time in different locations. In other words to very firmly locate their owners in time and space. Their acquisition may have been more to do with their value as display items, intended to enhance the status of their owners. The most idiosyncratic of those portable sundials discovered so far, dating probably to the early first century AD, comes from Herculaneum and takes the form of a side of ham.

The many different types of geographical information produced and viewed in the Roman world together provide a multivalent impression of Rome's perceived place in the world. The processes of selection, composition, editing, production, and performing or display meant that this was not unmediated information but rather information mediated by culture and ideology. Issues of surveillance, control, access, and framing affected every single piece of information presented somehow as knowledge. The Roman empire moved outward into new geographic space as a way of moving forward in time. Language and metaphors created an ongoing discourse around foreigners that permeated Roman state art, not as a device but more as a demonstration of the omnipresence of the foreigner embedded deep in the Roman psyche. Place was essentially the ghost of all Roman lives. Yet it must be borne in mind that different people would have had different experiences and relationships to land and landscapes, possibly involving issues centred around ownership, exclusion, clearance, capital, segregation, and access. The pleasure of walking and observing was one layer of experience in the geology of the social city. But we must also consider the continual flux and change in a dynamic city such as Rome which probably meant that no path could ever be trod the same or views precisely replicated. Thus the idea of the city as an active space, as a concept, a lived environment, an experience, may have interacted with a kind of nostalgia, not for the past but for a lost future that was never to be. If the irregular vastness of Nature related to monotony and sameness, in the city, while Nature was still present, architecture, images, and decoration contributed towards a rhetoric of revelation, at the same time natural and unnatural, monolithic and fragmented, secret and public, pitiless and enveloping, sublime and beautiful.

---

[27] On Roman portable sundials see principally: Talbert 2017.

In many ways reading Strabo and the other Roman geographers there appears a narrative of control over the natural environment: the author or reader's eye commands what is placed within its gaze; mountains show or present themselves; the country opens up before the power of Rome; as indeed does the naked, indigenous bodyscape of conquered peoples.

In the next chapter discussion will turn to the contemporary significance of landscape art and representation and, in the chapter following that, to an examination of Egyptian landscapes.

Chapter Five

Staged Designs

In this chapter consideration will be given to Roman landscape art, especially painting, and, to a lesser extent, the geography of Rome's world as it related to the specificity of place.[1] The drive towards the creation and presentation in Rome of many different types of ideological landscapes was part of the broader use of geography products to define, or at least fix, Roman identity. Landscape art at this time clearly expressed the dialectic between humans and the natural world: life in the countryside with figures working was a carefully-managed constituent element of the farming year, rather than some natural ordering of the countryside which is what much Roman landscape painting aspired to.

Even in quite recent studies of the ancient world it is noticeable that there are often distinct splits between what might be termed environmental geographers and cultural historians. In this present book nature will be discussed as both a real material actor and a socially-constructed object. A Nature/culture divide will ultimately be eschewed. Of course, when considering Campanian landscape painting it is tempting to view Nature and landscape here as an object solely for human contemplation and controversy, even as a physical stage for otherwise quintessentially human dramas. But in seeing Nature as a historical actor at this time we can understand that landscape was being used in these paintings as an artistic way of seeing for the local Roman elite, not as recording. It was almost like the situation in eighteenth and nineteenth century western Europe where wealthy owners of grand houses and estates hired landscape designers and architects to redesign their estates to look like landscape paintings and once completed, hired artists to paint them. The idea that in western Europe landscape art was inextricably tied in to empire appeared later, in the nineteenth century when landscape painting can be viewed as inextricably bound up with projects of imperialism and the discourses around imperialism.

Short Distances and Definite Places

Roman landscape painting and other media for representing landscape and landscape elements such as triumphal paintings, easel paintings, stucco work, engraved gemstones, and relief sculpture could all be considered here as part of the same strategy of representation. Landscape painting was a specifically Roman genre of art, particularly in the southern Italian region of Campania where it reached some kind of artistic apotheosis in the decades before the eruption of Mount Vesuvius in AD 79. The destruction of towns such as Pompeii and Herculaneum in the aftermath of the eruption severely truncated and disrupted the history and technical development of wall or fresco painting in southern and central Italy, as most of

---

[1] On Roman and Campanian landscape painting see, for example: Bergmann 1991, 1992, 2002, 2018, and 2023; Clarke 1996; Conan 1986; Croissile 2015; Dawson 1944; Dietrich 2017; La Rocca 2008; Ling 1977; O'Sullivan 2023; Peters 1963; Platt 2009 and 2023; Schefold 1960; Silberberg 1980; Squire 2012; Thagaard Loft 2003; and Zarmakoupi 2014 and 2019. On the literary evocation of landscape see principally: Leach 1988; and Spencer 2011. On Roman gardens as private landscapes see, for example: Barrett 2017 and 2019; Bergmann 2008 and 2018; Boatwright 1998; Carroll 2018; Ciarallo 2001; Conan 1986; Hartswick 2004; Jashemski *et al.* 2018; Jones 2011 and 2013a; Kuttner 1999; Macaulay-Lewis 2009; Myers 2018; and Von Stakelberg 2009. On Romans and the broader environment see, for example: Hughes 2013 and 2014; Jeskins 1998; Kennedy and Jones-Lewis 2015; Lampinen 2022; McInerney and Sluiter 2016; Secord 2015; Shipley and Salmon 1996; and Thommen 2009.

its major practitioners of the AD 60s and 70s presumably died in the destruction of the towns and villas of the region.

It is interesting to consider whether these Campanian landscape paintings were viewed as being somehow quintessentially Campanian or Roman by their commissioners and viewers, in the same way that Constable's landscapes are now viewed as quintessentially English, Caspar David Friedrich's as quintessentially German, and Edvard Munch as quintessentially Scandinavian in capturing both inner and outer Norwegian landscapes.

A great deal of attention has been paid to the Roman penchant for landscape painting in various thematic genres. It is open to debate whether Campanian landscape painting should strictly be subsumed within the overall category of 'Roman' landscape painting, as it would appear to have been a *bona fide* regional phenomenon in Roman Italy and one within a specific chronological frame.

Roman landscape painting was not a universal or indeed universalised genre of art, but nor was it a series of distinct sub-genres based on technique, artistic practices, and styles of representation. If post-Renaissance western European landscape painting can be endlessly sub-divided into sub-genres, often chronologically-ordered, such as Ideal, Heroic, Pastoral, Beautiful, Classicising, Sublime, and Picturesque, Roman landscape painting would seem to have been about subject matter predominantly: (idealised) Campanian landscapes; garden landscapes; mythological landscapes; sacro-idyllic landscapes; other types of imagined landscapes; Egyptian or Egyptianising landscapes; and architectural landscapes or urbanised landscapes.

In a number of recently-published discussions of Roman landscape painting there have been curious analyses of style and (lack of) perspective in Roman pictures,[2] yet there is a certain circularity in the argument here as the impact and influences of the discovery of Pompeian landscapes by excavations at the town on eighteenth and nineteenth century landscape painting is a story still waiting to be written itself.

The archive of ancient Roman landscape paintings of which we are currently aware is dominated by Campanian works, alongside a much smaller number of landscape representations from Rome itself, and finally a few landscape representations in wall paintings outside of Italy. For good reason there is little or no provincial landscape painting, probably because of the heavily-freighted and ideologically-underpinned nature of the Roman source material and the sub-genre of the medium in particular. There can be no doubt that landscape was absolutely central to the Roman imagination: it was the part of most people's daily lives that imprinted personal, public, and collective fantasies onto places and scenes through their representation (Figures 49-53). Landscapes, particularly of the great villa estates, were sites of perpetual surveillance. Just as the genre of Roman still-life painting was part of broader Roman discourses on food and luxury as much as about resemblance or representation, so landscape painting was similar in many respects by reason of illusion and subterfuge. Idealised landscapes, sometimes called sacro-idyllic, also appeared in other media.

---

[2] I am thinking particularly of: Riggsby 2019: 130-153, but not criticising the study from that perspective.

STAGED DESIGNS

Figure 49. Roman/Campanian landscape wall painting from Pompeii. Early first century AD to AD 79. *Museo Archeologico Nazionale*, Naples. (Photo: Author).

Figure 50. Roman/Campanian landscape wall painting from Pompeii. Early first century AD to AD 79. *Museo Archeologico Nazionale*, Naples. (Photo: Author).

Pliny tells us that it was a painter called Studius 'who first instituted that most delightful technique of painting walls with representations of villas, porticos and landscape gardens, woods, groves, hills, pools, channels, rivers, and coastlines'.[3]

Roman landscape pictures were like frames on the land: contemplation of these landscape images probably would have made the local seem both familiar and reassuring, and at the same time somehow alien. At all times there must have been a feeling in the viewer's mind that he or she was seeing an image of a landscape and not the countryside itself with its

---
[3] Pliny *Naturalis Historia* 35.116-117.

Figure 51. Roman/Campanian landscape wall painting from Pompeii. Early first century AD to AD 79. *Museo Archeologico Nazionale*, Naples. (Photo: Author).

Figure 52. Relief depicting a sacro-idyllic landscape, Rome. First to second century AD. *Palazzo Massimo Museo Nazionale Romano*, Rome. (Photo: Author)

Figure 53. Wall painting depicting an idealised landscape, Pompeii. *Museo Archeologico Nazionale*, Naples. Early first century AD to AD 79. (Photo: Author).

myriad detail, and the mud, guts, blood, and muck that constituted its essence when polluted by human activity. Pliny the Younger[4] in lauding his Tuscan estate, wrote: 'you would be delighted, if you looked out at the geography of this region from the mountain. For you would imagine that you were looking not at the land but at some image painted to the highest degree of beauty'.

Just as it has been said that the most privileged settings in American history were elements of the landscape-the wilderness, the farm, and the city-so can each of these favoured locations be seen to have also been connected with events or ideological positioning that turned them from being uncontaminated into sites of contention and dispute. The wilderness was once indigenous land: the farm depended for its existence on slavery; and the city was market-driven beyond reason. These three strands of illusion driven by ideological reasoning can surely equally be applied to ancient Roman artworks depicting the Italian landscape. Colonial landscapes were an altogether different matter, as will be discussed below and in Chapter Six in the case of Nilotic landscapes. Perhaps we should also categorise pictures of Greek landscapes, popular with Roman artists after the Roman takeover of *Magna Graecia* and later, as being depictions of yet another kind of colonial landscape.

There is an overwhelming impression to be gained from looking at large numbers of Campanian landscape paintings in one go, as can be achieved by a visit to the *Museo Archeologico Nazionale* in Naples but which would not have been possible to achieve in Roman times, of the landscape

---

[4] Pliny the Younger *Epistulae* 5.6.13.

Figure 54. Wall painting of rural estate and estate workers, *Palasgarten*, Trier. Second century AD. *Rheinisch Landesmuseum*, Trier. (Photo: Author).

here having become an object of nostalgia for its own time. A lack of seasonal change in nearly all paintings known places them in a perpetual Spring or Summer, a point in time where no past or future change could be contemplated.

The Campanian oeuvre can be considered as 'framed landscapes without people', a second category being landscapes filled with mythological figures, a type equally popular in Campania as it was in Rome. Many of these works carried no hint of any attempt to make the landscape hosting this activity Italian or somehow 'real'. One of the few landscape paintings that can be said to define a third category on its own actually comes from the provinces and is now in the *Rheinisches Landesmuseum* in Trier. Dating to the second century AD it was found in a Roman building of unknown function in the centre of Trier (Figure 54). This image located the viewer in something like the real contemporary countryside where the keeping-up of appearances was maintained by the employment of slave labour on many estates. A large villa building is depicted, a tree, bushes and shrubs grow in the foreground, while three human figures, two of them in short tunics are shown toiling away, hard at work.

Most of the landscape painting under discussion here came from private houses rather than public buildings: it was art for estate owners to show their fellow estate owners. In towns like Pompeii and Herculaneum it was used to bring the countryside into the town, sometimes in quite modest houses whose owners perhaps wanted to restate rural roots and grapple with issues of identity and geolocation. Such struggles with issues of identity must have been

Figure 55. Wall painting from the garden room of the Villa of Livia at Prima Porta. Second half of the first century BC. *Palazzo Massimo Museo Nazionale Romano*, Rome. (Photo: Author).

magnified in the cosmopolis that was the city of Rome. It is uncertain whether freedmen or slaves who borrowed the idea of landscape painting as a suitable medium for funerary decoration in communal tombs and on cremation urns did so to mimic elite styles and secure status, because obviously they were not invested in the ownership of this land or the system that allowed for its exploitation and control. The creation of the great *horti* or public gardens of Rome was itself not just a series of exercises in landscaping and the bringing in of the countryside to the city. Rather, the *horti* themselves were in some ways representations of landscape: idealised, edited, controlled, and surveilled: almost indeed Roman paintings brought to life. Other Roman gardens became microcosmic examples of selected landscape features and plants from around the empire: they had both an ideological and a homeopathic power (Figure 55). The issue of botanical imperialism is discussed more fully elsewhere in the book in relation to the gardens of the Temple of Peace in Rome.

A visual vocabulary for painting these landscape scenes included a great deal of repetition, but a strategy such as this was common in other Roman artistic media, most notably sculpture, and therefore will not be viewed here as restrictive or unimaginative. Original works still stun though with their originality of composition and scale of ambition. Vast landscape works such as the Praeneste/Palestrina Nile Mosaic and the wall painting generally known as the Odyssey landscape from Rome represented a wholly new concept, in that they marked a conscious break from the representation of landscape in Greek and Hellenistic art: they reflected the period in which Rome's aggressive territorial expansion was telescoping the then-known world. Roman expansionism resulted in the exponential growth of slavery which itself then allowed for rich absentee landlords to acquire and operate enormous farming estates operated by slaves and their overseers.

It has been suggested that Roman landscape painting was a managerial tool related to the control of the countryside and its surveillance. Estate owners were not employing skilled painters to record their land-holdings in some kind of pictorial census-they could employ professional surveyors to do this if required- nor were they commodifying their views to enhance their identity and status. Of course, if it is accepted that a painted landscape is a natural scene mediated by culture then this all makes sense. Landscape painting does represent a new way of seeing and at different points in history has often been associated with various countries' imperial projects and ambitions. There was a future in Rome's dreaming, and landscape art was but one manifestation of that. The creation of a bucolic peaceful landscape was a way of screening off both the violence perpetuated in the countryside itself and in lands overseas, as the lush plant-life on the *Ara Pacis Augustae* did to some extent, metaphorically representing the ushering in an era of peace after years of war. Roman landscapes were not mere background and decoration, nor were they musing digressions: rather, by being representations of something which was itself a representation they helped to naturalise and emancipate the world as a unified whole.

As with the late eighteenth to early twentieth century phenomenon of the Panorama, there might be said to have been a 'near-far' paradox in the scope and scaling of many Roman landscape paintings and other representations of landscape or topographical features on mosaics for instance. Rather than having been worried about the lack of perspective in many such images-surely a modern problem superimposed in hindsight on Roman viewers-the contemporary viewer was in all probability more likely to have been concerned with issues around scale and composition, aesthetics and control. In Roman landscape art the wall painters and easel painters were not painting 'landscapes by numbers', that is something that exists; rather they were painting only its reflection.

It is interesting to note that some of the staged displays of panoramas were mounted as multisensory assaults on the viewer, with the actual paintings being enhanced by the placement of fake terrain scenery, noises, and sometimes artificial aromas and breezes of air. In this way an experience was created in the same way that the Roman triumph was not just a victory parade but rather a multi-media spectacle. Sometimes there was surely a need to broaden or change one's vantage point to recapture the sensation of a journey undertaken across a landscape or to imagine oneself at some future time making that journey and experiencing the joy, hope, despair, or ennui that the journey will trigger.

To try and deconstruct the processes and intellectual ideologies of Roman landscape painting by analyses which foreground discussion of vanishing points and perspective, not to mention the presence or absence of horizon lines, seems to be missing the point to some extent. Again, making comparisons to artistic strategies and practices of landscape painting of other periods (for instance to Cézanne and Matisse) is perfectly acceptable and interesting, but by omitting consideration of the broader categories of landscape art or place-based art (that is, not just painting) and the work and ideas of non-western artists, a kind of visual sovereignty is being established, linked to a western salon, studio, or gallery canon. Since the 1960s an expanded field of land art, performance art, events, photo-conceptualisation, and site-specific installations have usurped the primacy of painting, but have not altogether replaced it. Ideas centred around ecology, sustainability, conservation, and restoration/destruction are engaged with, rather than simply representation. Landscape and abstraction, specificity and universality, physical and metaphysical states are all explored. Where human traces are presented more often than not the relationship of the figures and the landscape itself go unresolved.

Quite commonly, horizon lines put the body in Roman landscape paintings by creating an eye-level. It would seem that the principal difference between painted Campanian landscapes and imagined sacro-idyllic ones was a matter of scale. The former allowed for the containment of the action and the scene while in the latter an overwhelming sense of scale acted against containment, and the larger-scale shapes and denser textures of such landscapes therefore seemed somehow oppressive. These repeating representations of villas, porticos, and landscape gardens spoke of architecture and management, of ownership or aspiration and envy. Images of woods, groves, and hills evoked more spiritual matters, though still mediated by ownership of land. The recurring images of pools, channels, rivers, coastlines, and massive coastal villa complexes aptly illustrated a more general Roman obsession with water management and, most importantly, control over Nature and its resources: beyond the ends of land the seas facilitated Rome's acquisition of empire and the wealth derived from that journeying away.

The *Odyssey* landscape wall paintings, from a portico-like structure on the Esquiline Hill in Rome along what is today *Via Cavour* and now in the *Musei Vaticani* (and a single panel in *Museo Nazionale Palazzo Massimo*), date to the mid-first century BC and represent the earliest example of Roman landscape painting that has come down to us.[5] The eight sequential scenes represent the companions of Ulysses meeting the Laestrygonians and an assault on the Greek ships, the destruction of most of the fleet, the escape of Ulysses to the island of Circe, Ulysses in her palace, his companions transformed back from pigs to men, and finally Ulysses in the underworld meeting the dead, Orion, Sisyphus, Tityus, and the Danaids. The landscape settings are open and expansive, panoramic even: a large rocky outcrop and tall trees, along with craggy coastal vistas, are particularly striking individual elements. Of great interest is the use of descriptive labelling, including *KRENE* (spring), and *AKTAI* (headlands), for example, to aid the viewer's understanding. These are highly literary works, probably more-or-less interpretable only by those with prior knowledge of Homer's epic work, and indeed of only a very specific part of the overall *Odyssey* narrative. The themes represented here are complex and diverse. In an era of great political transformation in Rome the idea of journeying and of heroic exploration might have seemed particularly relevant. These scenes created and evoked a sense of jeopardy in places to arrive at, and to depart from quickly. These were not so much individual places as points along a route, records of a journey and encounters with the other and unfamiliar, which in the architectural context here must have seemed altogether appropriate. The contemplation and consideration of epic journeying represented a way to return to the source of the signal, something that would also be reflected somewhat later in

---

[5] On the *Odyssey* landscapes see principally: Lowenstam 1995; and O'Sullivan 2019 and 2023.\

Figure 56. Christian sarcophagus with scene of Jesus preaching in Holy Land landscape defined by palm trees, Ravenna. Fourth century AD. *Museo Nazionale Romano*, Ravenna. (Photo: Author).

the choice of the *Odyssey* text for inspiration in the design of the programme of sculptures in the grotto at the Villa of Tiberius at Sperlonga, as will be discussed in Chapter Nine.

Nilotic landscapes such as the great Palestrina/Praeneste Nile Mosaic tell an altogether different story in ideological terms, though not one that places the two major genres of Roman landscape art in any sort of home/away opposition. Of course, there must have been an element of fashion in the choice of a Nilotic scene for wall decoration or on a mosaic floor. However, it would appear that fantasy and control apparent in these narratives would have formed a potent, perhaps intoxicating mix for their commissioners. Nile flooding, irrigation scenes, lush vegetation, sometimes scenes of sexual encounter alluded to Egypt's abundance and fertility, and to the psycho-sexual elements of Roman male imperialism which are discussed more fully in Chapter Eight. A scene of sexual encounter involving two pygmy figures appears in a mid-first century AD wall painting at the House of the Upper Floor in Pompeii and forcefully illustrates the links between the personal and political at this time. There is also a story to be told about how Middle East landscapes in which Jesus and other figures from the Bible came to be portrayed in later Roman art, and about how the familiarisation of such alien settings came to be part of the Roman artistic mainstream in a way in which other alien landscapes did not, with the exception of Egypt in earlier imperial times (Figure 56). Just as there was a Romanisation of Egyptian and Nilotic landscapes through artistic practice, so the same happened with a Romanisation of the Holy Land, a topic which will be considered during a discussion of pilgrimage in Chapter Nine.

The Inconstant Ones

Perhaps surprisingly, the images of natural landscape elements of Dacia presented on the frieze on Trajan's Column in Rome would seem to represent the most conspicuous and detailed depictions of any ancient landscape in Roman times, with the exceptions of Campania and perhaps Thessaly. Other images of natural environments or individual landscape elements are known though and will be considered briefly below: depictions of the German landscape on the frieze around the Column of Marcus Aurelius; landscape elements on the battle friezes from the Baths of Caracalla in Rome; the landscapes of northern Roman Britain as manifested on a significant number of artworks from the region; and landscape elements depicted on mosaics from the North African provinces.

As is so often the case with Roman imperial art exceptional visual schemes can be found on certain monuments which move narrative conventions forward in an unusual and unprecedented way, as is the case of Trajan's Column in Rome in terms of its depictions of 'alien' peoples and places. Of course, the column's helical frieze is principally about so much more; Trajan as great emperor and soldier-emperor, the exploits and political significance of the Roman army, conquest and imperialism, and about the benefits of *Romanitas*. Yet other sub-themes are presented and readable, relating to women and children, provincial life, and so on. However, I will concentrate on the geographical sub-theme here to the exclusion of the others (Figures 57-58). But before beginning this discussion some mention must be made about the question of the visibility of parts of the frieze, particularly in its upper registers. My account is based on the examination of the portfolio of numbered photographs of casts of the frieze first prepared by Conrad Cichorius in the late nineteenth century and subsequently re-presented and analysed scene by scene by Frank Lepper and Sheppard Frere in 1988.[6] I will start the analysis here by concentrating on depictions of natural scenery both as a setting for action on the column frieze and as a background to that action. Cicorius, Lepper, and Frere attempted to relate many of the key scenes and topographical elements of the frieze to recorded events in the historical record of Trajan's Dacian Wars and to specific geographical locations, but with obvious exceptions, like the death of the Dacian king Decebalus at the fortress of Sarmizegetusa, I will eschew such attempts at matching up here.

The helical frieze depicts an imagistic narrative account of Trajan's two Dacian Wars of AD 101-102 and 105-106, with the two distinct actions separated by the figure of Victory. Scene I, the so-called 'Watch on the Danube' shows Roman military watchtowers and frontier-works along the banks of the flowing river. The river's movement is captured by drilling that makes the susurration of the waters resemble rippling locks of wavy hair. So almost instantly the viewer in Rome would have been transported by this image to the very frontier of the Roman empire along the Danube. However, the military buildings and the alert sentries portrayed would have acted to reassure the viewer that they have not been set adrift in entirely alien territory and, indeed, in Scene II we are presented with images of boats being loaded with barrels for transport along the river. In Scene III walled settlements cluster on the banks, a

---
[6] Lepper and Frere 1988.

Figure 57. Scenes I–II, Trajan's Column, Rome. AD 113. (Photo: Author).

Figure 58. Scene XX, Trajan's Column, Rome. AD 113. (Photo: Author).

few trees are depicted. The scene is overseen by the mighty presence of the great river god of the Danube. This image very neatly introduces at a very early stage in the monument's visual narrative the idea that conquest, empire, and expansion was inextricably linked in the Roman mind with the exploitation of new natural resources and products, and that conquest and trade went hand-in-hand at this time, as with virtually every project of empire-building in every era of history.

With the geographical setting now having been established as being in distant lands, in Scene IV the viewer would probably have been jolted out of their reverie by the depiction of massed Roman forces on the march over the mighty Danube bridge. The army is then depicted marching on, holding a council of war and staging a sacrifice. Scenes involving the depiction of landscape or natural features such as trees, groves, and forests do not occur until Scene VII, and after that point they become more common in an almost rhythmic and inexorable kind of way, often used as in Scene VII/VIII to contrast with Roman building work in a Nature/culture contrast of sorts. I will return to the significance of the depiction of trees on the column below. Again, the Romans' power over Nature is more pointedly conveyed in Scene XIII in the back of which two Roman soldiers carry large pieces of recently-cut timber logs to aid camp construction and in Scene XV where a whole unit of Roman troops swinging axes is engaged in cutting down a grove of trees, either for use in building or to remove covert shelter for enemy forces or to provide a firebreak around their camp, or indeed both. The tree-fellers on the relief might now be missing their miniature accessory bronze axes once affixed in their hands, but even without their flailing arms create a blurred vision of frantic activity that brings to mind the flailing limbs of the Trojan priest Laocoon and his two sons as they tried to hold off the rapacious and unstoppable sea serpents sent against him by Poseidon. The destruction of this ancient woodland might also have been an attack on a natural location which had some local or regional symbolic value. The same point is made in a slightly different way in Scene XXIII where the Roman troops are depicted destroying part of a forest to route a new road through it, symbolising access, progress, and conquest. The uprooting of these trees appears almost as a harbinger of the regime change that the Romans were seeking here in Dacia.

All of the scenes so far discussed acted as the lead-up and prelude to the first battle of Trajan's Dacian Wars. As the Roman cavalry and infantry forces move into battle the emperor and senior officers are shown looking on as combat commences and trophy enemy heads are taken and paraded before them in a grotesque display of Roman power. The Dacian soldiers in this battle can be seen to emerge from deep forest cover represented by crowded trees in Scene XXIV and retreat back into the tree-cover like wounded wolves or boars. Roman victory here is achieved with the aid of Jupiter who from the sky hurls a fiery thunderbolt at the Dacian soldiers. The artists of the column frieze were also proficient in showing that the landscape had changed in other ways, and that there was not only dense, wild forest for the soldiers of the Roman army to contend with, but that much of the landscape they found themselves camping in, traversing, or fighting in was rocky and indeed mountainous in places. In Scene XIV it can be clearly made out that the Roman fort being visited by Trajan is built on a prominent rocky bluff.

The crossing of the great River Danube marked a hugely symbolic event, as well as a purely military one, but other smaller rivers also needed to be bridged to allow the army to cross quickly and in safety, and we see one such river and crossing in use in Scene XVII. Bridge-building becomes a trope for the inexorable forward motion of the Roman advance and bridge-building occurs again in Scene XIX, and a small, completed bridge is shown in Scene XXI. A long Roman pontoon bridge of boats appears in Scene XLVIII, with Roman forces crossing it. However, the army did not always have to call upon its skilled engineers to progress forward in all situations, and in Scene XXVI we see a unit of the army wading across a small river or stream. One Roman soldier, stripped naked down to the waist, carries his equipment balanced on his upturned shield on his head in order to keep it dry. In complete contrast to images of Roman bridge crossings and the scene of orderly wading of Roman soldiers in Scene XXXI the viewer is confronted with a vibrant image of Dacian cavalry attempting to cross a deep and seemingly-hazardous stretch of wide river, presumably the Danube or one of its larger tributaries, while being harassed or pursued by Roman forces. The crossing appears chaotic and disordered, with horses plunging deep into the churning waters, as their riders struggle to control their mounts and to stem their panic and alarm. Some swimming riders seem to have lost their mounts.

Substantial Roman port facilities on the River Danube appear as images in Scene XXXIII, as the emperor can be seen waiting to embark from this riverside harbour. In Scene XXXIV we see Trajan progressing along the river by boat, to disembark in Scene XXXV at a smaller military landing stage. After the next major battle depicted in Scene XL the defeated Dacian troops are depicted fleeing into a landscape defined by trees and rocks, in Scene XLI suggesting that they might now be fleeing from the river floodplains uphill into the foothills of the mountains, there to regroup and rearm. Dacian soldiers captured during this battle are later shown in Scene XLIII claustrophobically contained in a walled stockade, guarded by a Roman gaoler. Their stillness and lack of movement, the sense of physical confinement in this image contrasts brilliantly with those images of Dacian men moving freely around and across their ancestral landscapes.

Further river embarkation/disembarkation scenes occur in Scene XLVII. After further advances, the reception of Dacian emissaries, and a sacrifice, the Roman forces now appear to be advancing into upland territory themselves, as suggested by the rocky outcrops depicted in Scenes LV and LVI. Once more, this difficult terrain requires the building of further stretches of usable road (Scene LVI). In Scenes LVII and LVIII-LVIX the Romans are shown traversing and circling around mountain ridges and bluffs, preparing to assault Dacian villages high up in this hilly region. In Scene LVIII the bridge that the emperor is depicted crossing must presumably now therefore be a land bridge thrown across a gorge or chasm, with water flowing some distance below. More steeply-hilly terrain appears in Scenes LXIII-LXIV. Once more, after another pitched battle and Roman victory the Dacians flee to a large forest. In Scene LXVII the Dacians cut down trees to refortify one of their fortresses, while just two scenes later we see Roman troops also clearing woodland and trees in preparation for their own assault. The assault then follows, with crowded scenes of battle, the storming of a citadel by Roman troops advancing in *testudo* formation, guarded by their upraised and interlinked shields. The emperor is brought news and is presented with trophy severed heads. The final pitched battle of the First Dacian War is now depicted, after which Trajan addresses his victorious army (Scene LXXIII). Scene LXXIV is particularly curious, in that its place in the narrative appears to make little or no sense. Here we see Roman soldiers visiting a spring. The waters are shown contained in an elongated basin structure or natural basin which has been upended in perspective to give it the impression of being vertical. Cichorius and Lepper and

Frere suggest that this is the site of the hot springs at Calan, equated with the site *Ad Aquas* named in the Peutinger Map and other ancient geographical works. *Ad Aquas* was one of three sites of natural hot springs in Dacia which became the focus for religious activity, healing, and votive dedication. The other two hot springs, *Ad Mediam* and *Germisara*, are also named on the Peutinger Map but not marked with a topographic image as was the case with *Ad Aquas*. It is highly likely that this curious image of Roman soldiers at a highly significant Dacian religious site was intended to emphasise the role that religious syncretism played in the pursuance of Roman imperial domination and empire, through the emperor's control of the rites of the state religion. In subsequent scenes we are shown various images of the subjugation of the Dacian peoples and the forced removal of families from their land. Finally, Trajan once more addresses his army before Victory is depicted, writing on a shield, flanked by trophies, this image breaking up the depiction of the two wars on the column.

It was some three years or so later that the emperor launched a second war against the Dacians, the defeated Dacian king Decebalus now agitating against Rome once more. The frieze begins the account of the Second Dacian War with Scene LXXIX, a spectacular depiction of Trajan's sea journey to the theatre of war. The imperial fleet can be seen leaving a great, built-up port city in Italy, quite possibly Ancona on the Adriatic coast, and making its way across to strike land at a port city somewhere along the Dalmatian coast. Two dolphins follow the armada. Great emphasis has been placed here on the emperor's journeying to the war which, given the great metaphorical significance of the journey as a conceptual trope in Greco-Roman literature and art, seems unsurprising. After reaching land the emperor is greeted by the provincial elite, there are sacrifices and so on, before the emperor and army set out by land to Dacia. The same landscape and topographical tropes as appeared in the scenes narrating the first war now start to appear: the appearance of distinctive trees and woodland; the clearance of forest to build roads, Scene XCII being a particularly well-realised example of just such a Roman taming of a foreign or alien wilderness, with the zigzag course of the road suggesting ascent into the mountainous territories. Scene CX is centred on the depiction of Roman troops foraging for food supplies, cutting cereal crops with small handheld sickles or falxes. This scene is of particular interest in the present context because it acts to bring the viewer up short, having been led to believe that this is simply a land of war.

The army has already now moved up into the hills, and rocky outcrops appear in Scenes CXI-CXII. Scene CXVII shows preparations for the blockade of a major Dacian fortress, involving Roman soldiers yet again cutting down trees and preparing lengths of timber to build siegeworks or defensive structures. Scenes CXIX and CXX show the Dacians setting fire to their own fortress, and their military commanders committing suicide in order to evade the indignity of capture by the Romans. The action and landscape setting of Scene CXXVIII neatly encapsulates the ethos of Roman military expansionism. Here the viewer is presented with what at first appears to be a comfortable genre scene, breaking up the intense build-up to the final events of the second war. A number of helmetless Roman soldiers guide a baggage train of mules along a zigzaging mountain road, the beasts being heavily laden down by the weight of the open, bulging panniers strung across and secured around their bodies. The panniers would appear to be stashed with looted metal items, probably silver and gold plate, war booty in other words, making explicit the symbiotic link between conquest and commerce which the booty scene CXXVIII on the column frieze so eloquently and unashamedly portrays.

The great mountain stronghold of Decebalus at Sarmizegetusa now hoves into view for the Roman forces. It seems somehow logical and fitting that the action at the mountain eyrie of Sarmizegetusa takes place high up towards the column's top. Everything that had gone before seemed in any case to be leading to this point, to this place, to this event. It was as if it was all fore-written, somehow all meant to be. As the viewer looked at the scene of the Dacian stronghold set aflame and perhaps glanced up to then take in the bronze statue of Trajan once atop the column, they might well have thought about the consummation of the emperor's body on his funeral pyre and his subsequent apotheosis. It would seem to have been important to the designer of the frieze that the suicide of Decebalus was very specifically located in time and space for the viewer, and that this momentous event was presented as taking place in a defined location, defined by topographical indicators.

The topographical images on the column frieze are more about the idea of landscape, the foreignness of such landscapes, than they are depictions of actual landscape. The seemingly unchanging Dacian scenery of mountains and steep passes would seem to ultimately be topographically unspecific. Yet when the designer of the frieze wanted to be topographically specific sufficient detail was provided in the images, for instance at the site of the famous Danube bridge and at the Italian port city from which seaborne Roman forces departed for war, this town being identified as Ancona by reason of its hilltop temple and the triumphal arch down by the harbour. Pliny the Younger's *Panegyricus* to Trajan describes Dacia's mountains, rivers, and seas in such a way as to almost suggest to the reader that they were themselves active participants in the war, swayed by Trajan's power, and won over to cooperate with the Romans in aiding their campaigning.

It is worth returning briefly to discussion of the significance of the depiction of trees on the column frieze, a subject that had received very little attention until quite recently. While some of the trees depicted would seem to have been placed at intervals simply to break up scenes of action or to frame certain actions or individuals, most though were there to make very specific points about Dacia's natural landscape and about the Roman discourse surrounding imperialism and Nature. Two detailed studies centred specifically on the images of the trees on the column have counted 224 individual trees portrayed here, 222 being species that would have been native to Dacia and portrayed as being on Dacian territory on the frieze, with two trees, identified as non-Dacian species, appearing on the Roman side of the River Danube.[7] Thirty seven different leaf types were carved (and probably originally painted). It seems extraordinary that such great care was evidently taken to accurately depict specific tree species, identifiable by both their form and structure of trunk, branches, and canopy, and, most significantly from a botanical identification point of view, by leaf shape, density, and clustering. Both deciduous and a large group of coniferous, resinous trees are identifiable. The principal Dacian native trees are predominantly oaks (*quercus*), but the service tree (*Sorbus torminalis*), and sycamore (*Acer pseudoplatanus*) are also easily identifiable in the artwork.

In the same manner, on the Arch of Trajan at Benevento images of an oak tree appear as symbolic of Germany. Likewise, the fragmentary reliefs of battle scenes from the Baths of Caracalla in Rome used images of oaks to represent the landscape of either Britain or Germany

---

[7] On the trees on Trajan's Column see particularly: Fox 2019 and Fox 2023: 79-88; and Stoiculescu 1985.

(which is uncertain on present evidence) and images of palm trees were used to represent and evoke the Parthian landscape in the east.

There is considerable evidence for the veneration of certain sacred trees in the Roman world, and indeed mention has already been made with regard to the sacred fig trees of Rome in Chapter Three. Again, as has already been mentioned, living trees were sometimes carried as booty in Roman triumphs, emblematic and symbolic of their original geographic location, suggesting that the significance of the trees on the column should be carefully considered as somehow ideological. Andrew Fox has even suggested that following their parading in Trajan's Dacian triumph 'captive' Dacian trees were subsequently planted in pits which formed double colonnades in Trajan's forum, though there is no firm evidence for this from the archaeological excavations there.

The many scenes of deforestation on the column frieze could also have been highly significant in themselves. Possible motives behind military deforestation depicted here could have been strictly practical: to clear areas of woodland through which a new road was intended to be constructed; to clear woodland whose position was interrupting sight-lines for military signalling; to clear woodland whose position was interrupting clear 360 degrees views from inside a Roman camp or fortification; to clear woodland that either had provided shelter or which had the potential to do so, a hiding place, or ambush location for enemy forces; and to clear an area of woodland which had a non-strategic value to the enemy but which was in some way highly culturally significant, perhaps in terms of there being a historic or sacred tree or trees there. The latter of these acts of clearance and deliberate deforestation would have been intended to be particularly provocative, a goading violation of sorts. We can learn something about instances of deliberate tree violation in Virgil and learn there too about reactions and ambivalences towards the practice.

Another, but much more recent instance of Nature being viewed as necessary collateral damage in wartime is represented by the systematic and programmatic use of bombs carrying the chemical Agent Orange by the US army and air-force in Vietnam, Cambodia, and Laos in the 1970s Vietnam War. This tactic was viewed as being a practical way to firstly destroy enemy forces making use of the deep forest cover, to destroy the forest itself so that it could no longer be used as deep camouflage and a place of relative shelter by military forces, and, although the term was not used then, but was used in the later Iraq War of 2003, it was very much a tactic of 'shock and awe', clearly demonstrating the devastating firepower and resources of the United States and indeed its sway over even Nature itself. This was also a significant propaganda exercise, with this becoming as much a media, and particularly a televisual, event as it was a series of historical moments. I still vividly remember as a youth seeing on the TV news images of runs of bombs exploding sequentially one after another in startling balls of bright orange flame that immediately obliterated vast swathes of the dense, dark green forest canopy. One after another, after another, after another. The effect was dazzling, hypnotic, mesmeric.

If we think of the geographical and landscape elements presented as images on the column as almost akin to theatrical scenic backdrops or flats then we might be able to understand the thinking behind their creation and presentation. What was created on the frieze on Trajan's Column in terms of the scenic machinery wheeled into place was the creation of an otherspace, a foreign image space through which viewers could understand a foreign people, in this case the Dacians. The emperor's travel and the Roman army's transit through this otherspace dynamically organised and constituted the space around it. The constant recourse to the depiction of Roman movement through the landscape and the means of transport to facilitate that movement was constant and almost unrelenting. The emphasis on this was certain and insistent. Marching columns of men, riders on horseback, baggage mules, wagons, wheeled artillery pieces all figured heavily. Roads were built, rivers were forded, bridges were built, either as temporary pontoon structures, or in a more substantial and permanent form like the great bridge over the Danube. The sheer mobility of the Roman army on campaign was clearly stressed and enumerated.

To viewers of the frieze when first unveiled the distance between them and the images was strictly a matter of geography and space. The distance between the viewer and the images on the column frieze will have been removed with the passing of time. Contemporary events become part of the historical continuum. The further away in time viewers were situated from the events portrayed the further back in time would the images recede, time and space here then becoming one dimension.

The experience of looking at images of the physical world, in lieu of at that world itself directly, would have meant that the images became more and more elusive to define with the passing of time. A move towards a kind of pictorial abstraction here on the column, represented by the great simplification of form and light and the use of colour could perhaps have resulted in the images becoming perhaps too simple. Scenic backdrop elements had to be of a certain scale to be visible, something perhaps aided by the use of painting, as they had to be seen from a distance. The viewer needed to look at the frieze not as a series of individual scenes comprising a logical sequence and narrative, a history perhaps, but rather as a series which far from being repetitive was actually highly inventive throughout the work. This was what has been called by classicists a 'landscape of war', though much of the writing on this subject has been textually centred rather than considerate of images. There is no reason though why we cannot equally write of a Dacian or Trajanic landscape of war as of a Homeric one.

Most obviously viewers of the frieze were being presented not with a dispassionately-designed and independent series of images of war in a specific, real landscape. This was a Dacian landscape mediated by its being traversed and controlled at times by the Roman army. As invaders the landscape to the Romans was first and foremost strange and potentially hostile, while to the Dacians, whose viewpoint viewers were not intended to consider, it was their home environment, knowledge of which would have empowered them. However, war itself would have also changed the Dacians' own views of their home landscapes: suddenly their armed forces had to realign their perceptions of their native spaces and places, and consider them from a tactical point of view. Defendable positions, camouflaged cover, spots for an ambush, hidden hazards, clear lines of sight would have created a geometry of fear. Critical reimaginings of the landscape would have to have taken place on both sides. The feeling of the relentless forward momentum of the narrative on the column, the onwards march of the Roman forces, has already been commented on, and in literary narratives describing landscapes of war it was a common trope to convey the goal of moving towards the front. An objective like this led to the shrinking of the surrounding landscape and the telescoping of the land around the route of travel that led to the front. Like an itinerary, but here driven by military necessity, the army looked to move from one point to another with little or no concern with the broader landscape through which they were moving. Obstacles in the way were simply problems to be solved. The throwing up of defences, the repurposing and bolstering of natural positions, offensive war-works, clearing lines of sight, building roads and bridges, destroying strongholds: all these acts of war left marks on the landscape, war inscribing the landscape itself as rotting corpses polluted the ground, rivers, and watercourses, maybe conversely sometimes fertilising the soil, enriching its being.

The emphasis on the column of the scene involving the River Danube is interesting in that Roman thinking often saw rivers as natural frontiers, useful to Roman geo-political strategies, and not necessarily as fixed boundaries or barriers to movement. This scene, and the Tiber river scene on the Arch of Constantine, may have been designed with the Homeric story of the battle between Achilles and the Trojan river/river god Scamander in mind and might have been not simply ideologically tinged but also religious as well.

A small number of other artworks depicting the Roman army in combat overseas can also be considered from the same perspective. Depictions of the German landscape on the frieze around the Column of Marcus Aurelius are less numerous than on Trajan's Column and nowhere near as cleverly deployed. Landscape elements on the battle friezes from the Baths of Caracalla in Rome, in this case trees, were used in a contrastive way to signify whether the military actions shown and celebrated were in the east (Parthia) or in the west (Germany or northern Britain).

In my previous book *Visions of the Roman North. Art and Identity in Northern Roman Britain*[8] I proposed the thesis that the topography and landscapes of this northern frontier zone had a significant and profound effect on much of the art produced here and that referencing the landscape and almost recording journeys across it provided the peoples of the region with a way to come to terms with their environment. In other words, that in this region art as an expression of identity was almost symbiotically linked to the broader, physical world. These links could be manifested in something as simple as using local stone to produce a sculpture, thus making the artwork *of the landscape*, it could have involved the depiction of some element of the landscape in terms of its trees, flora and fauna, even in a sketchy form or as simplified motifs, or more allusively it could have involved recourse to the depiction of classical sacro-idyllic landscapes, landscapes inhabited by gods and other mythological figures, into which viewers could escape. Such interior landscapes reflected reactions to external ones and provided some comfort in harsh times and in harsh conditions. Again, filling the frontier zone with art and inscriptions quite clearly separated the sacred, forbidden lands beyond from the territory systematically mapped and traversed by the Roman army. With the establishment of formal frontier defences and other works in northern Britain the limits of the Roman empire literally became inscribed on the land like lines on a map. As frontiers changed, as they did most notably in northern Britain with the change from Hadrian's Wall to the Antonine Wall and then back again, so the lines became blurred or erased, like a ghostly palimpsest.

But if these various landscapes of war provided the Roman viewer with the vicarious thrill of virtually campaigning with the Roman army, marching along with the baggage trains, and seeing alien landscapes unspooling ahead, before experiencing the clamour of battle in

---

[8] Ferris 2021a.

strange, foreign locations, another foreign landscape, that of Parthia, came to be presented in Roman art almost as a 'landscape of defeat'.[9] Dacia and Britain might have been contested landscapes for a while but they soon became tamed spaces, with former battlefields returning to agriculture and the plough, familiar in their subservience. It has already been discussed in Chapter One how the decorated breastplate of the Prima Porta statue of Augustus carried images of captured Roman standards being returned to Roman envoys in a non-specific setting, years after the disastrous defeat of Crassus at the Battle of Carrhae. It has been suggested that Roman authors were unable to come to terms with Rome's antipathetic relationship with Parthia, following military setback and diplomatic intransigence in a way that did not occur in the case of any of its other opponents in war.

If Augustus' treaty with Parthia which saw the return of the captured Roman standards was hailed, at least by him, as a major victory over the Parthians, it would not be until the time of Trajan that an actual military victory was won there in the field, even if the advantages of that too to Rome were let slip away. If Parthia in Augustan times could be presented as merely being in Rome's 'sphere of influence' while still being an active enemy, the very definition of that relationship, its vagueness, would have meant that for those at Rome who wished to imaginatively conjure up the landscapes or peoples of Parthia there was very little in the way of conceptual frameworks from past experiences to allow them to do so successfully. It has even been claimed that this was some kind of vicious geo-political circle, that Rome encountered defeat and humiliation in its dealings with Parthia, up to the time of Augustus, and one could suggest for some time after that as well, because of a failure on behalf of Rome's rulers to fully comprehend and understand the geography of Parthia and the ethnography of its people. Indeed, among the late Republican and Augustan writers there developed a discourse that wrote the country off as largely being desert, with only a few known named cities, including of course Carrhae which would ever be associated with the name of Crassus. This desertified, open landscape, ideal for the famed mounted Parthian archers, was presented almost as if it was itself a weapon to be used against Rome, as if the landscape was itself an enemy. When other cities were named in these historical accounts, no further information about them was given, as presumably none was available or widely in circulation at Rome and in Italy. It has been suggested that an idea emerged of Parthia being in another world-*alter orbis*-from Rome which allowed Romans to still feel all-conquering and powerful in *their* world.

Later Roman encounters with the Parthians were successfully narrativised in Roman monumental art under the Severan emperors, with the Parthian landscape being tamed in the visual records of monuments to the dynasty's Parthian Wars of AD 195-217, such as the now severely-damaged Arch of Septimius Severus in the Roman Forum,[10] the greatest of these monuments, the Arch of Septimius Severus and Julia Domna in their North African home city of Lepcis Magna, and the Parthian monument of Lucius Verus in Ephesus.

Like scratching an itch Rome returned to conflict with Parthia once more in the third century, these wars being commemorated for the benefit of the Roman public on the Arch of Septimius Severus in the Roman Forum. Dedicated in AD 203 the arch bears a very lengthy inscription.

---

[9] On the idea of Parthia as 'a landscape of defeat' see: Reitz-Joosse 2021. On the Seleucids and ideology see, for example: Kosmin 2014. On landscapes of war and topographies of violence see, for example: Rietz-Joosse 2021; and Riess and Fagan 2016.

[10] On the Arch of Severus see principally: Brilliant 1967.

Just as Augustus dedicated the *Ara Pacis* in Rome to celebrate the Augustan peace, after bitter and protracted civil wars, so Septimius Severus sought also to present himself as a bringer of peace after civil war. The arch was seriously damaged by fire in antiquity and its now-compromised and partial relief decoration suffered in particular, and for this reason has often being neglected by academic study. As Richard Brilliant pointed out in his still definitive monograph study of the arch, its inscription, and the decoration on the major relief panels on the monument each centre on the capture of an urban centre by Severus' army, including depiction of the siege of Ctesiphon. River gods appear in the spandrels of the arch. The depiction of Severus' Parthian triumph in AD 202 consists of a parading of captured spoils on wagons before a large seated female personification of Parthia. Many of the images of Parthian prisoners consist of captured soldiers being led away in shackles.

But what were the Senate and people of Rome to make of this monument set up to help usher in a new imperial dynasty whose roots were very firmly and proudly in Roman North Africa? Did the Parthian War monuments built at other locations convey the self-same messages and in the same way? Septimius Severus never attempted to hide or disguise his origins.

In this chapter it has been argued that the formal presentation and use of images of nature, land, and landscape was highly important to the creation and maintenance of the self-identity of many ancient Romans. This occurred not just in the presentation of 'home' landscapes in Latium and Campania as so commonly represented in domestic wall paintings, but in the presentation of alien landscapes. Colonial landscapes of Egypt and Greece were common topics in Roman wall paintings and on mosaics. It has also been noted how the background landscape elements on the frieze around Trajan's Column and other Roman imperial monuments served to locate the action quite precisely in landscapes of war that Roman viewers could relate to. The forest, the primal place where everything was at once clear and yet at the same time fearfully confusing, often acted as a backdrop. The past too rewilded itself and oblivion sprung up out of it like a forest. We can identify certain alienating trends in the discourse around the visual dissemination of geographical knowledge at Rome and it seems that there were distinct and different orientalising trends, that is with regard to eastern peoples and places, and occidentalising ones, that is relating to western places and peoples. While much Roman monumental art was created to reflect the Roman's utopian visions of their city bringing visualisations of alien landscapes within the city limits somehow helped to stop what might otherwise have occurred there: the swift degeneration into characterless urban gridzones of despair and anomie.

The colonial landscapes of Egypt will be considered in detail in the next chapter.

Chapter Six

Landscape and Desire

In the case of Egypt and the Nile there is a vast scholarly literature on the inter-relationship between Rome and Egypt and, more recently, a growing literature on the interplay between images of Egyptian themes in Roman art and life and the reuse of Egyptian monuments and artefacts in Rome.[1] Some of this writing is too eager to accept the idea of a rampant 'Egyptomania' at Rome and in Italy more broadly almost at face value, without contextualising the quite profound engagement between the two societies that would appear to have underlain and to some extent explained this phenomenon but which did not necessarily control or direct it. I have previously blundered into this trap myself. Egyptomania really is a quite problematic category of action and reaction. The context of each appearance of Egyptian or Egyptianising images in Roman contexts is by definition unique to a particular time and a particular place. To lump all these contexts together as a unified phenomenon perhaps risks losing sight of factors such as contact and choice, motivation, emulation, aesthetic taste, ideology, and the interplay of culture and power. Ordering the transport of a massive obelisk from Egypt to Rome at huge expense was not the same as having a Nilotic scene painted on your dining room wall in your house at Pompeii.

There is a quite firm chronological framework for Rome's interest in Egypt. Egyptian motifs in wall paintings and on mosaics first became popular from the later second century BC onwards, reaching some sort of peak of popularity around the time of Augustus' formal annexation of Egypt in 30 BC.[2] While interest in all things Egyptian did not thereafter then simply fall away there can certainly be perceived a temporary loss of interest among Rome's elite until the time of Nero. This new enthusiasm was tempered with a form of nostalgia, and is best exemplified by the art of the cities of Pompeii and Herculaneum where Nilotic scenes in wall paintings and on mosaic floors were almost as common as paintings of idealised Campanian landscapes. Later Roman Nilotic scenes on mosaics have an altogether different resonance.

Possession

Personifications of the Nile in statuary form must have been relatively common in the Roman period, as over twenty of these are known today and many more must surely have been created at the time. Sometimes the Nile was twinned with the personified figure of *Tiberinus* or the Tiber, pendants not necessarily of partnership but perhaps of interdependence. There must have been so many images of the Nile in Rome it must have felt almost as if the distant river flowed through the city. It also seems to have flowed through the imperial imagination, as best personified by the paired Nile and Tiber statues at Hadrian's Villa at Tivoli, the Nile and Tigris or Tiber (it is uncertain which) in present-day Piazza Campidoglio, perhaps originally

---

[1] On Roman and Egyptian relationships and interactions, and on Roman Egypt see, for example: Blouin 2014; Bricault *et al.* 2007; Davies 2011; Greco *et al.* 2016; Leemreize 2014; Mazurek 2020 and 2022; Merrills 2017; Meyboom 1995; Poole 2016; Spier *et al.* 2018; Swetnam-Burland 2011; Van Aerde 2019; Vasunia 2001; Vecchi and Vecchi-Gomez 2002; Versluys 2010; and Vout 2003.

[2] On Egyptian and Nilotic landscapes see, for example: Awan 2003; Barrett 2013, 2017, 2018a, and 2018b; Gullini 1956; Hachlili 1998; Meyboom 1995; Schrijvers 2007; Swetnam-Burland 2009, 2012, and 2015; Versluys 2000; Walker 2003; and Whitehouse 1980. On Egypt and Roman cultural interaction see, for example: Pearson 2021; and Trimble 2017.

from the *Serapaeum* of Hadrian, and the Nile from Domitian's Villa up in the Alban Hills about twenty kilometres from Rome.

Io and the personified Nile appear before the Egyptian goddess Isis in a wall painting of the later first century AD from the *Casa del Duca di Aumale* or House of the Duke d'Aumale at Pompeii. As a literary image the Nile also appeared in the famous *ekphrasis* description of images on Aeneas' shield in Virgil's *Aeneid*: 'the cerulean Nile, grieving in his great body and spreading out his folds and with his entire robe calling his conquered people into his lap and into his furtive currents.'[3]

The evident popularity of Egyptian imagery and symbolism within Romanised society in Italy can perhaps be gauged by the fact that around one hundred and thirty Nilotic scenes on mosaics and wall paintings have been recorded at Pompeii, for example, a perhaps surprising number in what was a small provincial town. Nilotic scenes took many forms, often involving the perceived semi-humorous depiction of dwarves and pygmies,[4] but interest here will quite specifically be on those scenes that incorporated depictions of landscape and environment, including flora and fauna. An *opus sectile* mosaic pavement in the House of Publius Cornelius Tages, also known as the *Casa dell'Efebo* or House of the Ephebe, a wine merchant living in Pompeii, carries images of roses and lotus flowers, the latter not plants grown in the Pompeian landscape. These simple floral images were themselves geographical indicators, conjuring up the landscapes of Egypt. The majority of these Nilotic images occurred in private houses rather than in public buildings and contexts. It should also be noted here that Nilotic scenes were very popular on the so-called Campana Plaques or decorated tiles.

When Nilotic scenes appeared outside of Rome and Italy it must be asked why that might have been so and if the use of such images reflected the same motives of creation and consumption. Such scenes away from Italy have been found in Gaul, in Spain, in North Africa, which is probably not too surprising, and in the Near and Middle East. The fact that the Near Eastern group of twelve examples is chronologically distinct, belonging to the fifth and sixth centuries AD, placing them much later than the majority of Roman Nilotic images, and mostly associated with ecclesiastical buildings, suggests an altogether different use of the motif. It has been suggested that the image of the Nile in flood and creating fertile land was highly symbolic in a Christian context as reflecting the creation of God and a generalised conception of fertility, abundance, and prosperity. The Nile flood could perhaps have been an allusion to the gathering of the waters to create dry land. Jesus' arrival in Egypt could also have been alluded to by reference to the Nile as a signifier for Egypt more broadly.

The Nile was thought to be equated with the Gihon, one of the four great rivers of Paradise (Euphrates, Tigris, Gihon, and Pishon) and thus to have represented part of a broader Christian sacred geography than simply tying the river to the land of Egypt. The use of such scenes continued into Byzantine Christianity: the eighth century AD mosaic from the Church of St Stephen at Umm er-Rasas in Jordan has an inner frame that is formed by the flowing waters of the river, live with fish, birds, and vibrant water plants, afloat with boats, and its banks alive with fishers and hunters. The ten walled cities of the Nile Delta are spaced out along the

---

[3] Virgil *Aeneid* 8.711-713.
[4] On grotesque dwarfs and pygmies in Roman painting see, for example: Barrett 2018a; and Tybout 2003. On depictions of Nubians see: Gates-Foster 2021.

river's route. The same religious symbolism, that the Nile was thought to represent Paradise, in this case in the Koran, may also be found in the boat mosaic from the north colonnade of the courtyard of the Great Mosque at Damascus in Syria. Dating from the first phase of the mosque, AD 705-715, this places this example earlier than the Umm er-Rasas mosaic.

While it is argued below that many Nilotic scenes can be categorised as what are called colonial landscapes,[5] many pre-date, some from the second century BC by around a century, the Roman colonisation of Egypt and its subsequent incorporation as a province. Were these earlier images political statements of a kind: soft power depictions of a future target for exploitation and military and political takeover?

The most spectacular of the surviving Nilotic mosaics in Roman Italy is the huge pavement from the site of ancient Praeneste, modern Palestrina, to the south of Rome[6] (Figures 59-65). Dating to around the first two decades of the second century AD it was commissioned for a large basilican hall in the centre of Roman Praeneste, a public building of some kind, though whether a temple to Fortuna, a civic basilica, an *Iseum* or temple to Isis, or a building incorporating a nymphaeum, remains uncertain. Some authorities view the mosaic as being in the style of a Hellenistic artwork, even going so far as to suggest that it represents a copy of an Alexandrian court painting of the third century BC.

The mosaic today can be viewed in the town's museum, the *Museo Archeologico Nazionale* in *Palazzo Colonna Barberini*, where it is mounted on a wall on the uppermost floor and approached up flights of stairs from the museum's lobby. While the wall mounting allows a clear view of the design in its entirety to the modern viewer it perhaps forces consideration of its original context and the ancient act of viewing in that context to become obscured. It can also be forgotten that the mosaic as we see it today consists of only perhaps half of the original mosaic, with the rest restored to create a unified whole. While I do not necessarily accept the description of the restored mosaic as 'a pastiche', as it has been referred to by one academic commentator, discussion of the style of the mosaic must take the fact of its heavy restoration into account.

The Palestrina Nile Mosaic takes the form of a panorama or bird's-eye view of the course of the river from its mountain source to the Nile delta, and the annual flooding of the river, that great natural regenerative event. The peoples of the uplands, of Middle Egypt, and of the delta appear, as do rustic buildings and grand architectural complexes. Ritual or religious scenes are portrayed. Locals participate in animal hunts, most notably a hippopotamus hunt prominently positioned in the lower half of the pavement. Detailed and in most cases astonishingly-accurate representations of plants such as papyrus, the date palm, the Egyptian acacia, sycamore, and the Indian lotus acted as strong geographical and locational markers. The river and its banks can be seen to be teeming with wildlife. Here we see a virtual celebration of the fecundity of the country as represented by the great river, the environments through which it flowed, and, of course, its distinctive and abundant fauna and flora. The labelling of many creatures on the mosaic might at first sight suggest that this was to act as an aid to

---

[5] On the idea of colonial landscapes see, for example: Blunt and Rose 1994; Dang 2021; Desbiens 2003; Greene 2023; Leech and Leech 2021; McMillan 2019; Mukherjee 2020; Sluyter 2001 and 2002; and Zarobell 2009.
[6] On the Praeneste/Palestrina Mosaic see, for example: Ferris 2018: 18-21; Gullini 1956; Merrills 2017: 5065; Meyboom 1995; Schrijvers 2007; and Thomas 2021: 41-88.

LANDSCAPE AND DESIRE

Figure 59. The Praeneste/Palestrina Nile Mosaic. First quarter of the second century AD. *Museo Archeologico Nazionale di Palestrina*, Palestrina. (Photo: Author).

Figure 60. Detail of the Praeneste/Palestrina Nile Mosaic. First quarter of the second century AD. *Museo Archeologico Nazionale di Palestrina*, Palestrina. (Photo: Author).

Figures 61-62. Detail of the Praeneste/Palestrina Nile Mosaic. First quarter of the second century AD. *Museo Archeologico Nazionale di Palestrina*, Palestrina. (Photo: Author).

Figures 63-65. Detail of the Praeneste/Palestrina Nile Mosaic. First quarter of the second century AD. *Museo Archeologico Nazionale di Palestrina*, Palestrina. (Photo: Author).

some contemporary viewers who might have had no idea as to what these exotic creatures were. Interestingly, the artist has quite carefully distinguished between the characteristic fauna of the Upper Nile in Nubia-the lion, giraffe, boar, rhinoceros, hyena, and elephant- and that of the Lower Nile and delta of Egypt-the hippopotamus, the crocodile, and the ibis. Monkeys, birds, lizards, otters, turtles, bears, dromedaries, peacocks, and onagers are also depicted. Some fantastical, mythical creatures also appear. However, these taxonomic labels are all in Greek and were unlikely to have been decipherable to all local viewers/readers: these labels represented a manifestation of Hellenistic learning and knowledge as promoted at the Ptolemaic court at Alexandria, a process that allegorically stood for the regime's geopolitical power and territorial domination.

The Nile Mosaic displayed a totality that encompassed human life, experience, and endeavour, the coexistence of primitive, unsophisticated life with civilization, as represented by architectural images, and natural phenomena and the environment as represented here by indigenous, yet to the Romans exotic, flora and fauna. The Nile Mosaic from Palestrina is also particularly interesting in that it used its sheer size and scale, along with its bird's eye viewpoints, to present not just an image of the Nile and its surrounding landscapes but also a geography of the river, literally from source to delta, a kind of pictorial map in other words.

The Nile Mosaic was in fact one of two great tessellated pavements adorning the basilica at Praeneste, pendants to one another inside apsidal-ended rooms leading off either end of the great hall. The second mosaic pavement, a fish mosaic, is today still *in situ* inside the remains of the basilica and though very badly damaged, particularly in the centre, its interpretation alongside the Nile Mosaic helps to shed light on the meaning of the overall decorative scheme inside the building. It sits in a recessed basin cut into natural rock. The Fish Mosaic is likely to have depicted marine life, that is fish, squid, crustacea, and octopuses, across its whole surface, representing the great Mediterranean sea. A solitary human figure stands on the foreshore by the sea.

Each of these two mosaics can easily be interpreted on its own. Indeed the Nile Mosaic has been variously described as a religious allegory, a historical painting, a topographical scene, a representation of a province, and a genre piece. However, if intended to be considered together as a conceptual whole then further ideas could have been intended to be expressed here, principally the crossing over the sea by the Romans to conquer the distant land of Egypt, or that the Nile fed the great sea on which Roman power and prosperity depended. I cannot agree with a recent suggestion that the two mosaics presented competing conceptions of time and space.

The classical art historian John Clarke has suggested that there were probably three levels of meaning underlying the depiction of Nilotic scenes in Roman art.[7] Firstly, that they could have acted as 'scenic entertainment', as he terms it, celebrating the exotic nature of this highly significant Roman province. Secondly, just as Roman depictions of barbarian peoples often presented them as being 'other' to the Romans, so the same sense of difference, and thus of power differential, could also have been suggested by Nilotic scenes with people and animals. Thirdly, they could have acted as protective or apotropaic images, repelling evil spirits and

---

[7] Clarke 2007: 88-89.

negative forces. I believe that a fourth level of meaning pivots on the identification of Nilotic scenes as part of a geography of Roman-ness. The depiction of exotic animals in such scenes was part of a strategy for a Roman urban audience to come to terms with the modernity of their situation, that is with being part of a vast new world, as represented by the Roman empire, that they would otherwise find difficult to conceptually comprehend.

Crucially, no other Roman province was celebrated and depicted in this way, or indeed in any other way than as a conquered land, through personification or through images of war and decimation. In the case of Dacia on Trajan's Column and Germany on the Column of Marcus Aurelius we see depicted there foreign lands through their battling and ultimately defeated peoples, sometimes through their villages, huts, and strongholds, and sometimes their farm animals and flocks appear. We get no sense of the actual appearance of these foreign landscapes-their topography, their rivers (unless being bridged or crossed by Roman forces), their flora, and their wild fauna-as we do of Egypt in so many of the Nilotic mosaics and wall paintings that have come down to us. Painted topographical scenes known as battle paintings were carried in the procession at Roman triumphs, though none has survived. In contrast, it is worth noting here how the visual idealisation of Roman and Italian landscapes in bucolic scenes was a common trope on wall paintings.

There seems to be a reluctance among some archaeologists to interpret Nilotic scenes outside the established rubric that they represent either purely decorative scenes, of aesthetic value only, or that when they appear it must be assumed to be because they originally had some symbolic religious value to their commissioners, even if those commissioners were not necessarily adherents of the cult of Isis or of other eastern deities. The either/or dichotomy does not leave much room for nuance or for other frameworks of interpretation that would allow for the the works to have been created and perhaps viewed through a colonial lens. Many of these Nilotic scenes were not simply genre portrayals of exotic and exoticised landscapes, they were what are called depictions of a colonial landscape, with the Roman political and military annexation of Egypt quite significantly changing once and for all whatever benign views of Egyptian culture and thought might have been inherited by Roman intellectuals and political thinkers from Greek forebears. There was a highly-significant earlier Greek discourse around Egyptian culture, philosophy, and religion that both actively involved and encouraged Egyptian contributions to the debate and which was used as a sometimes distorted mirror to reflect on the very notion of Hellenic identity.

The recent observation by Joshua Thomas that dark red and dark green oval shapes appearing in the mountainous terrain on the Palestrina mosaic are images representing deposits of precious or semi-precious gemstones is extremely interesting in terms of what it might allude to.[8] Roman colonialism and imperialism came to be defined economically by a remarkably efficient system of locating and exploiting local and regional raw materials for Roman gain, though more often than not this was often just a sophisticated form of asset stripping, before giving way to licensed commercial exploitation of these resources.

---

[8] Thomas 2021: 83-85.

(Dis)Possession

As depictions of a colonial landscape the corpus of Nilotic scenes is of course highly idealised, and represented only one of many individual and distinct landscapes in the country: there are no depictions of the quarried landscapes and the porphyry quarries in the western desert, economic centres driven by enslaved labour and sometimes the labour of condemned criminals. There are no equivalent working, enslaved or indentured landscapes to those of present-day Brazilian photographer Sebastião Salgado. Richness and abundance, growth and regeneration, jeopardy and plenty, were the main themes presented by the recurring tropes of Roman Nilotic representations. The production of Nilotic scenes in Roman art both pre-dates and post-dates the formal Roman annexation of Egypt. As already noted, it is difficult to gauge whether there was a discernible break or rupture in the way the Nile Valley and its hinterlands were presented as images before and after annexation, though it may well be that this was indeed the case. The before/after contrast would appear to centre on a pre-annexation concentration on Nilotic flora and fauna, with the number of landscape depictions being fewer in total number, and with fewer numbers of human inhabitants shown before, in contrast to the trends after annexation. These changes are not strictly quantifiable, but certainly they would appear to have marked changing trends, if not necessarily a break followed by change.

Yet the asset-stripping of the wealth of Egypt in the form of its artworks, its agricultural crops and other foodstuffs, its precious metals, and its other raw materials was taking place at the same time. Pliny the Elder noted that: '…with the entire world having been united by the majesty of the Roman empire, who thinks that life has not benefited from the commerce of things and the partnership of solemn peace….For even those things, which before had been inaccessible, are made of use without distinction'.[9] Sorcha Carey's recent study views Pliny's discussions of imported marbles and other stones in the *Naturalis Historia* as being part of a broad narrative of conquest in the work.[10] Indeed his metaphor of different coloured marbles representing different regions and peoples conquered by Rome was extremely powerful, in that it suggests that to many in the city parts of the very fabric of Rome provided a constant reminder of geographic spread and distance.

For instance, the great Egyptian stone obelisks brought to Rome and set up there can be seen certainly foremost as emblems of imperial power and munificence, as the money, resources, and organisation needed to bring them there was beyond the means of most private individuals even in a city and empire as powerful as Rome. Their choosing as suitable monumental stones reflected the expanse of that empire-its existence as a repository of goods and raw materials for exploitation by Rome almost at will. In addition to acts of quarrying, creation, transportation, and shipping-reflecting powerful infrastructures-the placing and erection of the obelisks was testament to the skills and knowledge of Roman architects, engineers, and builders.[11]

These Nilotic landscapes were landscapes of colonialism in a way, though much more subtle in intent than the phrase might at first glance suggest. The consumption and viewing of

---

[9] Pliny *Naturalis Historia* 14.2.
[10] Carey 2003.
[11] On Egyptian obelisks at Rome see, for example: Iversen 1968-1972; and Swetnam-Burland 2010.

these Nilotic landscapes produced their own historical transformations of space, place, and perception, independent of the political reality that led to their popularity in the first place. In other words their reception in Roman times had the effect of recreating ideological information each time they were viewed. Again, replication and copying acted to create a series whose power as a group in terms of impact was greater than each single work. If reference is made to the results of a number of recent groundbreaking studies of the creation of what have been dubbed colonial landscapes it can be seen that the images of these overseas landscapes are often notable for what they omitted and not necessarily for what they portrayed. The impact and reception of Jean-Charles Langlois's massive painted *Panorama of Algiers* of 1833 on French society and its acceptance of colonial adventuring presents an instructive model.

The landscape of Roman Egypt was the only colonised landscape regularly and consistently portrayed by Roman artists for the delectation of a Roman and Roman-Italian audience of patrons and viewers. Other conquered lands were certainly portrayed in paintings that were carried in formal triumphal processions, but these works were one-offs, indeed really quite ephemeral works, and not part of an ongoing series with all that this implies. While foreign landscape elements were stressed in the portrayal of Dacia on the frieze around Trajan's Column, to a lesser degree in the depiction of German landscapes on the Column of Marcus Aurelius, and much less again on other monuments depicting battles between Romans and foreign peoples, all of these depictions are more properly categorisible as landscapes of war-an altogether different genre-rather than landscapes of colonialism.

Andrew Stewart sees the Palestrina Mosaic as representing 'racist, imperialist and exploitative' themes,[12] in part because of the contrasting depiction of different peoples and types of people in the different geographic registers of the pavement, indeed in its contrasting of urbanised environments and monumental architecture with wild, untamed upland landscapes. The colonisers here he sees as Greco-Macedonians and their Ptolemaic dynasty who exploited the land along with cooperative elite Egyptians. It can easily be imagined that an ideological meaning such as this could have resonated in a Roman Italian context. The list of lists known as the *Laterculi Alexandrini* that appears on a papyrus roll fragment of the second to first century BC also indicated the Ptolemaic era's fascination with geographical information as ideological content, listing the seven wonders of the world, the seven greatest islands, the highest mountains, the longest rivers, the most lovely springs, and the largest lakes.

Architecture was commonly portrayed in Nilotic paintings in particular, and occurred occasionally on mosaics. Perhaps surprisingly this is not a topic that has attracted much academic research in its own right, even though extremely interesting insights can and have been be gleaned from the study of the meaning of architectural depictions on Roman monuments such as Trajan's Column and the Column of Marcus Aurelius, establishing a discourse between the appearance of Roman and non-Roman building types on each monument and between the two monuments themselves in this regard.

It is instructive to review standard interpretations of the Nilotic landscape genre against a number of other artworks of later periods, and from other countries, representing colonial landscapes. The first such portrayal is a landscape painting of the Bogotá Savanna-*The Painting*

---

[12] Stewart 2014: 227-233.

of the Lands, Marshes and Swamps of the Town of Bogotá- made in 1614, arising out of a legal dispute over land ownership between the Spanish New Kingdom of Granada and Francisco Maldonado y Mendoza.[13] This is thus both a legal document-suggesting accuracy of information-but it is also a painting and not a map, so while one can appreciate the spatial ordering of information here it was never intended to be accurate in terms of measurements and distances. In its style, form, and use of colour it was obviously to also be aesthetically appealing. The research project '*Redrawing Andean Territories in the Seventeenth Century*' set out to examine through this painting, other artworks, maps and plans, documentary evidence, and material culture how colonisation by Europeans transformed the social and environmental worlds of the native peoples and landscapes of what is today Colombia in the sixteenth and seventeenth centuries. The Bogotá Savanna map caught a landscape in transition, as the lands and field systems (ridge and canal) of the indigenous Muisca peoples were transformed by European settlement, town foundation, and agricultural practices, beginning quite soon after the conquest of the region by the Spanish military expedition led by Gonzalo Jiménez de Quesada in 1536. As the project's webpage points out, this painting to all intents and purposes is 'essentially a map of land dispossession', as European farming regimes, especially land seizure and the resettlement of Muisca peoples into villages, clearance, parcelling out, and extensive cattle ranching on large new estates impacted on the very land itself-images of many cattle are drawn on the picture, as well as its peoples. Other animals such as horses, sheep, and pigs are also shown here. The villages, each with a church, were nodal points for the introduction of Catholicism.

The study of colonial landscapes is a fast-growing sub-discipline of geography, history, and postcolonial studies and many other examples based on recent research could also have been presented here, in order to draw out some further interpretative schemes that might have helped elucidate the meaning and help contextualise the much earlier Roman Nilotic mosaics and paintings. British colonial landscapes in India and the Caribbean, and colonial landscapes in the American southern states would have made ideal comparative landscapes, but their deployment in such an analysis as being presented here really needs to be the subject of other, future studies.

A highly significant, perhaps metaphorical representation of Roman-Egyptian relationships appears on a fragmentary marble relief from Rome, now in the British Museum, London (Figure 66). The relief was acquired by the museum in the mid-nineteenth century and is unfortunately without a secure provenance. Dated to 30 BC-AD 30 in the museum wall caption and 100 BC to AD 100 on its website the image carved here is of a high-prowed boat progressing down the River Nile, the setting confirmed by the elaborate, pointed eastern headgear of the boatman at the tiller/rudder and the presence in the river of a large basking hippopotamus. A pair of dolphins guide the boat forward. On board the barge are a man and woman engaged in coitus, the man entering the woman from behind as she rests and steadies herself with her hands on the deck, this sexual act taking place under the shade of the boat's sail. What is to be made of this explicit scene? It could simply have been an example of the broader strain of Roman erotic art which was so common in elite household contexts in Rome and Italy, usually to cater for, satisfy, or stimulate the male gaze, establishing a gender-based power differential for its viewers. As the caption for the displayed item in the British Museum suggests it could be that the Nilotic setting married with the sexual image of the copulating couple could

---

[13] Website *Redrawing Andean Territories in the Seventeenth Century*: colonial-landscapes.com .

Figure 66. Nilotic relief with erotic scene, Rome. Date 30 BC-AD 100. British Museum, London. (Photo: Copyright Trustees of the British Museum).

metaphorically have alluded to the mingling of the great rivers Tiber and the Nile, of Rome and Egypt. Merging waters and exchanges of bodily fluids could be somehow conflated here. The submissive sexual position adopted by the woman could also have been significant in terms of Egypt now being subservient to the power of Rome, now being dominated by Rome. Another possibility is that the copulating couple were intended to represent two specific individuals, the British Museum captioner noting that it could have been intended to portray Antony and Cleopatra, in which case we are seeing literally the conquest of Egypt in the form of a scene of sexual domination.

Who would have commissioned and owned such a piece must remain unknown, though obviously it must have been an elite Roman male. The well-attested popularity and presence of Nilotic scenes in different media and of Egyptian or Egyptianising objects in many private houses in Rome, Pompeii, Herculaneum, and doubtless many other Italian towns and cities, in the late first century BC to the first century AD, along with a penchant for erotic artworks, means that speculation should in this instance certainly be directed towards elite Roman men. Also in the course of research for this book I have seen numerous terracotta Campana plaques of the period AD 1-100, bearing profuse but standard repertoires of Nilotic images. Images of the inhabitants of the country could also be used to evoke Egypt in the Roman mind (Figures 67-68).

Egypt was 'captured' in other ways too, and not just through frozen images of its landscapes. Four huge obelisks of Egyptian stone were transported to Rome and erected there. Two stand

Figure 67. Black and white mosaic, Nilotic scene with pygmies, Rome. First to second century AD. *Palazzo Massimo Museo Nazionale Romano*, Rome. (Photo: Author).

Figure 68. Marble statue of a black youth on a crocodile, Rome. First century BC to first century AD. British Museum, London. (Photo: Copyright Trustees of the British Museum).

LANDSCAPE AND DESIRE

Figure 69. Statue of the personified Nile carved in dark basanite, Rome. Flavian. *Musei Vaticani*, Rome. (Photo: Author).

in front of the Mausoleum of Augustus (in *Palazzo Esquilino* and *Palazzo del Quirinale*), a third brought here by Domitian now stands in *Piazza Navona*, and the fourth, the so-called Barberini Obelisk, now can be seen in the park on the Pincian Hill but originally was set up near *Porta Maggiore*. The latter seems to have marked the mausoleum of the deified Antinous as related by its hieroglyphic inscription, marrying Egyptian script with Roman ideological concerns.

It is also interesting how Egyptian stone was sometimes used for imperial portrait statues, again making a very specific point about Roman and Egyptian inter-relationships. These statues made very specific points about the emperor's geographical reach of power. A stone like Egyptian porphyry was only a luxury material because the Roman state designated it as such. In imperial contexts it acted as a signifier of Rome's command of raw materials from right across its empire.

Another such material would appear to have been basanite, another hard, coloured Egyptian stone from Wadi Hammamat and the Allaki Wadi. As far as can be ascertained it would seem that its use was reserved for statues of deities or of imperial figures, including the so-called Berlin Caesar, in the *Altes Museum* there, the Capitoline Augustus in the *Musei Capitolini*, Livia and Agrippa Postumus in the *Musée du Louvre* in Paris, Germanicus in the British Museum in London, and the Capitoline Hadrian in the *Musei Capitolini*. Other statues in Egyptian stone include the Flavian Nile statue in the *Musei Vaticani* (Figure 69) and the statues of Hercules and Dionysus in the *Aula Regia* in the *Domus Flavius* on the Palatine Hill, now in *Palazzo Farnese* in Parma.

Figure 70. Statue bust of Antinous, Rome. Hadrianic. *Musei Capitolini Centrale Montemartini*, Rome. (Photo:Author).

Figure 71. The *Canopus*. Hadrian's Villa at Tivoli. AD 133-138. (Photo: Author).

Ovid, in his *Metamorphoses*, wrote that the Egyptian goddess Isis answered a woman's pleas by turning her daughter into a boy.[14] It must be asked whether this miraculous transformation was meant to be a metaphor for Rome's transformation of Egypt after its annexation and incorporation as the Roman province of *Aegyptus*? A change of gender here was perhaps being conflated with the strength of Rome's male power structures. Similarly, with regard to the relationship between the emperor of Hadrian and the Bithynian youth Antinous[15] (Figure 70) the latter perhaps could be said to have had the lands of Egypt written on his body, his body eventually dissolving there, back into the land as part of a greater regenerative force, after his drowning in the Nile.[16] The so-called *Canopus* at Hadrian's Villa at Tivoli must also have sparked memories of this gilded, doomed youth (Figure 71).

Fraught dynamics were at play here, creating a complex web of secrecy, shame, pride, guilt, frustration, and desire, and the art came to stand for this. Rome's power was here, displayed through the process of intertwining the individual and the world, the detail and the environment, the visual and the sensual. The series of mosaics and paintings, plaques, and figures featuring images of Egypt and its peoples created doubling movements, of separation and creation, division and participation, deconnection and contagion: these folding, unfolding, and refolding processes of objectification created a kind of Corbusian frame for surveillance and colonisation. When imposed on the material culture produced in Rome and Roman Italy this participated in the maintenance of colonial possessions as a source of fantasy, as well as power. We can also think about the geographies of collaboration, usually from the provincial elites, that enabled and helped Roman imperial expansion into new territories. Roman Nilotic art took the margins of its empire and placed it at its centre.

The conceived difference and (dis)placement here marked a state in which marginality would become the condition of the centre, the illusion of continuity that they provided concealed the violence that remained out of frame. Roman Nilotic art was one strand of a number of projects of 'worldmaking' undertaken by the Romans.

If Hellenistic, Roman, and later Ottoman colonisation of the physical land of Egypt and of Nilotic landscapes in art appeared to represent a historical stranglehold on the indigenous voice and vision of the country then we can consider the imagination of Egypt to have been somehow decolonised and freed to a very great degree by the art movement represented by the Ramses Wissa Wassef School of tapestry weaving established at Harrania in the early 1950s. The vibrant tapestries of Nilotic scenes, flora and fauna, and local landscapes produced there by untrained but instinctive artist-weavers such as Ashour Messelhi, Karima Ali, Chehata Hamza, and Maryam Hermina provide visual statements that help rewrite the scripts of the past through the manifestation of the independence of imagination.

In the next chapter attention will be paid to Roman mapping and its ideological links with Roman imperial power and control.

---

[14] Ovid *Metamorphoses* 9.688-690.
[15] On Egypt and Hadrian's Villa at Tivoli see, for example: Mari 2008.
[16] On Antinuous see principally: Vout 2005 and 2006.

Chapter Seven

An Unseen Ruler

In the last two decades there has been a remarkable flowering of the study of ancient Greek and Roman geography and maps, best represented by the numerous writings on the subject by Richard Talbert and Duane Roller in particular.[1] This does not mean that there is nothing new to say on the subject, but rather that what will be discussed here in this chapter is more reliant on secondary sources and interpretations than the material discussed in the book's other chapters.

The subjects to be considered here include the work of professional Roman surveyors and the history of the Roman creation and use of maps, as far as the latter can be gauged from a very small database of evidence indeed.

In certain circumstances in the Roman world there were calls for accurate surveying of land or of routes through the landscape, and a large amount of information about the *agrimensores*, the Roman land surveyors, has come down to us, including items of surveying equipment, a survey handbook, and funerary monuments of named surveyors. To these surveyors we can ascribe the setting out of urban street grids and boundaries, the grids and footprints of Roman military camps, forts, and fortresses, road lines, cadastral plots of centuriated land, groundplans for the construction of buildings for both architects and builders, and plans of buildings or land required for official records or for use as base evidence in legal disputes.

Only a few items of surveying equipment have come down to us, including pieces of a *groma*,[2] a survey instrument, from the *Officina di Vero Gromatico* or Workshop of Verus in Pompeii and now in the *Museo Archeologico Nazionale* in Naples. The possible remains of a *groma* come from the German *limes*, at Pfünz near Eichstätt in Bavaria, and from the Fayum in Egypt, the latter find dated to the second to first century BC and perhaps representing an early prototype of the machine.

Recently two unusually decorated floors in a house in Pompeii-*Casa di Orione* or the House of Orion- have been suggested to carry images of surveying symbols and a stylised image that might represent a *groma* rather than a *groma* in use, suggesting that the property could have belonged to a professional surveyor or a retired surveyor. One image is of a square inscribed inside a circle, the circle being cut by two perpendicular lines. The second is a circle with an orthogonal cross inside it and another orthogonal symbol close by: it has been suggested that this could be a schematic image of a *groma*. The third is of a baseline with five equally-spaced transects radiating from it.

---

[1] On ancient maps and mapping, specifically Greek and Roman maps see, for example: Albu 2005 and 2008; Arnaud 2008 and 2016; Avi-Yonah 1954; Bellori 2021; Bosio 1983; Bowman 1998; Brodersen 2004; Clarke 2008; Coarelli 1991; Diederich 2021/2022; Dilke 1985, 1987a, and 1987b; Evans 2005; Fodorean 2011; Gee 2020; Hălmagi 2015; Haselberger *et al.* 2002; Hawes 2017; Irby 2012; Janni 1984; Klotz 1931; Krebs 2018; Lanciani 1901; Lozovsky 2008; Meneghini and Valenzani 2006; Najbjerg and Trimble 2004 and 2006; Nasrallah 2005; Rathmann 2016 and 2022; Reynolds 1996; Rizzo 1994; Rodriguez-Almeida 1977, 1981, and 2002; Salway 2005 and 2012; Sherk 1974; Talbert 2004, 2008, 2010, 2012, and 2023b; Tierney 1963; Trimble 2006, 2007, and 2008; Trousset 1993; and Wiseman 1992.

[2] On the *groma*, surveys, and Roman surveyors see, for example: Campbell 1996 and 2000; Dilke 1962 and 1971; Lewis 2001; Schioler 1994; and Thulin 1971.

Funerary monuments of named surveyors include the first century BC stele of the freedman Lucius Aebutius Faustus, now in the *Museo Civico* in Ivrea in northern Italy. This is of particular interest because not only is Faustus referred to as a *mensor* in the dedicatory inscription but we learn that he had the tombstone set up for himself while still alive, for his wife Aria Aucta (a freedwoman), their children, and the freedwoman Zepyra. The face of the stele is decorated with an image of a *groma*, represented in quite clear detail, with the instrument's hanging plumb-bobs easily discernible. A *groma* and plumbline also appear on the first century AD stele of Popidius Nicostratus from the *Porta Nocera* necropolis at Pompeii and now in the *Antiquarium di Boscoreale* nearby. Military *gromatici* or *mensores* are known from the fortress at Aquincum in Hungary: Aelius Rufus of *Legio II Adiutrix* dedicated an altar to Jupiter there and Aurelius Deipas, along with his wife, commissioned and dedicated a small lidded sarcophagus to their young daughter.

Early surveying treatises by Sextus Frontinus and Hyginus Gromaticus in the time of Trajan are known about through references by other writers, but the works themselves have not survived. An illustrated survey handbook dating from the fifth century AD, the *Agrimensorum Romanorum* or *Gromatici Veteres,* probably includes elements from these earlier handbooks.

Carving and Paring

What then of the products of the surveyors, in terms of drawings and plans? As might be expected, little or nothing of this sort has survived, given that sketches and original drawings would have mostly been done on parchment of some kind or have been done on papyrus. Small survey drawings, sketches, and notes of measurements could also have sometimes been made on wax writing tablets, but again no such documentation relating to survey work has survived. However, some records relating to the practice of centuriation do survive. As a geography product these are of huge importance, being as they constitute part of the physical record of the Roman colonisation of certain landscapes, first in Italy and then beyond, beginning in the fourth century BC quite close to Rome and gradually being extended to other parts of Italy as Rome's Italian conquests accelerated. Around 268 BC in the Po Valley of northern Italy new colonies were founded, such foundations requiring the legality that was seen to come with the confiscation, measuring, and gridding of the land at each chosen site. The land reforms of Tiberius Gracchus in 133 BC led to further needs for surveys. This carving and paring of the land was methodical and almost scientific: beneath the rule a country hid. Interrupting the locals' trains of thought were lines like lines of longitude and latitude, defining and refining space. Virgil in the *Eclogae* or Eclogues complained quite bitterly about the centuriation of land around his home town of Mantua, modern Mantova, confiscated land then allotted to veterans of the Battle of Phillippi in 42 BC.

It is not certain how much centuriation was applied to the landscapes of conquered territories outside of Italy, but certainly there is a great deal of evidence for mass centuriation in parts of Gaul and Spain in particular. The outlines of many centuriation grids imposed over previously open landscapes are often revealed by aerial photography and more recently by online survey systems such as Googlemaps. These grids of land plots, of military camps, of towns and cities are like negative images, almost hauntological remains of the work of surveyors whose labours otherwise would be undetectable and unknown to us. The science of land measurement, allotment, and reallotment was used in the service of Roman imperial ideology and would usually have been associated with the confiscation and reassignment of land in terms of colonial imposition. These invisible grids, though once doubtless all marked out with pegs and rods, twine and rope remain active monuments to Roman colonialism and imperialism. Indeed, quite a large number of such inscribed field markers have been found, variously in Italy, Gaul, Macedonia, *Numidia*, and *Africa Proconsularis*.

Only a very few centuriation plans are known. The marble *Cadastre of Arausio*, from Orange in Provence, in southern France, is part of a large marble slab, now represented by three conjoining pieces, of a centuriated area between the veteran colony at Orange itself and the town of Nice. The grid drawing is heavily annotated. Dating to the second century AD this must have been created as a display plan, presumably using measured information from parchment or papyrus field drawings, notes, and calculations. Part of a centuriation plan can be seen inscribed on a fragment of a bronze plaque from Verona in northern Italy and now in the *Museo Archeologico Legnano*. It depicts gridded land, with inscribed captions, north of the River Adige. Bronze plans such as this must have been produced for every centuriation episode and probably every large-scale land survey in Italy, and probably in many provincial contexts. Where they were centrally archived and stored in Rome is unknown. The Hadrianic era historian Granius Licinianus relates how Publius Cornelius Lentulus was tasked by the Senate of Rome some time in the mid-first century BC to organise a survey of state land in the region around Capua in Campania and to have his results inscribed on a bronze tablet that would be publicly displayed in the *Atrium Libertatis* in Rome. Such displays must have been common practice, particularly when legal cases involving land ownership disputes would appear to have been extremely common in Roman times.

If Greek colonisation had been predicated on the right to return, new Roman cities were to be organised according to the Roman model: land was to be divided up into cadastral plots, and Roman *coloniae* were to be an altogether different kind of proposition. If Greek colonisation was a system based on a network of mother cities and satellite colonies, obviously Rome was itself the sole model for its many *coloniae*.

The survival of an annotated plan, possibly of the Tusculum Aqueduct known as the *Aqua Craba*, inscribed on a marble slab found in Rome[3] suggests that engineering and architectural schemes such as the one spatially represented on the fragmentary slab also regularly required plans such as this, presumably for evidence of both land ownership and access onto privately-owned land, or to back up cases for land requisition or confiscation.

Similarly, parts of two detailed but small-scale architectural groundplans survive carved onto display slabs for some reason, one provenanced to Rome but now in the collections of the *Galleria Nazionale delle Marche* in Palazzo Ducale in Urbino and another in Perugia and in the *Museo Archeologico Nazionale dell'Umbria*. The Perugia plan is of a tomb, recording layouts and dimensions, being the mausoluem of Claudia Peloris and Tiberius Claudius Eutychus (Figure 72). Andrew Riggsby has listed fifteen Roman building plans in total that have come down to us, including these two examples, though I would amend that total to fourteen myself, as the so-called *Via Anicia* marble plan fragment, discussed below, is surely part of a city plan rather than being strictly a one-off architectural plan. On his list are a number of less formal plans

---

[3] *CIL* VI, 1261.

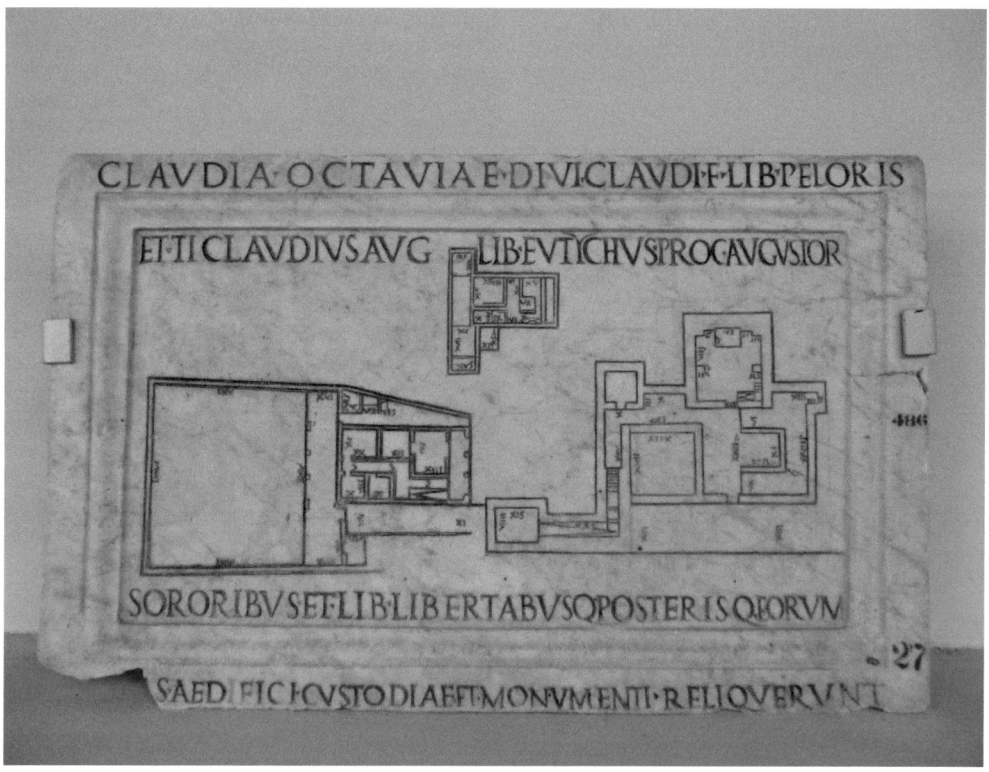

Figure 72. Marble ground plan of the tomb complex of Claudia Peloris and Tiberius Claudius Eutychus, Rome. Mid-first century AD. *Museo Archeologico Nazionale dell'Umbria*, Perugia. (Photo: Author).

than the Urbino and Perugia plans, including an elevation drawing of the Pantheon's facade inscribed on flagstones in front of the Mausoleum of Augustus in Rome, and a profile of a temple cornice scratched on wall plaster at the *Capitolium* in Brescia in northern Italy.

But there is a huge difference between the formal commissioning by an emperor of a map or plan in order to present geographical, topographical, or spatial information in a controlled manner, and to maybe even manipulate or distort that information, and the creation of a plan at the behest of a private individual. The so-called *Forma* (Map) of Via Marsala,[4] dating to the mid-third century AD, was found in Rome in 1872, and is unique in that it is a building ground-plan of a baths complex rendered as an image on a mosaic at a scale of approximately 1:16 (Figures 73-74). The walls of the building are in yellow edged with black, the interiors of each room appear as white *tesserae*, and each room contains Roman numerals delineated in red *tesserae*, perhaps representing the internal dimensions of each room.

The Roman measuring and mapping of space and the gathering of spatial information should be seen as having been a managerial tool in some circumstances, and in others a method of control, adjudication, or surveillance.[5] It could have served an economic and political

---

[4] On the *Via Marsala* mosaic map/plan see: Bouet 1998. On labyrinth mosaics see, for example: Molholt 2011.
[5] On the politics of mapping in more recent times see, for example: Albano 2001; Barber 2004; Barber and Harper

A Map of the Body, a Map of the Mind

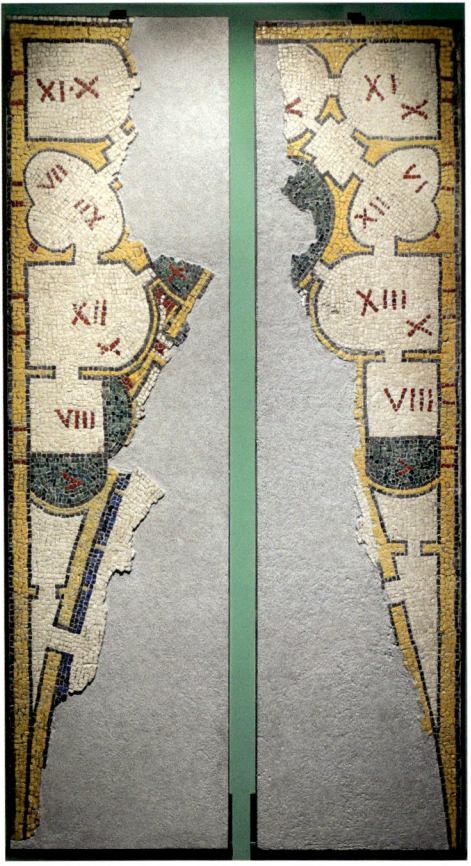

Figure 73. The *Via Marsala* mosaic map. Late second to early third century AD. *Musei Capitolini Centrale Montemartini*, Rome. (Photo: Professor Lynne Lancaster, by permission of the *Musei Capitolini*).

Figure 74. Detail of the *Via Marsala* mosaic map. Late second to early third century AD. *Musei Capitolini Centrale Montemartini*, Rome. (Photo: Professor Lynne Lancaster, by permission of the *Musei Capitolini*).

purpose, as well as an ideological one. The idea of accuracy in the Roman world has been much discussed recently by a number of academics who have suggested that of course there was a need to make weights and measures standardised and for methodologies of recording to be in place to ensure that systems operated across a wide swathe of territory in as unified a manner as possible.[6] In terms of dividing up and allocating land it seems as if an unseen ruler defined with geometry an otherwise unruleable expanse of geography. With these bird's-eye views the Roman administrators, like an aerial photographer over-exposed to the cartologist's two dimensional images, knew the area where the waters flowed: straining eyes must have tried to understand the works incessantly in hand.

2010; Black 1997; Gordon and Klein 2001; Harvey 1993 and 2010; Lemke 2002; Leuenberger and Schnell 2020; Minor 1999; Mukherjee 2020; Rapoport 2021; Short 2003; Taylor and Gregory 2022; and Van Den Hoonaard 2014.
[6] Particularly Riggsby 2019: 154-202.

Figure 75. The Mosaic of the Islands, Haidra, Tunisia. Third or fourth century AD. *Musée National du Bardo*, Tunis. (Photo: Author).

More public works of geographical art took the form of actual maps displayed in various locations in Rome which will be discussed here next, followed by a consideration of other types of maps that were drawn or created for more private consumption.

Livy in *Ab Urbe Condita*[7] describes a triumphal tablet made for Tiberius Gracchus, and presumably originally a prop carried in his triumphal procession in 175 BC, and which subsequently found a more permanent home on display in the Temple of *Mater Matuta* in the *Forum Boarium*. Described as being cut to the outline shape of the conquered island of Sardinia its surface was painted over with images of battles. This object then was both a map and a historical chronicle at the same time. Were maps carried at other triumphs? We do not know, but it is surely likely that they were. Lucius Aemilius Paullus earlier in 181 BC had a painting displayed of his victory over the Ligurians which again might have included some kind of map. It would appear then that the Temple of *Mater Matuta* became one of the first buildings in Rome where maps of different kinds were displayed for public delectation. Varro writing in

---

[7] Livy *Ab Urbe Condita* 41.28.8-10.

De Re Rustica[8] describes looking at what he calls 'pictam Italiam'-painted Italy-a wall-painting in the Temple of *Tellus* on the Esquiline Hill. Whether this was a map as such or an image of a personified *Italia* is not altogether clear. The temple was first dedicated in 268 BC and Varro's visit must have been some time around 45 BC, meaning that the wall-painting was made some time between these two dates. Pliny refers to a map of Armenia and the Caspian Sea drawn up by Corbulo's staff and sent to Rome in AD 58.[9]

Two extraordinarily-detailed mosaic maps are known from the Roman world, the so-called Mosaic of the Islands and the Madābā Map. The Mosaic of the Islands dates to the third or fourth century AD and comes from Haidra (Ammaedara) in western Tunisia[10] (Figure 75). It is now in the *Musée National du Bardo* in Tunis. It represents an overview of the Mediterranean sea, showing and naming fifteen of the important islands and some of their cities in a purely pictorial manner which took no account of distances and relative locations. Fish swim around in the sea between each island. It represents an informed geographical work that has been manipulated and transformed for purely aesthetic purposes. The Madābā Map appears on a mosaic of the sixth century AD in the Church of St George in Madābā, Jordan.[11] It is much more topographically accurate than the Mosaic of the Islands and, though today damaged, shows a huge swathe of the lands from Lebanon in the north to the Nile Delta in the south, and from the Mediterranean in the west to the eastern desert, including therefore part of the Holy Land and the walled city of Jerusalem, packed with buildings within, some identifiable to actual structures. Folded perspectives, bird's-eye, almost aerial, views, and detailed captioning in Greek of over one hundred and fifty towns and settlements, and the presence of topographical geo-locators like the Dead Sea, make this a remarkable public work which must have been designed using pilgrimage maps and itineraries.

Both the Sardinian outline map and the map-like painting of Italy were public works of geographical art whose production would have been dictated by political motives and which would have contributed towards the discourse around the achievements and status of their commissioners. The fact that both subsequently became display items in public and religious spaces is extremely important. Were public maps simply a manifestation of an interest in maps among some private elite Roman citizens? A particularly extraordinary story concerns the senator Mettius Pompusianus who was first exiled to Corsica by the emperor Domitian and later executed in AD 91, allegedly for a string of compound offences according to the historian Suetonius,[12] including possessing an imperial horoscope and having a map of the world painted on his bedchamber wall, slaves named after Carthaginian generals, and copies of extracts of speeches from Livy. Cassius Dio suggested that Pompusianus' map was on a roll of vellum rather than in the form of a wall painting. Whichever the case, it would appear that commissioning and owning maps at this time might have been considered an exclusively imperial prerogative.

One of the most widely-discussed passages in Roman elegy occurs in the Augustan poet Propertius' *Elegiae*, principally concerning the fictionalised correspondence between Arethusa,

---
[8] Varro *De Re Rustica* 1.2.1. On Varro's *picta Italia* see, for example: Roth 2007.
[9] Pliny *Naturalis Historia* 6.15.
[10] On the Mosaic of the Islands see principally: Bejaoui 1999.
[11] On the Madaba Map Mosaic see principally: Avi-Yonah 1954; and Piccirillo 1989.
[12] Suetonius *Domitianus* 10.3 and Dio *Historia Romana* 67.12.2-4.

a woman at home in the city of Rome, and her husband Lycotas who is away overseas serving in the Roman military.[13] Probably dating to around 16 BC the character Arethusa mentions that she is tracking her husband's travels and movements by reference to 'tabula pictos..mundos', that is 'worlds painted on a wooden tablet' which has to be a reference to a map of some kind. Propertius' readers would presumably not need to have been told what this object was that Arethusa was consulting, and the fact that a private individual, an elite Roman woman, albeit a fictional creation, had this item in her home and could consult it, could 'read' it and understand how to read it, would also have probably not occasioned surprise. The contrasting of a letter and a map in this poem is of great interest, in that it would appear that text and image were here being naturally marshalled together. Arethusa's attempts in the poem to body forth Parthia, its landscapes and peoples, in her imagination suggests that the considerable latent power of a geography product such as her *tabula pictos* need not necessarily have always been linked to Roman state ideologies. That we have here a Roman male author using the fictional creation of an elite woman to conceptualise geographical information is more-or-less unique.

Two other privately-created Roman maps are known, in the form of the Artemidorus papyrus and the Dura Map. The Artemidorus papyrus, now owned by the *Banco di San Paolo* in Turin, might in fact eventually turn out to be a forgery: it includes text thought to represent part of a geographical work by Artemidorus of Ephesus who wrote around 100 BC, and a very rough, unannotated outline drawing once considered to be the earliest map of the Iberian Peninsula. The Dura Map, from Dura-Europos in Syria and now in the *Bibliothèque Nationale de France* in Paris, dates to AD 230-235 and is drawn on a piece of leather, thought by some to have once been used as a shield cover.[14] It appears to be a route map of some sort, perhaps representing part of the route along the western and northern coastline of the Black Sea taken by the *Cohors XX Palmyrenorum* en route to be stationed at Dura at this time. Twelve stopping-off stations are named, each site represented by a simple stylised image of a building.

If Jupiter's prophecy in Virgil's *Aeneid* that Rome would one day become 'an empire without limits'-*imperium sine fine*-[15]was meant to relate to the contemporary geopolitical situation under Augustus then questions certainly need to be asked about the timing and motives for the creation and display of Marcus Agrippa's *orbis terrarum* or 'world map', as it is called by Pliny.[16] We do not have any firm evidence for the form of this map, other than that it was on display in the *Porticus Vipsania* in the *Campus Martius* where Augustus 'wrote' much of the new physical cartography of Rome through his various building projects there. Completed as a project by Augustus some time after Agrippa's death in 12 BC, possibly between 7-2 BC, this building might not originally have been intended to house this great map, but the original intentions of Agrippa are sadly not recorded.

Within a short walking distance of the presumed location of the *Porticus Vipsania* were seven of what Mary Boatwright has described as 'geographical monuments',[17] that is monuments somehow referencing aspects of Rome's empire, suggesting to her that the map display site

---
[13] Propertius *Elegiae* 4.3.
[14] On the Dura Map see principally: Hălmagi 2015. Professor Simon James does not believe that the leather on which the map is drawn comes from a shield.
[15] Virgil *Aeneid* 1.278-279.
[16] Pliny *Naturalis Historia* 3.17. On Agrippa's world map see, for example: Arnaud 2008 and 2016; Klotz 1931; Rizzo 1994; Tierney 1963; and Troussel 1993.
[17] Boatwright 2015: 235.

Figure 76. Portrait bust of Marcus Agrippa, Rome. 25-10 BC. British Museum, London. (Photo: Copyright Trustees of the British Museum).

might have become part of a conscious attempt to create a conceptual zone in Rome dedicated to the presentation and experiencing of codified and managed geographical information. The last documented reference to Agrippa's map, no fragments of which have yet been recovered by excavation or chance discovery, is from the fourth century AD. It is uncertain whether the map was painted on a wall or carved in marble. Obviously huge in size, it presumably placed Rome and Italy firmly at the centre of the known world, indicating coastlines, major rivers, and presumably also major roads. There may well have been some captions or textual annotations on or by the map to help orientate viewers. Its accuracy must have been considered important, and Pliny even discusses the issue of the incorrect representation of Baetica on the map as first set up. Anything beyond this constitutes pure speculation about the actual size, appearance, and display of the map. In the Greco-Roman world geographical knowledge could be presented in a factual way, using exact data, or in an allegorical or multivalent manner.

As well as a displayed map Agrippa (Figure 76) had also compiled or had had compiled an accompanying commentary or a database of a kind. This commentary/database, like the map, does not survive, nor was it ever described in antiquity in a way that might allow for it to be reconstructed today. Fragments of this text might survive in the geographical sections of Pliny the Elder's *Naturalis Historia* where he cites Agrippa as a source no less than thirty two times. It might be that Agrippa's map and commentary were very firmly in the Greek geographical tradition based on natural features, but novel in terms of a new Roman slant that ordered some part of the works on administrative districts, thus politicising the project. It is possible that the *Porticus Vipsania* was the headquarters of the *Cursus Publicus*, the official Roman state body overseeing a courier and transportation network around Italy and across the empire, which would have made the display of the map here highly appropriate, if correct.

It has been suggested that the creation of Agrippa's map was the very point at which the Roman world-view could be said to have become cartographic. The displayed map certainly caught the attention of contemporary Roman writers, and, as already noted, Pliny made around thirty references to geographical points or distances on the map, while Strabo referred to it at least seven times. Infuriatingly, neither of these authors described the map, but both presented petty criticisms of specific measurements on the map, at least presenting us with some idea of the incredible level of detail included on the map. Indeed, Pliny at one stage in his description wonders how a man as careful and meticulous as Marcus Agrippa could have made such serious errors of measurement and presentation. The very fact that Pliny has these measurements for discussion has suggested to some authorities that he might have had access to a written commentary accompanying the display of the map, but this seems unlikely. Perhaps more likely is that the drawn, painted, or inscribed map, on marble perhaps, either bore a number of measurements of distances on it or that there was an itinerary-like inscription of distances between major towns and transport nodes on the wall alongside the map figure. The last of these suggestions seems to me the most likely.

A number of attempts have been made by classicists and ancient historians to reconstruct the appearance of Agrippa's map: as a linear Itinerary map, in other words a sequential list of major transport nodes arranged in routes, with indications of distances between each node and thus along the length of each route; as an oblong map; or as a circular scheme. However, Pol Trousset's suggested reconstruction of the map in the Portico is to all intents and purposes immersive in terms of the viewing experience, comprising a triptych-like arrangement of the map over three wall surfaces, each with one of the three parts of the world, oriented respectively north, south, and east, with the viewer stood as it were in the Mediterranean Sea, looking straight ahead to Asia and the east, with Africa and the south on his or her right, and Europe and the north, including Italy and Rome, on their left.

The map must have been extraordinarily detailed in whole or in part, given that one of Pliny's descriptions refers to the appearance on the map of the relatively obscure town of Charax in Parthia, suggesting a level of specific detail altogether new in Roman geographical presentation. So it would not appear that this was simply a map of the extent of the Roman empire as it was at the time under Augustus, and it may indeed have been a map of the entire known world. Maps, like all images, have to be framed or to end somewhere and at some point, even if that boundary or constraint in this case was the height and width of the walls in the Portico.

Figure 77. Wall of the *Templum Pacis*/Temple of Peace, Rome. Now part of the Church of *SS. Cosma e Damiano*. (Photo: Author).

In summary, there would appear to be contradictory evidence and academic opinions on the form of Marcus Agrippa's great world map-whether it was a visual or written map or a mixture of image and text. The Severan marble plan of the city of Rome, although produced many years later than Agrippa's world map, seems to have been a companion piece to Agrippa's map project.

Augustus to all intents and purposes remapped Rome, not only in its appearance-the city of brick being transformed into a city of marble-but in many other ways. He extended the *pomerium*, that sacred but invisible boundary that marked a historical boundary and juncture but which stymied and stifled ambition, and he created the system of dividing up the city into *regiones,* allegedly for administrative and bureaucratic convenience. Placing Agrippa's world map in a prominent building in Rome made the city the centre of the known world. Maps have edges: thoughts are always with what might lie beyond. We need only to imagine the impact on the Roman crowds of Augustus celebrating three great triumphs in the streets of Rome over the course of just three days: his Illyrian triumph being followed the next day by the Actium

Figure 78. Wall of the *Templum Pacis*/Temple of Peace, Rome. Now part of the Church of SS. *Cosma e Damiano*. (Photo: Author).

Triumph, and that in turn being followed the very next day by his grand Egyptian triumph. It was at the Egyptian triumph that an effigy of Cleopatra, laying dead on her couch, was paraded. In 20 BC Augustus had the so-called *milliarium aureum* or 'golden milestone' set up in the Roman Forum, close to the Temple of Saturn. The possible marble base of the monument is all that survives today, though various reconstructions of its original form-as an exaggeratedly large milestone- and build have been offered by archaeologists and ancient historians. On it may have been inscribed in gilded bronze letters the names of the major towns and cities of the Roman empire and measured distances between them. The *milliarium aureum* was meant to symbolise the central nodal point of the empire's road system, quite simply that all roads led to and originated from Rome. Again, it was an act of centreing Rome in its contemporary world. The named towns, cities, and road routes it recorded and displayed to viewers and readers already existed, perhaps floating in unlocatable space in the imagination of many at Rome. Now suddenly a centreing, a grand gesture of recalibration, was taking place, with the *milliarium aureum* as a geographic mnemonic device to allow the Roman imagination to travel these roads, to visit these towns, and to understand principally their relationship with, and to, Rome, but also with each other in terms of distances apart. The display of Agrippa's map, the administrative reorganisation of the city's districts, the geographical elements of the triumphs, and the building of the *milliarium aureum*, the display of statuary in his *Porticus ad Nationes*, along with a number of other initiatives in his reign, suggest that Augustus saw geography products such as these as very useful tools in world-making in Rome.

In the instance of the great Severan map of the city of Rome set up in the *Templum Pacis* or Temple of Peace[18] it is Rome that is then to all intents and purposes the world to its rulers (Figures 77-78). As I have noted in a number of places in this book it is a useful exercise to consider how many of Rome's great geographic data monuments were extant at the same time and what their spatial relationships to one another might have been. In AD 211 could one in a short space of time have taken a walk to visit Pompey's Portico with its figures of various nations, the Augustan *Portico Nationes* and *milliarium aureum*, view the world map of Agrippa, and finish by viewing the Severan plan of the city of Rome?

The Temple of Peace was built on a plot of land on which once stood a market hall destroyed in the Neronian fire. Today it is part of a church, situated just a short distance south of Augustus' forum and east of the *Forum Romanum*. The complex was finally finished and remodelled under Domitian. The presence of a library on the site stressed the intellectual and ideological coherence of its conception, and it also became almost a museum, a repository of looted goods and war booty. Although much of the complex was destroyed by fire in AD 192 the Severan rebuilding still included a library. But this was not the first such place, even if it is the best known to us today. The *Porticus Metelli/Octaviae* brazenly housed a display of war booty in the form of looted statuary, eliding captor and captive, captivating viewers with stolen images now domesticated in a new Roman setting. Placing things from there here.

The Severan map of the city of Rome, known today as the *Forma Urbis Romae*, was created and set up some time between AD 203-211 on an end wall in the *aula* or hall of the *Templum Pacis* or Temple of Peace that now forms part of the exterior back wall of the Basilica Church of *Santi Cosma e Damiano*. This high wall is still quite noticeably scarred by the attachment holes for the metal fittings which would have held the individual marble blocks that made up the map in place. The installation of the map formed part of the Severan refurbishment of the complex. Detailed analysis of this wall face has allowed certain conclusions to be made about the size and form of the original map as mounted here. Probably originally 18 metres wide and 13 metres high, the map consisted of 150 blocks or slabs of Proconnesian marble onto which had been carved in plan the buildings and streets of the city at a scale of 1:240. The map was mounted with its lower border around 4 metres above floor level, meaning that the uppermost edge was c. 17 metres above floor level. Today, around 1194 fragments of the map have been recovered by various excavations in the area, comprising somewhere between just 10-15% of the original map surface.

The orientation of the map was such that south-southeast was at the top. The map had no topographical features on it, meaning that the seven hills that Roman poets believed gave Rome its unique character, did not feature. Likewise the River Tiber was both present and

---

[18] On the *Forma Urbis Romae* see, for example: Bellori 2021; Coarelli 1991; Lanciani 1901; Meneghini and Valenzani 2006; Muzzioli 2007; Najbjerg and Trimble 2004 and 2006; Reynolds 1996; Rodriguez-Almeida 1977, 1981, and 2002; Trimble 2006, 2007, and 2008; and Tucci 2004, 2007, 2013, and 2018.

absent, its position and course marked as a negative, that is an area marked principally by the absence of streets and buildings. There were no geographical or political boundaries marked on the map and the city as defined by the map was simply portrayed as all that area framed by the map's edges, rather than delineated by walls or boundaries. No measurements were inscribed on the map and relatively few individual buildings were accompanied by their inscribed names. No names of building owners were inscribed, unlike on some other known maps, as discussed below.

It has been calculated that 13.5 square kilometres of the city were recorded on the map. Who commissioned the map? Was it was a planning or tax-collecting tool? How accessible was it for viewing by the public? And did it become a focus for the visualisation of the city for its inhabitants? These are questions that all need to be asked but which cannot necessarily all be satisfactorily answered here. The choice of marble used is interesting. Proconnesian marble comes from quarries on the Turkish island of Marmara and is typically coarse-grained, with blue veins.

Some buildings, including a number of temples, the *Septizodium*, and the *Porticus Aemilia*, would seem to have been highlighted on the map by the use of double incised wall lines and probably further enhanced with paint. Thus, it could be that the map's designers intended the viewer's eye to be drawn to certain parts of the map rather than the whole map having equal visual character and value. The fact that some structures were depicted in an earlier form that had been subsequently altered by Severan times simply suggests that the Severan map makers made use of both archived older and contemporary cadastral plans, probably on parchment or papyrus, in creating their city-wide plan. There may though have been some ideological motive for this, but that sadly remains obscure.

Its role and purpose in promoting the dissemination of geographical knowledge both about Rome itself and about certain foreign lands must be considered in the context of the other great Severan monuments of Rome such as the Arch of Septimius Severus in the *Forum Romanum* and the *Septizodium*. As an integrated programme they allowed viewers to consider both Rome and its place in the wider world. Why such a reminder was necessary at this time is open to question, and it may be that the coming to power of the North African Severans after an extended period of civil war meant that the issue of *being* Roman had replaced debates relating to *becoming* Roman that had so dominated earlier times.

Surprisingly little attention has been paid to the question of the longevity of the map as a display item in Rome. How long was it on show? What occurred to make the map redundant as a monument? How was it decommissioned? Was it pulled down, destroyed in other words, and disposed of? Did it decay *in situ*? After all, it is generally agreed that the wall face to which the map panels were attached is now still largely standing, indeed with the lines of clamp holes for attaching the slabs still discernible and measureable, as was noted above. It is thought that in the early fifth century, around AD 530, a passageway door was cut through the wall on which the plan was mounted, damaging a large part. But why? The wall on which the map was mounted now became part of a newly-built church. There is some suggestion that differential damage to certain parts of the map allow for it to be inferred that the bottom parts of the map were broken off first, with slab removal then moving upwards, leaving only a triangular portion of the marble map *in situ* in the central, upper part of the wall until it was finally destroyed by fire in the Medieval period.

It is interesting that today discussions about the viewing of the map centre on issues of illegibility and unreadability. It has been argued that the map was never intended to be viewable in its totality, that is that a viewer on the ground could read off details of individual streets and buildings high up towards the map's top. Its illegibility was just the point it is said by some. Rather, some academics argue that the map was intended to overwhelm the viewer by its sheer size, a visual essay to the grandeur and scale of Severan Rome. Maybe it should be considered not as a map as we understand it today but rather as some kind of exhibit, even a monument. The map might have been intended just to be an image of the city of Rome and not an archival record of its precise geography. But who would and could have seen it? Might colour have been used on the map to improve legibility? While such interpretations seem initially persuasive, doubt can creep in. If the map was not only meant to be seen but to be used as well, the question arises as to what it might have been used for, and how it might have been used from a practical point of view. It is not beyond the realms of possibility that the curators of the map could have made use of it by having on hand sets of movable wooden steps-like library steps-which would have allowed access to, and close viewing of, any particular part of the map for whatever reason.

But the map needs to be considered as the end product of an enormous and laborious undertaking in terms of its teams of surveyors gathering data out in the city's streets, indeed gaining access to and measuring every ground floor room in every building in Rome. The raw field data and information then needed producing as scale drawings, and then for those drawings to be carved and incised on the map's individual marble blocks. The size of the finished map must have been dictated by the room in which it was intended for display. The manpower and resources to undertake this task seems almost out of all proportion to the utility of the city map, to its impracticality as a useable map.

Along with the great Severan City Plan of Rome it can be suggested that the Severan period was a time when grand gestures expressed through monumental architecture and grandiose architectural projects reflected the need to recentre Rome in its contemporary world following the harrowing period of civil strife that brought Septimius Severus to power. Rome was to gain the great Arch of Severus and the *Septizodium*, and numerous pre-existing buildings were to be refurbished or repaired. The city map should be seen in this context, though there were earlier precedents. The Severans were the first non-Italic dynasty to rule at Rome. Caracalla's citizenship policy and the continued growth of Christianity which seems to have been particularly marked at this time might have produced doubts about the exact nature of Roman identity. Again, one can see in Severan literature a trend towards large synthetic works which captured and defined certain intellectual traditions.

A recent innovative study of images of deities on the reverses of Severan coin issues suggests that the African origins of Septimius Severus were expressly presented in terms of affiliation with certain deities, such as *Liber Pater* with Hercules, whose cult worship was tied to local provincial centres as well as to Rome.[19] As this study makes clear, the use of images

---

[19] Harl 1987.

An Unseen Ruler

Figure 79. The *Via Anicia* marble map fragment. Augustan or later. *Musei Capitolini*, Rome. (Photo: Author).

of geographically-specific deities had a long history and was not necessarily a Severan innovation. Indeed, the practice went back to the Republican period when, for instance, the mint at Tusculum produced coins with reverses carrying images relating to the local cult of the Dioscuri and the one at Lanuvium issued some reverses with the local goddess Juno on. There is no quantified evidence available about the distribution area and reach of locally-struck coin issues such as these. If circulating purely within their local region the deities depicted would have marked local pride and celebrated the specific geographical locale, while if circulating outside the immediate area they would have carried information about the home geographical locale to others, if such messages were understood by outsiders. The idea here of two concise layers of information-insider and outsider-is intriguing. As an aside, many coins issued at the numerous mints throughout the Roman empire in the later third to fifth centuries AD carried mint marks, informing those in the know about the origins of literally every coin in their pockets. With there being twenty one separate mints operating between AD 313-498 this allowed records to be kept of mint production, so that the geographical indicators here in the form of mint marks were principally bureaucratic marks, allowing surveillance and control, such a need having been identified as necessary during the sweeping financial reforms instigated by Diocletian. Careful choices of coin images allowed for the visual expression of the *patria* or native city: for instance Spanish born Trajan and Hadrian produced coin issues including depictions of *Hercules Gaditanus*, a syncretised Roman and Spanish god. Lanuvium-born Antoninus Pius had *Juno Sospita* put on coins, and so on.

It is possible that the Severan city map was simply the last and the very grandest in a line of city maps of Rome or of certain areas of the city. We simply do not know. Six marble fragments from large plans have been found in and around Rome: one at the *Necropoli di Porto*, Ostia; one from Amelia in Umbria; one from *Via Anicia* in Trastevere (Figure 79); one from *Colle Oppio/Via della Polveriera*; one from the Forum of Nerva; and one from the Temple of Peace itself. That these fragments did not derive from part of the Severan marble plan can be discerned either by the examination of the marble used, or its thickness, or by the style of representation, which presumably related to earlier display maps. The most significant of these eight fragments are the so-called *Via Anicia* fragment, the fragment from the Forum of Nerva, and the one from the Temple of Peace. On the *Via Anicia* fragment can be seen the Temple of Castor and Pollux in the *Circus Flaminius* and what appear to be commercial buildings or depots on the right bank of the River Tiber. Owners' names are inscribed by each building and a series of units of measurement run along the river bank. The early fragment from the Temple of Peace shows part of an unfinished plan of the Forum of Augustus in greater detail than on the equivalent self-same area on the Severan map.

These fragments of earlier marble maps probably were not from maps of the size of the Severan plan: indeed, they were in all probability from much smaller, perhaps localised maps of specific areas of the city and not of its whole vast, messy expanse. Again, that one of these pre-Severan fragments also came from the site of the Temple of Peace[20] might suggest that an earlier map was displayed there, setting the precedent for the Severan plan being installed here, but as an altogether grander gesture that reflected other trends in Severan culture. The Temple of Peace appears to have acted as a depository/repository for a curious mixture of objects which each in their own way said something about Rome's geographical reach, and in this context it would not seem altogether strange to suggest that the temple was therefore an apposite place for a unique map of the city of Rome itself to be displayed. The fifty feet high temple column shafts were monolithic shafts of Aswan granite, brought here both at phenomenal expense and by long and difficult routes over land, sea, and up river: mnemonic markers of Rome's dominion over Egypt. Among the disparate items on display inside were the Menorah and other spoils looted from the Temple at Jerusalem, and famously portrayed in forensic detail on one of the reliefs on the Arch of Titus, and the original bronze cow of the great Greek sculptor Myron. Also here, according to Pliny, was displayed or stored a statue of the Nile, presumably in the form of a conventional personification but carved from dark or black Ethiopian basanite, the colour choice of the stone being as equally important as its source. Collected or commissioned by Vespasian it was brought here in the early 70s AD. It has recently been suggested that the temple also somehow acted as the home for an imperial collection of plants, that it was for a time a kind of botanical garden displaying specimens collected from various locations around the empire: a particular kind of artifice, growing and creating landscape plants from distant lands in symbolic rebirth of a kind.

---

[20] On the *Templum Pacis* see, for example: Luke 2010: Moorman 2022; Noreña 2003; and Pollard 2009. On the *Porticus Pompeiana* see, for example: Gleason 1994; and Kuttner 1999. On this and other sites of public display see: Macaulay-Lewis 2009. On Rome as a museum see: Rutledge 2012.

Two later Roman maps are known to us, though there must have surely been many more, one (the Autun Map) known simply by a reference to its existence and the other (the Peutinger Map) by its copying in the medieval period.

Writing about the need for restoration work at the school of rhetoric at Autun in Gaul, the Roman writer Eumenius who taught there, writing probably in the later third to early fourth century AD, describes at length a map on display at the school which he evidently viewed as important and in need of conservation. 'Let the boys and girls see on the colonnades all the lands and all the seas……the points where rivers rise and where they have their mouths, and the extent of bays'[21] This building has not been located or excavated, so what form this map took remains uncertain: presumably it took the form of a wall painting.

The Peutinger Map or *Tabula Peutingeriana*, now in the Österreichische Nationalbibliothek in Vienna and circulated widely across Europe through copy engravings and modifications by Marco Velsero in 1598, Franz Christof von Scheyb in 1825, and Konrad Miller in 1887, is believed to be a copy of the oldest map to have come down to us from Roman times[22] (Figures 80-81). Dated to around AD 1200-1300 this medieval map, named after Konrad Peutinger of Augsburg who obtained the map in 1508, depicts the extent of the Roman empire and territory beyond to the east as far as India, and is generally accepted by archaeologists and ancient historians to be a copy of a Roman map dating to some time between the early second century AD and the early fourth century: indeed, one expert very precisely dates the original on which the map is based to AD 300 using place-name evidence and other information from close-reading of the map. The various arguments about dating will not be reiterated again here: suffice it to say, the Peutinger Map can safely be discussed as a document based on an ancient Roman prototype and thus can tell us much about the systematising of geographical and spatial knowledge in the Roman world. It has been suggested by Richard Talbert that the original map on which this was based might well have been a display map in the emperor Diocletian's palace at Split in Croatia at the end of the third century. Of course, this display element would place the Peutinger Map very much in the tradition of Agrippa's world map and the *Forma Urbis Romae* display maps, taking it away from the world of practical itineraries and relocating it in a (semi-) public space where viewing was more significant and expected than referencing and use.

The Peutinger Map consists of eleven sheets of parchment which when assembled together to form a single, long and narrow continuous diagram is around twenty two feet in length. On it are portrayed over 500 cities and 3500 named places, connected by roads that are marked in red. Distances between each node are recorded on the map, linking the concept to the idea of the written Itinerary and being necessary because of the compositional liberties taken with scale on the map. In other words, it is clear that the knowledge hierarchy that dictated the form and design of the map dictated that the cities and other named places were the nodes in the transportation system, with the road links between the nodes. Some landmarks such as mountains are shown, as are other natural obstacles to overland travel such as rivers, with established crossings also clearly marked. The basic design elements which dictated the form of the map were themselves dictated by the requirement for the map to support the planning

---

[21] Eumenius *Panegyrici* 21.1.
[22] On the Peutinger Map see, for example: Albu 2005 and 2008; Bosio 1983; Diederich 2021/2022; Fodorean 2011; Rathmann 2016 and 2022; Salway 2005; and Talbert 2004 and 2010.

A Map of the Body, a Map of the Mind

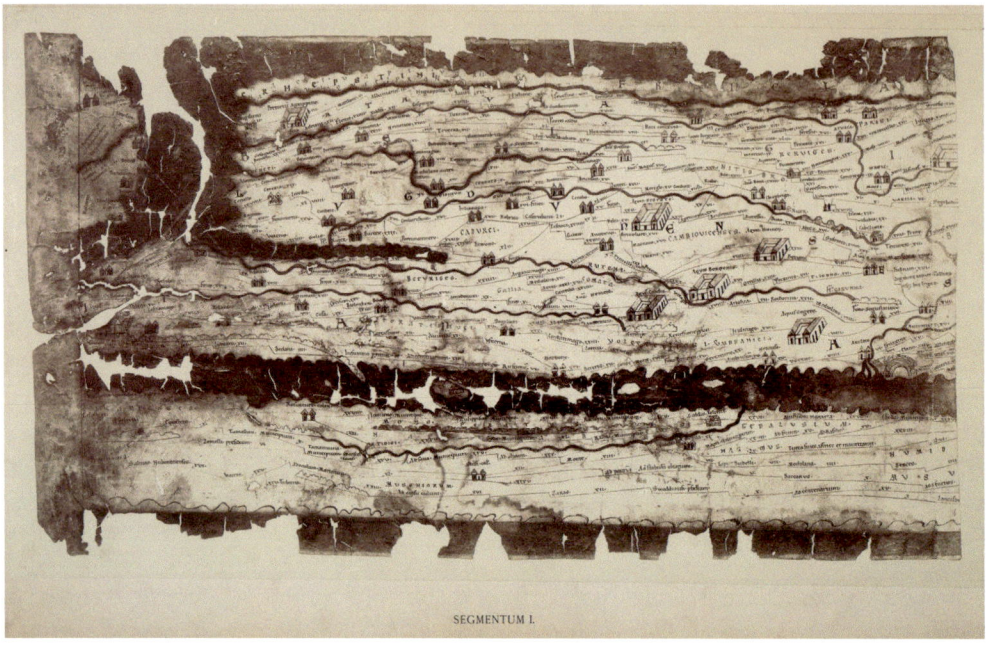

Figure 80. Detail of part of the Peutinger Map. Original early second century to early fourth century AD. (Photo of 1888: downloaded from cambridge.org/us/talbert/mapb.html TP1888seg1).

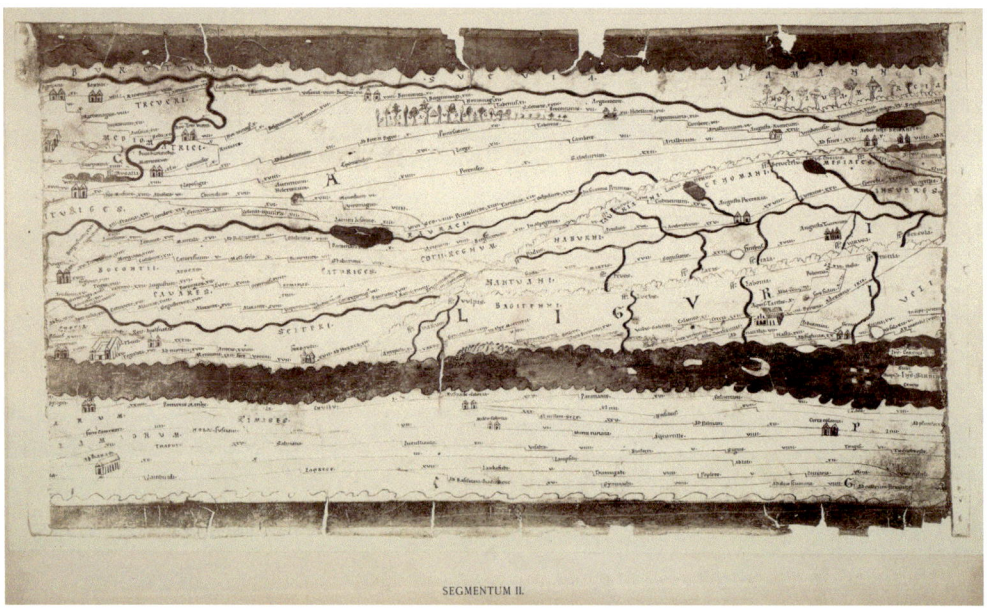

Figure 81. Detail of part of the Peutinger Map. Original early second century to early fourth century AD. (Photo of 1888: downloaded from cambridge.org/us/talbert/mapb.html TP1888seg2).

of overland travel. North is at the top of the map, with the East-West axis tipped to the South: Italy is shown in a lateral orientation, rather than North-South. A number of systematic distortions were used: North-South distances are highly compressed; areas between towns which were not connected by roads were severely compressed; the area covered by water is strictly minimised; and towns are positioned on the map often in relation to the rivers which constrained travel. The Atlantic coast is folded in, while the Mediterranean, the Black Sea, and the Persian Gulf are substantially reduced, as portraying their extent was evidently not part of the schema and purpose of the cartographers in this case. The greatest amount of detailed information and the densest occur on those sheets mapping Italy itself.

As an example of how telescoped and brilliantly-schematised detail is on the map Sheet 1 includes the south-east coast of Britain, France, and the North African littoral and hinterland together. The conceptual world view which envisaged visualising nodal overland travel in this way is both inventive and interesting. The map favours the sequence of stops along a route-toponomy-over the geographical relationships between the stops-topography. This is a transportation diagram and not a map as such, because on it geography is subservient to the visualisation of routes of movement, and indeed could not only be deemed to be secondary to the purpose of the map but actually quite unnecessary and in some ways problematic. A transportation network to operate efficiently needs the network's connections to be favoured over accurate geographical data.

Many natural features are shown. Seas border much of the map, but by reason of this being a land map feature in a much-reduced way. Rivers, lakes, and mountains appear, generally spatially vague. No geo-political frontiers are shown as such, with no marked boundaries between peoples, countries, or provinces. Rivers are simply things to be crossed, mountains are inconvenient blockages or barriers to travel, in contrast to the conception in later Christian theology that mountains were virtual meeting points between heaven and earth, sites of revelation.

It is not altogether surprising that the Peutinger Map is very much structured in a way that reflected the importance of the written itinerary in Roman military and bureaucratic planning, a subject which will be discussed more fully below. The question has been asked as to who in Rome or in the empire might have commissioned the map and why? How was it intended to be used? Was it displayed somewhere publically or was access to it restricted? Was it intended to be duplicated and circulated somehow?

Crossing the Line

This leads on to the need to briefly discuss the use of itineraries in the Roman world, practical aids to travel that to some extent represent the quintessential Roman geography product.[23] With the exception of the Peutinger Map most ordering of strategic travel information in the Roman world was in the form of itineraries, that is written lists of towns and cities in order along a route, with distances in miles between each nodal centre recorded. One could therefore plan and arrange to travel from A to F, via B,C, D, and E, knowing the distance between each successive node and thus being able to calculate the total mileage of the journey by adding up

---
[23] On Roman itineraries see, for example: Bowman 1998; Brodersen 2001; Douglass 1996; Elsner 2000; Fodorean 2011 and 2012; Koeppel 1980; Salway 2007 and 2011; and Talbert 2023a.

Figure 82. The Vicarello Itinerary Cups, Vicarello. First century AD. *Palazzo Massimo Museo Nazionale Romano*, Rome. (Photo: Copyright Ryan Baumann).

the individual figures. If tied in with the *Cursus Publicus* it would also allow breaks in travel at *mansiones* or staging posts at each node, and the possibility of changes of mount or carriage pair and so on. As the main Roman roads only went from A to B and B to C directly there was no need for the itineraries to record or dwell upon otiose extraneous geographical information or topographical data. To a regular or seasoned traveller perusal of an itinerary could aid recall of journeys made, as well as aid planning for future journeys taking new routes. Yet in many ways these static lists were also interactive, almost visual aids, allowing the imagination of their reader or viewer to visualise journeys yet to be undertaken. In hauntological terms viewing them was to be looking into the future.

In Chapter Nine there can be found discussion of the four inscribed silver vessels known as the Vicarello Cups[24] (Figure 82), souvenir vessels whose inscribed texts record itinerary routes from Gades (Cádiz) to Rome. Roman itineraries would more usually have been documents on parchment or papyrus whose regular and widespread use in the Roman world is well-attested by numerous Roman writers. The most significant itinerary document to have come down to us, preserved through copying in the fourth century AD, is the so-called Antonine Itinerary- the *Itinerarium Antonini Augusti*- an official imperial document which lists fifteen separate itineraries in different parts of the empire, including Britain. Within each itinerary there are sometimes numerous road routes given, with there being 34 routes in the Itinerary covering Spain for example. Several thousand individual place names appear on the itineraries listed.

---

[24] On the Vicarello Cups see: Popkin 2022b.

It is likely that there existed similar documents covering itineraries across all parts of the empire.

There is no question that though doubtless corrupted in many places through its copying in medieval times the text of the document known as the *Notitia Dignitatum* is recognised as being transcribed in the first instance of its copying from a genuine late Roman original, now lost. The *Notitia* is a long, highly-detailed document listing the administrative organisation and offices of both the western and eastern Roman empire. The amount of geographical information in the body of the text is phenomenal and one could draw all kinds of maps based upon the data collected and presented here. However, the illustrative 'maps' derived from the various copies based upon the so-called *Codex Spirensis*, along with the almost heraldic designs presented of Roman military shields and equipment, are obviously fourteenth-sixteenth century creations. The 'maps' of the *Notitia* will therefore not be considered further here.

In his recent book *Mosaics of Knowledge: Representing Information in the Roman World*[25] Andrew Riggsby set out to discuss how what he terms 'information technology' was deployed in five distinct and disparate fields, that is listing, compiling tables, weighing and measuring, using perspective in painting, and mapping. The visualising of data as a constant in the practices of Roman art is not explicitly discussed, nor is the evident use on occasions of mnemonic devices in the interpretation, ordering, and processing of data generated by each of his five technologies, including the kind of visual mnemonics discussed in this present study. With the exception of the use of artistic perspective these technologies each had their roots in the necessity of bureaucratic practices, but each was hierarchical in that its employment could filter down the social classes and be adapted to different circumstances, situations, and contexts. If the American space programme of the 1960s could be said to have given us all non-stick teflon pans, then Roman imperial authority's use and manipulation of geography products helped bolster the identity of Rome's citizens and define their place in the wider world.

We learn from the geographer Strabo and a number of other sources that terrestrial globes existed at the time, but none has yet been found by archaeologists. Such globes represented another category of geography product that the Roman elite could use in order to enhance their status,[26] as we have seen elsewhere in the book also occurred with relation to the ownership of portable sundials. Such globes can be seen as part of an erotics of knowledge at the time. However, one Greek and three Roman celestial globes are known. The first is held by Atlas in a sculptural work now in the *Museo Archeologico Nazionale* in Naples and known today as the *Farnese Atlas* (Figure 83). In the collections of the *Römisch-Germanisches Zentralmuseum* in Mainz is the bronze Mainz Celestial Globe. Finally, there is the so-called Paris-Kugel Celestial Globe. Dating to the second century AD the Farnese Atlas is probably a copy of a Hellenistic original. On the globe is to all intents and purposes a map of the night sky in low relief, with 41 or 42 of the classical Greek constellations named by Ptolemy depicted.

Cartography can be seen to have been a literary genre in the Roman period, utilised for political ends, in a remarkably similar way to examples from much later contexts, in Renaissance Italy for example, in Elizabethan England, and in the Fascist Italy of the twentieth century.

---

[25] Riggsby 2019.
[26] On the globe as a Roman image see, for example: Arnaud 1984; and Dekker 2009.

Figure 83. The Farnese Atlas, Rome. Second century AD. *Museo Archeologico Nazionale*, Naples. (Photo: Author).

I would now briefly like to consider the phenomenon of 'the map room' in sixteenth century Italy, the examples being discussed here dating to the mid-1560s to the mid-1580s. The *Galleria delle carte geografiche* in the Vatican, the *Sala delle carte geografiche* in Palazzo Vecchio, Florence, and the *Sala del mappamondo, Palazzo Farnese*, Caprarola represent the most developed examples of such a room, intended to enhance the status of their owners through the conflation of geographical knowledge and information with political power and ideology. Here a map was a very particular type of decoration, transforming the walls of the rooms, almost turning them into permeable surfaces that could be entered and travelled through. Control of access to the rooms, the use of the rooms for greeting guests and clients, and possibly surveillance of others inspecting the rooms as guests would have set the maps in a broader context of political manipulation.

In one of the rooms in the *Gallerie degli Uffizi* in Florence are three wall paintings of maps commissioned by Ferdinando I de Medici in the late 1500s, following the creation of the unified Grand Duchy of Tuscany, after the Republic of Florence's defeat of its bitter rival Siena. This historic context for the creation of the maps is crucial to an understanding of their meaning and significance, and reveals the link at this time between cartographic realisation and political power.

The three maps were designed by the court cartographer Stefano Bonsignori and painted by Ludovico Buti: the detailed maps are of the Florentine lands, Siena, and the Isle of Elba which lies off the Tuscan coast. More than 1200 cities and towns are named in gold lettering on the maps. A huge picture window looks out onto the contemporary Florentine cityscape, including *Palazzo Vecchio*, the basilica *Santa Croce*, and *San Miniato al Monte*. The view out of the window helps to make an explicit conceptual link between the city of Florence and the territories displayed as images on the walls of the room, mapped, framed, and controlled.

Mapping and the deployment of a range of geography products linked to the display of royal power found one of its most interesting manifestations in Elizabethan England.[27] The so-called Ditchley portrait of Queen Elizabeth I of England by Marcus Gheeraerts the Younger in 1592 shows the aloof monarch standing on a scrumbled and tattered map of the English Home Counties, probably Christopher Saxton's map of 1579. As depicted here she is quite literally dominating the country. A Dutch portrait of 1598 presented her as Europe, while her image appeared on the frontispieces of several contemporary atlases. While this woman-as-land trope was very historically specific, nevertheless it suggests that geography and power can often be viewed as being symbiotically linked within certain political systems. Discourses of mapping were not just always discourses on power: they could also be discourses on gender, as in the case of Elizabeth. I have argued elsewhere in this book that geography could also influence genealogy and pedigree, as it did in the Roman period and the Elizabethan era.

Wistful and lustful comparisons between women's bodies and foreign lands can also be found in some of the poetry of the Elizabethan era and in the succeeding decades. John Donne's *To His Mistress Going to Bed* of 1593-1596 includes the exclamatory lines during seduction': 'Licence my roving hands, and let them go/Before, behind, between, above, below./O my America! My new-found land,/My kingdom, safeliest when with one man mann'd/My Mine of precious stones, my Empirie,/How blessed am I in this discovering thee!'. Meanwhile Andrew Marvell in *To His Coy Mistress* of the early 1650s pursued a similar geographical theme: 'Thou by the Indian Ganges' Side,/Shouldst rubies find; I by the tide of/Humber would complain.'

The emergence of cartography as a tool of government in the early modern period in western Europe is a complex tale but one which has certain resonances in terms of informing our understanding of the relationship between geographical knowledge and power in Roman times, and of the ways in which the visualisation of geographical knowledge can be presented and sometimes manipulated to ideological ends.

When in 1934 Mussolini had four large (4.6 metres square) coloured marble maps installed at the Basilica of Maxentius in Rome, affixed to the outer wall faces facing onto his new *Via dell'Impero*, the series was intended to present to the Italian people the chronological development of the ancient Roman empire in the first four maps and the extent of Fascist Italy's new empire in a fifth added in 1936.[28] Viewers were expected to marvel at both the

---

[27] On Elizabethan maps and geography products see particularly: Pilhuj 2019.
[28] On maps, symbolic landscapes, and Rome under Fascist Italy see, for example: Baxa 2010; Hecker and Bedarida 2023; Minor 1999; and Painter 2005.

territorial extent of the first Roman empire and at the creation and planned expansion of the second Roman empire under Mussolini. The fifth map was taken down and put into storage in 1946: the other four were left in place and remained *in situ* until relatively recently when they too were removed into storage during the building works associated with the *Metropolitana* station at *Colosseo* today.

The first four maps were unveiled in a formal ceremony on April 21st, a day whose significance in the Fascist calendar was doubly meaningful, being both the anniversary day of the founding of the city of Rome and the *Festa del Lavoro* or Fascist Labour Day. Subsequent annual marches past the site of the maps on the same day made them a *locus* for both commemoration and recreation for the Italian Fascist party.

It almost goes without saying that the Fascist use of large display maps such as these owed a great deal to the ancient Romans' own use of public cartography such as the Severan marble city map, the *Forma Urbis Romae*, and to the earlier World Map of Agrippa in the *Porticus Vipsania* as didactic tools to direct the definition of Roman identity. The geography theme in Fascist ideology was further enhanced by the imagery of the floor mosaics designed by architect Roberto Narducci at *Stazione Ostiense* laid for Hitler's visit to Rome in 1938, including an image of Augustus in a Nilotic landscape with a crocodile.

In conclusion, and returning to ancient Rome, it would seem that Rome found a number of ways to use maps and mapping, itineraries and lists to enhance state power, allied to the use and display of other types of geographical products. On display in the Roman Forum, possibly on the walls of the *Regia*, inscribed on a series of marble tablets, was the *Fasti Triumphales*, also known as the *Acta Triumphorum*. Dating to c. 12 BC this monumental inscribed list recorded around 200 triumphs celebrated in Rome up to the time of Augustus. While some of these were mythical events, the vast majority were actual historic triumphs recorded here by year, the name of the triumphal commander, the name of the peoples or countries conquered or defeated, and the day on which the triumph was held. Peoples/countries defeated and listed were in Italy (45 names), Greece (eight names), elsewhere in Europe (24 names), Asia Minor (five names), Asia (five names) and Africa (three names). To those stopping to read these lists, or being read the lists, a vivid sense of the chronological development of Roman power across Italy and the gradual creation of the Roman empire across three continents would have been experienced, while the geographical span of that empire would have been received through the iteration of the names of conquered tribes or territories, even if the specific locations alluded to by these names were not necessarily all familiar or widely known. This was like a form of historical atlas, and it must be borne in mind that it was just one of many such geographical indicators which would have co-existed together on display in the centre of Rome, though unfortunately we do not know how long the *Fasti* were on display before their eventual removal, so firm evidence for which markers co-existed with each other cannot be presented. If we suppose that the *Fasti Triumphales* were on display to the Roman public from 12 BC onwards, then it is likely that a viewer/reader of these having taken in the names of many conquered peoples and lands could walk to the nearby Augustan *Porticus ad Nationes* or Portico of Nations and look at personifications of conquered nations there, perhaps also doing the same at the Theatre of Pompey. Some years later, when Vespasian built the Temple of Peace c. AD 71 this too could have been visited in the same perambulation, to take in the wonders from around the world brought there for display as a symbolic marker of Roman geographical power and reach.

A trigonometry of journeying defined like deracinated snapshots Rome potentially imprisoned in its own legends and perhaps unable to come to terms with an age in which it could so easily have become a stranger to itself in a climax of lacerating cruelty. Yet the production of images of other lands and peoples alleviated matters: broke the tension. We should view this accreted archive of images not as a window onto the past but as fragments from it, wild atavistic refractions, representing a collective cathexis, leaving a kind of morphic resonance. In Rome empire was presented as not only the conquest of land and peoples, but also of the discontinuity between individuals, as the city reconstituted itself as an oneiric location. Rome both hymned the *domus* and the complex places of transit that helped ensure its wealth-the Tiberside docks, the markets, the ports and so on-and displayed the Map of Italy in the Temple of *Tellus* on the Esquiline Hill, setting some kind of precedent for future displays of maps of different kinds.

Like a digital nomad, the true identity of Rome under the empire came to be in its absence: the simple early lifestyle of its inhabitants was now largely absent but perhaps continued somewhere else. People were able to inhabit the same space, to be physically proximate, and yet to live in different worlds. The spatial uses characteristic of the late Republic and early Empire were determined by its potential not just for sociability but also for secrecy and security. In terms of reimagining something as big as the Roman empire at its greatest expanse artworks needed to both catch the attention of the viewer and at the same time to restrict the field of vision.

Maps are records of learning and follow experience, often embodying a narrative voyage, an itinerary of emotions which was either positive or negative, or even sometimes neutral. It made a world of affects visible-the exterior world conveyed an interior landscape. Emotion was materialised as a moving topography. To traverse such a world was to visit the ebb and flow of a personal, yet social, psychogeography. Conceptual mapping such as employed by the Roman state was used to unify the diverse perspectives presented, so that the viewers' input and engagement were directed to certain conclusions, or at least to questions relating to the meaning of the output. Roman maps were of a place, about ownership or control of that place, and not about how to get there.

In the next chapter discussion will turn to representations of barbarian peoples and lands.

Chapter Eight

Maps of the Body

In this chapter I intend to survey the widespread Roman practice of making and displaying images of non-Roman peoples: that is non-Roman Italic peoples, barbarian enemies of Rome, and barbarian peoples outside the empire's boundaries, and in a very few instances provincial peoples within the empire. In order to try and understand how Roman or Romanised viewers might have reacted to such images it is first necessary to consider whether or not geographical or ethnic origins actually might have been an issue that this audience cared about. In other words would the Roman viewers themselves have used or identified with locational tags and how would they have described or perceived their own geographical origins. Once more, this is to some extent a return to the question of the definition, creation, and maintenance of personal identity at this time. A broad-brush analysis of Roman funerary monuments and dedicatory inscriptions suggests that the Roman elite more often than not identified with family rather than locale, though sometimes both, and that family lineage was the overall defining structure for the public presentation of their identity. From the first century BC onwards Rome's growing population of freedmen and freedwomen also staked out a public position for expressing their identities, and this sometimes involved their geotagging of their locale in inscriptions.

In my first book *Enemies of Rome. Barbarians Through Roman Eyes*, published in 2000,[1] I examined in detail the many different contexts in which images of barbarians were used in Roman art (Figures 84-85). Most commonly these images appeared in contexts linked to the celebration of military victories, on imperial monuments in Rome and in provincial locations, and on military art. However, many other motivations for using such images were explored, from pure provocation to quite nuanced needs to define Roman-ness through contrast with 'the other'. There were also contexts in which a gendered motive might have been paramount in the images' usage, in terms of both defining what made a Roman man through public expressions of hyper-masculine display and how sexualised depictions of female barbarians related to Roman male power. The study concluded that images of barbarians were often as much about Roman self-definition as they were about actual barbarian peoples themselves. Images of barbarian peoples in Roman imperial art also sometimes acted as geographical markers for the definition of *Romanitas* and it is that particular category of barbarian image that will be further considered here.

This is not simply an opportunity to replay my greatest hits, pleasing though that might be. Rather, it presents a moment for reflection, for perhaps an editorial intervention in my own past. The main thesis of *Enemies of Rome* was that such art was both reflective and relexive, that it said much, if not more, about the Romans themselves than it did about the barbarian peoples portrayed. While I have not moved too far away from this stance since I adopted it almost a quarter of a century ago, I look back at it as perhaps being too atomised a position. Here I am repositioning my former argument in a broader study of Roman geographical

---

[1] Ferris 2000 and see the bibliography there for sources up to c. 2000 on images of barbarians in Roman art.

MAPS OF THE BODY

Figure 84-85. The Dying Gaul. Roman copy of a Hellenistic original. *Musei Capitolini*, Rome. (Photo: Author).

engagement. I do not feel this negates my previous analysis: however, it certainly would seem to locate it within broader trends than I had perhaps realised at the time.

Images of bodies, even in the form of personifications, were in some ways archives of empire in Roman times. The scrutiny which non-Roman bodies were under in the urban choreography of the city and the lack of agency many foreigners had over the public narrative and perception of them added an intimacy to the art's anonymity. They represented the kind of 'spatial stories' championed by the French theorist Michel de Certeau in *The Practice of Everyday Life*. Maps and tours aim to produce legible figures and results while spatial stories extend beyond numerical and quantifiable constraints, lending a humanity to places.

There is no doubt that foreign female bodies increasingly became objects of desire and part of a broader discourse of possession and desire in Rome and elsewhere in its empire. The dense intertextual spaces within which these images were situated might have varied over time but nonetheless a linear trajectory of male potency can be discerned throughout. When it came down to it authenticism in such cases may have involved a different set of intertextual relations than did a process like bricolage, but the former was no less contrived or arbitrary than the latter.

Such images were of peoples away from home and yet at home, at the very centre of the world and yet unseen of the world. In some cases there was a danger of these images of the other falling into the emptiness of an abstract and indefinable beauty, possessed but somehow for ever unobtainable, present yet always absent. Shallow knowledge and capricious attachments must have contrasted with claims of expertise and posturings of reverence. These figures evoked vulgarity, aggression, danger, transgression, the exotic, the alien, yet still were paradigmatic images as objects. The practices of producing and consuming such works reflected, bolstered, and in turn helped to configure the discourse around Roman imperialism. A grouping of concepts, distinctions, oppositions, rhetorical ploys, and allowable inferences such as was centred on the use of such images as a whole fixed the limits within which they operated. The viewer must often have been left with an inchoate and abstract residue of geographic information in his or her mind, shorn of any historical specificity.

Roman culture raised particular problems in the case of the representation of non-Roman women, as has been seen throughout this book, this being part of a broader discourse around representation in general in a masculinist culture. It is still very much open to debate and formulation how Roman power in some ways depended on aestheticised politics in the frequency, intensity, and omnipresence of such practices. Here the foreign body and the body politic revealed themselves as part of a joint ideological history, the contrast between the male/masculinised cityscape of Rome and these personifications of young women there being surely jarring and disconcerting to many.

But it is impossible to fully and confidently reconstruct the field of associations these images might have invoked. The clarity and completeness of these personifications found expression in metaphors which were intended to encourage the transfer of feelings about the body to the state. It provided ideological embeddedness for the artworks thus employed which were often remarkable both for their mass-psychological effectiveness and for their revelation of the ideological concerns and anxieties of the Roman state itself.

This urban montage of images of places (and of places by association) produced a new sense of reality, of place, of home, collapsing not only space but also time: it was not a case of nothing connecting with nothing, in an Eliotian sense, but rather enough connections being suggested as to bring a distinct consciousness to the city.

The Widening Gyre

While much of the material discussed here resists easy explanation, an impossible archaeology of all this nostalgia for distant lands and peoples is discernible in the painful desire to return to an authentic and singular notion of what it meant to be Roman. Contemporary viewers of these images would have been simultaneously gazing upon their own city through these tight close-ups of images of foreign bodies, fragmenting and unveiling the gaps within the urban landscape, the spaces around the places, which would have been loaded with political and racial poignancy. The harm of Roman power and cruelty would have rippled so violently across the hierarchies of gender, class, and ethnicity that even Rome's cultural bubble would hardly have contained its latent and potential destructiveness. It cannot all have been negativity though. The mundanity of serial repetition would have underscored anticipation, as some works tried to elevate empathy and understanding together as a narrative principle: intention was only one facet of their meaning. The effect must have been sometimes elusive, somehow at once swift and languorous, bearing a premonitory feeling of some impending return or encounter. If away from Rome, elsewhere on the fringes of the empire, landscapes were unknowable and treacherous, sources of unease and even vertigo to the Romans who had claimed sovereignty over them: finding a peaceful sense of home remained an option for each viewer.

As each image receded in time from the moment of its creation, the motive behind its creation, so it became an object in an archive. Ultimately the archive as a whole was part of a broader Roman discourse on borders and limits, of a certain *outside*, of certain borders *between* insides and outsides. It was a psychic archive, not a spontaneous memory. The memory traces of external events were there in the fiction of its suspense. All kinds of symptoms, signs, figures, metaphors, and metonyms attested to and contributed towards an archival documentation project. In the absence of linking memory it might have been thought that there was not one to identify and analyse. There was a process, rather than an end point.

Thinking about Rome and geography together is now much more complex than it may once have been. In an age responsive to postcolonial studies, transnational histories, and subaltern studies identity is now accepted as seldom being fixed or static, a situation compounded by increased mobility and migration. The city of Rome itself provided a way for its inhabitants to explore the alienation that inevitably attended living in a cosmopolis. This sense of alienation and longing, not necessarily experienced together or simultaneously, created an urban geography of certainty and not one of doubt.

This society of consumerism and erotic illusion used images to create loops of meaning that interconnected and interpenetrated. They represented what was both overt and hidden in the urban maze as power was transformed through the workings of the city: the biological processes of life and death as encapsulated in the human body had moved from the realm of Nature to the realm of the city. To paraphrase Hamlet, these were discovered countries to which urban travellers returned again and again without leaving their city home.

The city was a changing rather than a fixed realm: it made those at the centre aware of the peripheries, of the remote and of the frontiers, and constantly called them back to a nothingness. Every system of organisation necessarily contains a principle of disorganisation. Rome became the first of a new type of colonial city where there was no *there* any more. Rome both contained and organised vital energy. Rome as a city was not a personal, isolated experience, with each inhabitant caught in his or her own subjectivity. Rome made itself outside of time, a relic of the past but toward which all future life would paradoxically flow.

While certain Roman historians detailed the incremental manner in which Rome conquered, colonised, and incorporated the lands of first their neighbours of central and southern Italy and then the more distant Italian peoples north of the River Po up to the Alps,[2] this would not appear to have been a narrative that found much expression in Roman visual art. Indeed, in pre-imperial Rome it is perhaps only a now-fragmentary wall painting from a tomb on the Esquiline Hill, now in the *Museo Capitoline Centrale Montemartini*, that appears to do so. The painting appears to represent not war and battle exclusively, though some fighting is depicted, but rather treaty and negotiation, those vital steps on the way towards Roman hegemony over most of the Italian peninsula south of the Po. Dating to the third or second century BC the two principal protagonists depicted are labelled as Marcus Fannius and Quintus Fabius, perhaps the historically-attested Roman general Quintus Fabius Maximus Rullianus, who may have been opposed commanders in Rome's Samnite Wars of the later fourth century BC.

As was noted in Chapter One Augustus' completion of the extended project of conquering all Italy (as we understand it today as an entity) found expression both in the text of his *Res Gestae* and physical form in the great Alpine Trophy monument of 6 BC, the *Tropaeum Alpium*, at La Turbie, near Monaco, with its figures of dejected, chained male and female captives and its unusually long dedicatory inscription listing the names of all forty five conquered Italian Alpine tribes. This was a hugely provocative monument in terms of presenting the unity of Roman Italy in the style of a victory monument over barbarian peoples. The presence of such a large number of names of tribes in the inscription suggests that each named tribe had such distinct characteristics to need qualifying by the citing of their name, even though they all lived in quite a small area. There must have been a clear link between each tribe and a very specific tranche of home territory, and the careful curation of culture and memory in each specific place.

It was also in the Augustan period that in Rome the rape of the Sabine women, to provide sons for Rome's future army, and the subsequent punishment of the Roman woman Tarpeia for treason in letting the vengeful Sabine men into the city, one of the founding historical myths of Rome, was portrayed in artworks commissioned for the renovation of the *Basilica Aemilia* in the *Forum Romanum* in the Augustan period. As I noted when describing these artworks in *Enemies of Rome* 'conquest and sexual penetration here were construed as being one and the same, while the fate of Tarpeia was also seen as somehow justifiable in that her betrayal of the men of Rome was little different to sexual betrayal, and hence deserving of the grim death she suffered in consequence.'

---

[2] On early Roman and Italian relationships see Chapter One Note 10.

Tarpeia seems to have been seen as both simultaneously a (deliberate) betrayer of Rome and (unintentionally) a saviour of the city. She was situated in the very fabric and landscape of the city, both in the various tellings of her myth and in the reception of those accounts. She was both comprehensible and containable within the city of Rome. If as has been noted elsewhere in this book, Vitruvian man had the form of the buildings of Rome imprinted on his body, so Tarpeia herself was a monument of a sort, breachable like the walls of Rome. Her story contributed towards what has been called 'a topography of punishment' in Rome, encompassing the Tarpeian Rock, the Gemmonian Steps or Stair, and the *carcer Tullianum* or prison.

It is worth taking a moment here to consider Anne-Marie Leander Touati's interesting observation that a number of Roman imperial monuments bearing images of barbarian prisoners in chains or captives being led along by Roman soldiers were sited close to the Mamertine prison or *Tullianum*.[3] She has suggested that Parthian caryatids in the Augustan *Basilica Aemilia*, Dacian prisoners in and around Trajan's Forum, and, much later, captives on the Arch of Septimius Severus were together a geo-locational reference to the existence and role of the prison where there most famously died the Gaulish king Vercingetorix, Adiatorix of *Galatia*, and Jugurtha of *Numidia*.

If the act of rape is writing conquest on the body, the fate of Tarpeia was very different, with her story very much stressing in some versions her intact virginity. It has been noted that the story of Tarpeia was always presented as a moralising tale: it stressed the dangers inherent in looking outwards, away from home, and on a broader scale that a state such as Rome expanding its borders and horizons needed to guard against pollution and moral contagion, that is to know and fully understand its place in the world.

Celine Lillie has written about 'Roman imperial discourses of rape'[4] and sees a continuum between: the promotion of the myth of Mars's rape of Rhea Silvia that led to the birth of Rome's mythical founders Romulus and Remus; the Rape of the Sabine Women; the rape of Lucretia by Sextus Tarquinius; the attempted rape and killing of Verginia; Apollo's rape of Daphne as recounted in Ovid's *Metamorphoses*; and the Rape of Eve narratives derived from early Christian retellings of *Genesis* in the three texts known to academics today as *The Secret Revelation of John*, *The Reality of the Rulers*, and *On the Origin of the World*. Lillie's highly nuanced argument is that in the Roman world there was legal protection against rape of a citizen but that non-citizens, slaves, and some others did not fall under this legal protection, making rape in her mind in the case of the Sabine Women 'both a tool and a "spoil" of Roman conquest'. Subjugation, gender, and rape intersected in many Roman narratives and in many Roman imperial personifications of 'conquered', captive, or enslaved women, but these women were actually victims of Roman rule and not simply symbols of it.

The versions of these stories that have come down to us generally date to the Augustan era, though some of these narratives recur in later writers such as Plutarch. The emperor Augustus constructed a labyrinthine ancestry that through the Julian line of Caesar linked him to the Trojan Aeneas (and to the goddess Venus, mother of Aeneas by Anchises) and to Rhea Silvia, a

---

[3] Touati 2015.
[4] Lillie 2017: 5-6.

Trojan ancestor of Aeneas. Convolutedly, by also claiming ancestry with Romulus, son of Rhea Silvia and Mars he could link himself with Rome's god of war.

Ovid's *Metamorphoses* was a consciously Augustan proposition of a poetic suite, relating tales of human metamorphosis and change at this time of great political and social change and upheaval at Rome. The Republic had changed to an imperial system, Rome the city was changing in its appearance, and the very idea of what Rome was and who the Romans themselves were was in a state of flux. Ovid's work can be seen as a skewed mirror-image of this world, a 'chaosmos'[5] to help the Romans locate themselves and redefine their identity in the wider world.

In the context of the invention of a fictional Roman flaneur here it is interesting to note that in his *Ars Amor* Ovid suggested that walking around Rome was a good way to meet and pick up women.[6] As Celine Lillie noted, one of his recommended routes was 'loaded with political overtones', as it took in Pompey's Theatre, where were displayed female personifications of fourteen conquered territories, and the *Porticus Octaviae* where again could be seen female caryatids representing individual personifications of conquered countries (*simulcra gentium*), and the Portico of the Danaids in the Temple of Apollo on the Palatine Hill.[7] This route represented a veritable geography of conquest, of Roman male dominance over women; but did not Ovid himself not suggest that 'love itself is a kind of warfare'-*militiae species amor est*.[8] I cannot help but think of the work of William Burroughs where sometimes narrative was almost superseded by something altogether different, creating an autonomous zone of text and place built out of images and the atmosphere created around them. The Sabine women 'become conduits for the continuance and expansion of Rome' and indeed from their bodies could be said to emerge the Roman state and subsequent empire.

Threading a Dream

Anyone with only a passing interest in the Roman world will probably be aware of the fact that erotic and sexual images abounded in Roman art and that much of this art was at one time deemed too shocking and explicit to be openly displayed to the public. Here I will consider the nature and context of eroticised and sexualised images of barbarian or non-Roman women in Roman art in terms of their viewing through both the male and female gaze and the meaning and reception of such images in different contexts.[9] Laura Mulvey's theory of the 'male gaze' is of some interest here, as it may well have been that some images of non-Roman women in Roman art were intended to be filtered or processed through the Roman male gaze or indeed will have been created purely and simply for male viewing, delectation, or gratification. Sometimes erotic or sexual images seem to have been deliberately politicised and the relationship between male imperial power and female (sexualised) subjugation will need to be considered first.

---

[5] Lively 2002; 17.
[6] Ovid *Ars Amatoria* 1.67-68.
[7] Lillie 2017; 77-79.
[8] Ovid *Ars Amatoria* 2.233.
[9] Ferris 2000: 55-60.

Though erotic or sexualised images abounded in the Roman world Roman sexuality was very different in character from twenty first century sexuality. That vision was an instrument and symbol of Roman male sexual power is without doubt, but viewers of many Roman artworks could be both male and female. To some extent the demonstration of restricted or limited access to certain erotic or sexualised images at this time might suggest the operation of power on behalf of the work's owner, especially if access was allowed or denied on gender grounds alone. The penetrability of space was a crucial factor in the display of sometimes explicit images. That most of the images of sexualised or eroticised non-Roman women were placed on public monuments in Rome and elsewhere was therefore highly significant.

The politics of sexual desire and of male dominance in the Roman world are two separate issues altogether. Erotic images which were politicised and public were very different indeed from those which were public but linked to the commercialisation of sex, and again from those which were private rather than public erotic images. Much has been made of the theory that the Augustan city of Rome was to some extent an architectural model of phallocentricity, even down to the groundplan of its forum. Whether this was indeed the case is open to debate. How would a woman feel living in a phallocentrically-designed city and moving through its public spaces? How might the form of such a city have affected the display and reception of works of art there that bore images of women?

There is an undoubted link between eroticism and power in certain works of Roman imperial art, most significantly on a number of the marble reliefs found in the *Sebasteion* at Aphrodisias in modern day Turkey, a temple to the imperial cult at the time of the Julio-Claudian emperors. Panels depicting Claudius conquering *Britannia* and Nero conquering *Armenia* could be read as object-lessons in the visualisation of male imperial potency, though other, more benign interpretations of these scenes are also equally possible.

The religious context of the complex may have been of primary significant here. Roman state power was further codified and strengthened through religious rites and control of 'foreign' goddesses such as Cybele, the *Magna Mater*. The highly-overt female sexuality of the goddess as originally conceived was turned around on itself by Roman usurpation and control over her story and the religious rituals of her cult. This sexuality almost became a metaphor for Roman imperial power and its expansionist nature and became more explicit in terms of ritual symbolism as the religious vision was further distorted over time in new territories and in new contexts. The construction and dedication in 191 BC of a temple to Cybele on the Palatine Hill in Rome was crowned by the housing there of an image of the goddess or representative of her carved of a meteoric stone and looted from Pessinus, in Phrygia in Asia Minor and brought to Rome in 207 BC. Like a captive or slave this iconic stone was conquered, then possessed, by the Romans and displayed for their delectation[10] (Figure 86). The same applied to other female deities, including *Artemis Ephesia* (Figure 87).

The first of the two Aphrodisias *Sebasteion* reliefs,[11] Claudius and *Britannia,* as named in the inscription on a plinth below, is the most complete (Figure 88). The figure of Claudius dominates the scene and is portrayed in what is known as heroic nudity. He towers over the

---

[10] On the capturing and appropriation of oriental gods see, for example: Alvar 2007; Burton 1996; Mazurek 2022; and Orlin 2010.
[11] On the Aphrodisias *Sebasteion* reliefs see principally: Smith 1988 and 2013; and Taussig 2012.

Figure 86. Cybele/*Magna Mater*, Rome. AD 250-275. *Museo Archeologico Nazionale*, Naples. (Photo: Author).

female figure, a personification of the province of *Britannia*, and with his left hand grasps or pulls her hair at the back of her head, thus pulling up her head ready to receive a blow from a weapon held in his right hand, though both the weapon and the emperor's right forearm are now unfortunately missing, so that his original stance and attitude cannot be fully reconstructed. One of the emperor's legs is hidden behind *Britannia's* body, but his stance suggests that he could be bracing her body against this leg which he is driving into the small of her back. With his other leg he is kneeling on her thigh in an attempt to pinion her body down. The expression on the emperor's face is perhaps both determined and cruel, one could almost say demented and lustful. *Britannia* lies partially sprawled on the ground, apparently stunned and overwhelmed by the ferocity of the onslaught, but still appears to be trying to fend off Claudius with her raised right arm. Her right breast is exposed. On her face is an expression of evident pain and an almost stunned resignation to her fate.

It seems inconceivable that the role-playing depicted here was not obvious to the contemporary viewer: that the Roman emperor in the guise of a Greek hero or god was a man for all that, and that the personification of a conquered province in the guise of a mythological Amazon was a woman underneath all this pictorial, metaphorical subterfuge. I would suggest that no matter how many layers of allegorical and symbolic meaning are heaped upon this work in order to understand and interpret its message, once they are stripped away there remains what is a profoundly disturbing event caught *in stasis*. Without doubt, the Claudius and *Britannia* panel

Figure 87. *Artemis Ephesia*, probably from Rome. Second century AD. *Museo Archeologico Nazionale*, Naples. (Photo: Author).

from Aphrodisias depicted a virtual rape caught *in stasis* which says a great deal about Roman male imperial power.

If simply territorial conquest was shown here by the action of a male figure battering a female figure, then this must surely have been a statement of some kind, at once both startlingly direct and yet deliberately allusive and oblique, about attitudes to Roman imperial actions and policies and to conquered lands and peoples, and finally, of course, about contemporary male attitudes to women. The scene might also be considered to have been the depiction of the prelude to a sexual assault, in addition to a physical assault, and to the rape of the woman. Certainly the hair pulling, the pinioning with the knee, and the striking of a blow to the head could suggest such a culmination to this act of 'conquest'.

The second Aphrodisias relief, of Nero and *Armenia* is more damaged and therefore less complete (Figure 89). Again, the emperor appears in heroic nudity, looming over the female personification of the conquered province. She is virtually naked and only the fold of a thin cloak is draped around her neck. Both of her breasts are bared. She is slumped either in a faint

Figure 88. Relief of Claudius and *Britannia*, from the *Sebasteion* at Aphrodisias. Julio-Claudian. (Photo: courtesy of New York University Excavations at Aphrodisias. Photographer G. Petruccioli).

Figure 89. Relief of Nero and *Armenia*, from the *Sebasteion* at Aphrodisias. Julio-Claudian. (Photo: courtesy of New York University Excavations at Aphrodisias. Photographer G. Petruccioli).

or at the point of death, her head lolling to one side and her body partially supported by the hands of the emperor.

Arguing to some extent against my interpretation is the broader geographical and specific architectural contexts of the Aphrodisias reliefs, produced and commissioned in a locale far from the centre of Roman power, in the Greek east of the empire where a marrying of Roman themes and concerns with a Greek sensibility might have dictated their form. Sponsored by two local elite families the *Sebasteion* was profusely furnished with artworks. The two individual reliefs discussed here were but two out of over two hundred reliefs from the building complex, comprising three main series of representations: *ethnos* reliefs or peoples of the empire; gods and emperors; and myths and heroes. There were also honorific statues inside the building as well. Thus it might be possible to see the overall artistic scheme as representing the deployment of innovative strategies of imperial portrayal almost in a grand Hellenistic style that located the Roman emperors in a continuum of Greek mythological heroes. For instance, it has been suggested that the scene of the emperor Claudius overcoming and conquering *Britannia* could have represented a variation on the common image trope of Achilles slaying Penthesilea or indeed have mirrored another depiction elsewhere in the *Sebasteion* of a Greek warrior grasping the hair of a falling Amazon.

But, of course, these two reliefs are not altogether unique and can indeed also be seen in the broader context of depictions in Roman imperial art of violence against women and of the commonly employed trope of female abduction and rape in Greek and Roman art.[12] Again, the fact that it is Roman emperors themselves depicted committing acts of violence would seem highly significant. Of course, no Roman emperor ever was engaged in hand to hand combat or fought directly in battle. Only one other instance of such an image is known in sculpture, interestingly again from Asia Minor, suggesting that this was somehow an acceptable visual trope in parts of the Greek east of the Roman empire in a certain period. Martial kingship had of course been commonly celebrated in the earlier arts of Egypt, Assyria, and Akkadia: pulling the enemy's hair or quite literally crushing them underfoot. Greek images of martial kingship of course went back to the time of Alexander the Great, and we know that the Alexander Mosaic from the *Casa del Fauno* or House of the Faun in Pompeii probably owed its origins in terms of subject, composition, and style to a lost late fourth or early third century BC painting of the Battle of Issus by Philoxenus of Eretria or Apelles. It was in the Flavian period that the image of the Roman martial emperor appeared, most commonly on certain coin issues: mounted on horseback and spearing an enemy or standing in a dominant position with one foot holding down a submissive foe. The latter, though not strictly a martial act of violence, still placed the emperor directly and personally in a position where his own physical force was brought into play against others.

The sculpture from the city of Maionia, near modern Koula, in Lydia, now in the *Museo d'Antichità J.J. Winckelmann* in Trieste in Italy, is both damaged and fragmentary. However, an inscription in Greek on the relief allows us to identify the two protagonists depicted as Caligula, given the name and title Gaius Germanicus Imperator Caesar, mounted on horseback, and *Germania*, depicted as was standard practice as a female personification. Although not in direct combat, the image captures the emperor charging on horseback towards *Germania*. He holds his spear or lance ready, in a stabbing position, as his horse rears up in front of the female figure, its hooves potentially about to trample her. *Germania* stands upright, her hands on her hips, calm, and ready for the imminent onslaught, seemingly accepting of her fate. The composition of the scene on the Koula relief is very much in the style of some of the coin portrayals of the Julio-Claudian emperors, and presumably must have represented a local response to such coin images, providing inspiration for a new and different way to depict Roman imperial authority in sculpture for a Greek-speaking audience, perhaps an audience which was also used to the Greek tradition of the portrayal of riders on funerary monuments.

A stone relief, possibly from Egypt and now in the World Museum, Liverpool, shows a bearded emperor, perhaps Caracalla, dressed in military attire and holding a seated bearded male barbarian by the hair, while at the same time menacing the prisoner with a sword. If the relief

---

[12] On emperors and violence see, for example: Davenport 2020.

did indeed originate from Egypt it could have presented an overt reference to Caracalla's brutal and violent response to unrest in Alexandria in AD 215.

Not quite in the same category are a small number of second and third century AD terracotta genre figures of Roman emperors with captives, produced in Egypt, probably in Alexandria. These are interesting in that a number of variations in composition subtly suggest the need to elicit different responses from different viewers of the objects. In all cases the emperor is shown bearded and asserting dominance over a barbarian captive, holding the captive by the hair and with the other hand holding a sword blade to their throat. While some degree of official sanction probably allowed for the commissioning, creation, dedication, and display of the Koula relief and the Aphrodisias artworks from the *Sebasteion*, and images on coins would have gone through some form of approval for use or at least a vetting of some kind, these terracottas represent images created outside official systems. It has been suggested that like many terracottas produced throughout the Roman world these Egyptian examples were probably votive items, for dedication at a temple or shrine or for display in a household shrine. They may have been produced to commemorate a specific event, perhaps Hadrian's crushing of the Jewish revolts, or may have been generic pieces celebrating Roman imperial power tied in to local traditions of warrior kings.

While no definitive catalogue of images of the city's presiding protective goddess *Roma* and of Roman foundational myths found outside of Italy exists, nevertheless it is apparent that such images were relatively common in Greece. Why this should have been so is not altogether easy to interpret. If the empire of Rome was born from a cradle of violence then explicit reference to such a torrid birth might have been expected. That rape was also one of the foundational Roman tropes is true, in the case of the Rape of the Sabine Women, so perhaps the image of a Roman emperor, representing all Roman men, forcibly overpowering a non-Roman woman, representing all sexual conquests, can be seen in this broader context without discussion glossing the sheer malignancy of the image.

While there are some admiring accounts of the history of the Roman empire by certain Greek writers and historians such as Polybius, Dionysius of Halicarnassus, and Appian of Alexandria, this admiration was tinged with a narrative twist which often seemed to focus the apparent admiration not primarily on the forceable winning of that empire or on its extent, but rather on the continued duration of the empire and the length of time that it had survived. Maybe some elements of surprise and incredulity can be discerned here, though none of them disputed the historical precedents set by Roman expansion and empire building.

The most common mortal woman portrayed in Roman imperial art in all media was the generic barbarian woman. Images of chained or captive or dying and dead barbarian men were also common and regular tropes on Roman imperial monuments, mainly in Rome but also quite widely throughout the empire where their meaning and context would have been very different.[13] Fighting barbarian men also appeared in huge numbers on Roman monuments such as Trajan's Column and the Column of Marcus Aurelius. In the earlier Roman empire,

---

[13] On relationships between Romans and non-Romans see, for example: Balsdon 1979; Bowerstock 2005; Broadhead 2002; Castro-Páez 2023; Edwards; Farney 2007; Favro 2005; Gruen 2010 and 2011; Hingley 2005; Isaac 2004 and 2011; Mathisen and Shanzer 2011; Meinecke 2020; Nippel 2007; Noy 2000 and 2010; Pandey 2021; Price *et al.* 2021; Purcell 2005; Thomas 1982; Thompson 1989; Versluys 2015; Pitts and Versluys 2015; and Woolf 2011.

before what could be called the Aurelian watershed, that time around which attitudes to barbarian peoples seem to have hardened, there may have been more of an emphasis, in terms of numbers of appearances on public monuments, on the use of more mixed-gender images of barbarian groups and peoples. It can be suggested that in some depictions of barbarian couples, the fact that the man was bound or in chains, and that the woman remained untied, in a mournful or dejected pose, indicated a definite strategy of suggesting barbarian male aggression and female passivity, for the woman evidently posed no threat and was therefore not constrained. All the more extraordinary then was the isolated but highly significant depiction of Dacian women torturing Roman prisoners in one scene of the frieze on Trajan's Column.

When barbarian couples were depicted, this would seem to have been a way of staking-out some element of common ground, of a common humanity between the Romans and the barbarian peoples portrayed. An absence of such pairings could be construed as the opposite- a declaration of a state of complete otherness. As well as sharing common humanity with the Roman couple, the barbarian couple, though defeated and often shown dejected and distraught, could then be transformed by the incorporation of their land into the empire and subject territory of Rome. While they themselves may not have become citizens, their children and their children's children may have achieved such a status, perhaps through the service of a son in the auxiliary units of the Roman army. A people defeated by Rome was not necessarily a despised people in terms of the design and presentation of images to viewers in Rome. However, the wars fought by Marcus Aurelius were not wars of incorporation as such, though he may have hoped to eventually create new provinces in captured lands, and therefore it is this spirit of civilising *Romanitas* that is missing from the depictions of warfare and of violence against barbarian women on Marcus' column. There may have been a sense to the male gaze that these women were somehow colluding in their own subjugation and objectification.

Of the individual images of barbarian women suffering violence found in Roman art, the most noteworthy individual violent motifs pre-dating the Column of Marcus Aurelius occur on the *Gemma Augustea*, an Augustan court cameo, where the so-called hair-pulling motif is employed in the depiction of a barbarian woman being pulled along by the hair by a Roman soldier and on the two reliefs from the Aphrodisias *Sebasteion* discussed above. The harrowing scenes of desperate warfare and depictions of violence against women on the Column of Marcus Aurelius, including another hair pulling scene, represented an altogether different conception of the boundaries of taste in visual representation of imperial policies. That there is always sexual tension and sexual violence between the men of an invading or victorious force and the women of a conquered area is unfortunately and sadly true for all conflicts, both ancient and modern. While the hair-pulling scenes on the *Gemma Augustea*, the Claudius and *Britannia* relief, and the Column of Marcus Aurelius were isolated examples in Roman art, they may have derived from the hair-pulling motif in a mythological battle on the Altar of Zeus at Pergamon and could thus could be construed as having been stock Greco-Roman images. Nevertheless they were used in contexts which, in the case of the Augustan example, denoted a generalised state of mistreatment of a captive and, in that of the Antonine image, a more-focused use in the context of the fate of female victims in the German campaigns of the time. The Aphrodisias depiction of Claudius physically overcoming and conquering *Britannia* was an image which denoted one version of the nature of the relationship of the male conquerors with the female conquered, between whom sexual and social boundaries were perhaps deemed not to exist.

The most intensive use of images of barbarian women suffering from Roman military violence occurs on the Column of Marcus Aurelius where it perhaps provides an insight into the Roman male imperial psyche. Many of these barbarian women portrayed were mothers and the interaction between these mothers and their children and between some of these mothers and Roman soldiers was often portrayed in quite fraught images of violence and of the threat of impending violence. These scenes can be divided up into four categories, the first being scenes of actual physical violence against barbarian women, including the killing of some women, the second being a scene of the aftermath of a possible rape or sexual assault, the third being scenes involving the physical manhandling of women and children or their mistreatment as they are taken prisoner, and the fourth and final category being scenes of women in captivity and their transporting away from the area, presumably into slavery in Rome and elsewhere.

There are two definite instances of the depiction of barbarian women being killed in Scene XCVII on the column, following a battle. A third possible killing appears in Scene XLIII, although in this instance the actual killer blow is not depicted as it happens; rather, it would appear to be about to be struck by a Roman horseman during the sacking of a German village. The depiction of the terrified woman in an almost diaphanous robe which clings to her legs and hips as she runs for her life, the material billowing out with her movements, is like an image of an eroticised dancer, perhaps a Maenad, rather than a victim of war. The killing of these barbarian women is shocking in its ferocity and understatement and perhaps also surprising, given that captured women would have been valuable to the Romans as slaves.

The scene that perhaps shows the aftermath of a possible rape occurs on the column in Scene CII, a composition that is again centred on the portrayal of the destruction and burning of a German village. As huts are set ablaze by a Roman soldier holding a flaming torch other soldiers proceed to capture, bind, and lead away the men of the village. On the right hand side of the scene stands the solitary figure of a woman holding the shoulder of a small girl standing at her feet. The woman's dress is pulled completely off her shoulders and her upper body is exposed and naked. With her other hand she is holding up her ripped dress and has managed to bring it up sufficiently to cover her breasts and secure some modesty. It is difficult not to view this woman as the victim of a sexual assault by one of the Roman soldiers sacking the village. Here the rape of the woman was being equated with the rape of her people by the Romans and the rape of their village and homeland.

Depictions of abductions leading to rape were quite common in Roman art, as in the popularity of depictions of the abduction/rape of Persephone/Proserpina in funerary contexts for instance. While it might be true to say that rape was perhaps being used here as a metaphor for the violence with which death can carry us off, with a promise of some type of return from the Underworld as achieved by Persephone herself, the underlying violence of the action and its suggestion of a state of social and sexual disorder still remained at the heart of both the image and the metaphor.

When viewed in isolation individual images representing abduction or rape stories from Greco-Roman mythology or from Roman historico-mythology might be viewed as metaphorical.[14] However, when viewed in the broader context of Roman art, so often dominated as it was by depictions of warfare, battle, hunting, and the celebration and commemoration of male achievement and power, abduction or rape scenes might be thought of as having been part of an overall narrative of male dominance, of a cultural male hypersexuality.[15]

By usually implying that rape was about to take place in such images rather than depicting rape itself a boundary seems to have been deliberately drawn by male art commissioners. Indeed, it might have been considered by them that Roman women viewers would place some degree of positive spin on these at first sight wholly negative images, particularly as these were generally individual women portrayed and not anonymous signifiers of a generalised female powerlessness. The abduction and rape of Persephone did not define her: indeed her place in mythology relied as much on her escape and afterlife as it did on her terrible suffering. She was though never allowed to forget what had happened: indeed, she was forced to remember by being fated to return to the Underworld for part of each year for ever afterwards. Again, in the case of the Sabine women they eventually were expected to accept their fate.

Scenes involving the physical manhandling of women and children or their mistreatment on the column are more numerous than killing scenes, and include further hair-pulling scenes, as well as more general depictions of mistreatment and herding of captives. Taken together, these scenes of rape and violence against non-Roman women seem to be suggesting that war was inevitably something that affected all society, and that women and men were equally affected, even if victims rather than active protagonists. Portrayal of such events without the masking and filtering gauze of allegory or allusion was something quite new and revealing in Roman art. A comparison with the scenes involving women on Trajan's Column is instructive. There, barbarian women were captured or taken as hostages. They were present at scenes of siege and to the rear of battle, but nowhere were they shown being mistreated or killed as part of the general routine of war, as seems to have been the message on the Aurelian Column. The only jarring scene involving women on Trajan's Column was one showing Dacian women torturing captured Roman soldiers, a scene so extraordinary in that context that it can be interpreted as probably relating to a recorded, notorious incident from the Dacian wars or as an allusion to the inherent cruelty of Dacian and barbarian women in general. The kicking of the preserved head of the defeated Dacian king Decebalus down the Gemmonian Steps in Rome, following its parading around the streets in Trajan's Dacian Triumph, was a grotesque and cruel pantomime to further punish and forever urbanise his person within the walls of the city rather than in his Dacian homeland: a counterpoint in memory to the cruel behaviour of the women torturers memorialised on the column.

The scenes of violence against women on the Column of Marcus Aurelius do not represent the portrayal of soldiers somehow out of control, of the breakdown of military discipline, but rather they would appear to represent perhaps the first portrayal on a Roman imperial

---
[14] On rape, violence, and Roman power see, for example: Davenport 2020; Ferris 2000: 56-57; Lillie 2017; Pandey 2021; and Riess and Fagan 2016.
[15] On sexual violence, violence, and the power differential between the Romans and the enslaved see, for example: Bielfeldt 2018; Blake 2012; George 2003; Levin-Richardson 2021; Kamen and Marshall 2021; Pollini 2003; and Richlin 2017. On sex and exploitation, particularly involving the non-Roman and enslaved see, for example: Peakman 2019.

monument of what may be called the collateral damage of war, a more honest portrayal of the carnage and randomness of war. Was this a return to some earlier modes of the portrayal of Roman conquest by the use of the metaphor of sexual conquest as in the *Sebasteion* reliefs? The masculine sexual conquest of feminized space in both cases had a very real, non-metaphorical dimension. We are seeing here a political landscape that was also the setting for the human body, observed almost with a clinician's eye as it underwent trauma, as it was anatomised, penetrated, cut and crushed, and humiliated. This body landscape was also an image of itself, a mass media projection made up of female erotica and news footage of the wars of Roman conquest. Exploring the psychic fallout of all this horror and violence, a new logic which would explain all these events, with bloody apocalypse being not some kind of abolition but more a transfiguration. There is an underlying sense that a more primitive world was biding its time, that there was no past, no future, and a diminishing present in a transforming but still deeply-conservative society that did not know, but was perhaps coming to realise, that its past was all it had.

Finally on this topic, it became fashionable for first Roman imperial women and then presumably elite women at Rome to wear hair extensions or false hairpieces made from German barbarian women's hair, at least in the early empire. Whether the hair was cut off women forcibly or it was willingly sold to Roman merchants is uncertain.[16] This extraordinary phenomenon quite literally represented conquest and imperialism on a very personal micro-scale. Whether when worn by imperial women the visual and physical response of their menfolk was one of voyeuristic, vicarious excitement can only be guessed, though of course the imperial women could themselves have experienced similar feelings of power through adding such pieces to their costume repertoire.

In my study *Enemies of Rome* I did not view the presentation of images of barbarians in art at Rome as part of a broader phenomenon, but here I have taken the opportunity to do so, to consider such ubiquitous images as just one category of geography products created and displayed in the city in the form principally of statuary and sculpture and on coin issues that served to explain to viewers of these images how Rome related to the wider world, where Rome was, and who Romans were.

Bodies and Metaphor

Another way to depict people of another place, and indeed that place itself at the same time, was by using geographical personifications of countries or provinces which in Roman art were always in the form of female figures. These personified figures were usually completely passive, identifiable by accompanying inscriptions, clothing, attributes, or appearance. While often isolated figures, particularly in the form of statuary, equally often they appeared as captive figures in thrall to the person of the Roman emperors, figures such as *Britannia capta*, *Germania capta*, and so on, these images of subservience being particularly common on Roman coins.

As was mentioned in the Preface to this study, viewing the short documentary film of 1984 *Les Dites-Cariatides-The So-Called Caryatids* by the French film director Agnès Varda has helped

---

[16] On the ideology of hair see, for example: Tarlo 2016.

me to focus my thoughts on certain aspects of categories of gendered representation and to theorise around many ideas relating to place and home. The common use of architectural caryatids as supports in Greek and Roman buildings in lieu of supporting structural columns was described by the Roman architectural writer Vitruvius who contextualised and sought to normalise the use of such images of captured, subservient women, as he recorded it, in Greek and Roman building practice. His thinking, which might have been commonly shared by the Romans, that these women from Caryae were condemned to slavery after betraying Athens during the Persian Wars was mistaken, for in Greece the caryatids were simply seen as young women who danced with baskets on their heads at celebrations at Caryae in honour of Athena. Whatever the Roman interpretation of cayatids their use would seem to have been common in Rome's buildings and to have had an influence on the creation and use of other kinds of female figures as architectural supports and free-standing architectural decoration, including personifications of non-Roman lands and peoples and of Roman provinces.

Pompey's Theatre-Portico-Temple complex dedicated to *Venus Victrix* in 55 BC has often been called a museum of sorts.. As well as a mixture of classic Greek artworks, his own portrait was sited here, along with a collection of female personifications of all the *nationes* he had brought under Roman rule. These female images, by the sculptor Coponius, were altogether appropriate in a complex dedicated to martial Venus, and represented the first major commission of new artworks in Rome designed to transmit geographical information to their viewers.

It has already been noted that the content and form of the emperor Augustus' *Res Gestae* in many respects was like a Roman itinerary, listing 'places' along a route, mapping a journey, though in the *Res Gestae* these itineraries were not all geographical as such and mapped an itinerary of the emperor's life and significant achievements, including the listing of places visited, peoples conquered, buildings constructed and so on. In his famous comment that he found Rome a city of brick but had transformed it into a city of marble he was not only centreing Rome in its contemporary world but also as a staging post in his own life-course, that the city had been alchemically transformed from a thing of fired clay to one of hard stone at his instigation. The city had thus become Nature itself. Visitors to Augustus' Forum in Rome would have been able to view many architectural caryatids, a choice of image not only linked to classical Athens but also alluding directly to the original story of the capture and exiling of the women of Caryae, a story which somehow reflected the known and attested Augustan practice of demanding women hostages to secure the future good behaviour of their defeated tribes. In such a context the choice of the caryatid image seems very apposite. Parthians as caryatids also appeared on the reconstructed *Basilica Aemilia* and on the pedestal of the altar of Augustus' *Ara Pacis*. It would seem that the Augustan preference for images of subdued non-Roman women being displayed in the centre of Rome was as much about the punishment of deviance among women in general as it was part of the discourse centred around victory and submission.

As was discussed above, the Julio-Claudian period saw the use of images of female personifications in alarming active tableaux alongside the emperors of the dynasty at the *Sebasteion* at Aphrodisias in Asia Minor, but there also appeared here more conventional personifications in the form of the so-called *ethne* employed to present ideas about the geographical span and extent of the Roman empire to the local Greek elite here in a way

Figure 90. Relief depicting *Mauretania* from the *Hadrianeum* Rome. AD 145. *Palazzo dei Conservatori*, Rome. (Photo: Author).

that allowed them to marvel at the nature of the Roman empire in a manner similar to that displayed by the Greek historians Polybius, Dionysius, and Appian as discussed above earlier in this chapter. They also provided a reference back to the programme of artworks representing the provinces in the Augustan Forum back in Rome.

Later still, in Trajan's great new forum, the last and the grandest of all Rome's fora, numerous free-standing statues of defeated Dacian men contrasted greatly with the Augustan preference for images of subjugation to be female. The potency of these statues was so lasting that a number of them would be reused later as part of the accumulation of *spolia* used to construct the fourth century Arch of Constantine. If we can read a sexual subtext to the creation and display of many of the female geographical images presented as personifications, then perhaps a similar reading should be applied to images such as the Trajanic Dacian figures,

Figure 91. Relief depicting *Hispania* from the *Hadrianeum* Rome. AD 145. *Museo Archeologico Nazionale*, Naples. (Photo: Author).

variously described as being defeated and proud. Perhaps they were intended to be seen as both potent and impotent.

But the meaning of certain images displayed in Rome for many years must have changed considerably with time. Perhaps the best group of artworks to chose to demonstrate this large group of statues of Dacians originally commissioned by Trajan to adorn his new and spectacularly-large forum in Rome. At some time these statues, or some of them at least, were removed from the forum. Why and when might that have been? We really have no idea, though some potential dates and motives can be suggested. Some at least of these statues were put into storage, for them to re-emerge as an integral element of the design of the Arch of Constantine, dedicated to the emperor by the Senate in AD 312-315. One of the statues is inscribed under its base with the scratched instruction *ad arcum*, that is 'to the arch'. Why were these statues chosen and for what ideological purpose? Now this is something we can

Figure 92. Relief depicting *Gallia* from the *Hadrianeum* Rome. AD 145. *Palazzo dei Conservatori*, Rome. (Photo: Author).

never know. In their new architectural context did they mean the same as they meant in their original setting? It would seem highly unlikely that this could have been the case. Of course Dacia was inextricably linked with Trajan and as one of the so-called 'good emperors' cited on the monument by allusion as a role-model for Constantine, but so many decades later what did Dacia mean to an inhabitant of Rome in the first decades of the fourth century?

Such antagonisms now perhaps became simply archaicisms. The effect of older images of foreign peoples, or personifications of their countries, still being available to view many years after their newsworthy currency was devalued or indeed when they perhaps had become ideologically worthless could have created a 'pastness', through the display of ahistorical barbarian figures. Experiencing such a 'pastness' may have been akin to Jean Baudrillard's *simulcrum* which references other signs but not contemporary reality.

The personified provinces of the empire which appeared at the *Hadrianeum* (Figures 90-92) and on Hadrian's coin issues are discussed elsewhere in this book and in fact represent very different messages to those of images of captured or submissive barbarians. The Hadrianic nations, the so-called *Provinciae Fidelis* or faithful provinces, are almost shocking in their anodyneity.[17]

---

[17] On the *Hadrianeum* see, for example: Hughes 2009; and Sapelli 1999. On personifications of provinces see: Houghtalin 1996; Juhász 2016; Méthy 1992; Ostrowski 1990 and 1996; and Russell 2018.

Under Hadrian the empire was consolidated and no new conquests were undertaken, hence the gallery here of personifications could have been designed to be understood as a gathering of the provinces of the empire presided over by the benevolent emperor whose cautious and wise stewardship had helped guarantee an era of general peace and prosperity. Individually they stated, and in their seriality restated, a finality that suggested not an end to empire but an end to expansionism. They might have been viewed as reassuring in their presence, in a way that those confused and looking for a stable routine welcome stasis.

A Long Way from Here

It will have become apparent during the discussion of Egyptian imagery in Roman culture in Chapter Six that images of subservient Egyptians were never part of the repertoire of images used by the Roman state to impart geographical knowledge about Egypt to Roman citizens. Certainly there were some negative pictorial tropes relating to Egypt in Roman art, such as the use of images of pygmies, often in comical situations, but nothing comparable to the way in which say, for instance, Gauls, Germans, Britons, and Dacians were treated. However, in Chapter Six I have argued that most Nilotic landscapes depicted on Roman mosaics and in wall paintings were intended to represent colonial landscapes, not things of wonder but rather things of possession.

One unique item that raises multiple issues with regard to discussions of trade and contact, globalization and transculturality, aesthetics, and influences is an ivory carving from Pompeii now in the *Gabinetto Segreto* in the *Museo Archeologico Nazionale* in Naples[18] (Figure 93). Found during excavation work in the 1930s inside a wooden box in what was then called the *Casa dei Quattro Stili* (House of the Four Styles), a house believed to have been the residence of Maius Castricius, this 24 centimetres high carving takes the form of a long-haired, naked female figure adorned with heavy jewellery and attended by two female attendants or worshippers proffering toilet articles. The context dates the item to somewhere between the first century BC and the mid-first century AD. Without any doubt the carving is Indian in origin and in style: the ivory therefore must be Indian ivory. The figure is most commonly called a statuette, but as there is a hole at the back of the piece to take a fitting of some kind it has been suggested to have been a mirror handle or, less likely, a furniture leg. So significant was the find, the house in which it was found became more commonly known as the *Casa della Statuetta Indiana*.

The principal female protagonist portrayed has been suggested to be the Hindu goddess Lakshmi, goddess of female beauty and fertility. The idea put forward that she is in fact an easternised image of Venus and her attendants has few adherents and, if accepted, would have significant implications for discussions of appropriation and taste at this time. More recently academic concensus has moved towards, but not universally landed upon, the identification as being of a Yakshi or tree spirit. Stylistically the carving has been considered to parallel pieces from the north-western Indian regions of Begram, Ter, and Bhokardan, places with links to Gandhāra whose art of this period displayed Greco-Roman influences itself.[19]

---

[18] On the Indian statuette/figure from Pompeii see principally: D'Ancona 1950; Parker 2002; and Weinstein 2021.
[19] On Gandhāran connections displayed in art in India see principally: Rienjang and Stewart 2020.

Was the piece specially commissioned? Was it bought in Italy from an Indian Ocean trader?[20] Was it viewed simply as a luxury item by nature of the material from which it was made, its rarity value, and what would have been its considerable cost? Was it appreciated aesthetically, even though not in the favoured Greco-Roman style? Was the image of the naked woman something that placed it in the more general category of Roman erotica so common at this time in Italy? Was the naked image of a non-Roman woman of particular significance or did it provide a particular kind of sexual frisson defined by the Roman male gaze?

This is the sole item of Indian art known from the central Roman empire, despite the now well-attested busy trade between west and east. However, other finds have come from the margins of the Roman world, including a small carved marble Buddha head among the finds from the excavation of the Temple of Isis at Berenike in Egypt on the Red Sea. Also on the Red Sea, but somewhat distant from here at Adulis on the Gulf of Zula in Eritrea, was found a Gupta period fourth to sixth century AD terracotta figurine which must have had its origins in India. The recent re-evaluation of a Roman period ivory knife handle found at Wels in Austria in 1918 suggests that it represents another example of the traces of cultural encounters beyond the boundaries of the Roman empire. With a head and face carved on the end of the pommel the handle also carries an inscription in the Kharoshthi (Gandhārī) script, suggesting to the item's curators that it came from Niya, in the Chinese Xinjiang province along the Silk Road trading route. The inscribed dedication reads: 'Honouring gift to Master Tadara'. Trade between Rome and China appears to have been considerable, but though we know that Roman trade ambassadors visited Han China there is no manifestation of the trade reflected in Roman art.

Figure 93. Ivory Indian figure from the *Casa dei Quattro Stili* or House of the Four Styles, Pompeii. First century AD, before AD 79. *Museo Archeologico Nazionale*, Naples. (Photo: Author).

Of course, the Indian triumph of Bacchus was a popular theme on decorated sarcophagi, though in most cases the land of India was evoked by images of its wildlife, particularly Indian elephants. However, one of the fourth century AD mosaics at Villa del Casale, Piazza Armerina

---

[20] On Roman trade with India see, for example: Evers 2017; McClaughlin 2010; Parker 2002; and Young 2001.

Maps of the Body

Figure 94. Personification of India on a mosaic from Villa Casale, Piazza Armerina, Sicily. Fourth century AD. (Photo: slide collection of the former School of Continuing Studies, Birmingham University).

in Sicily carries an allegorical image of India as a semi-naked, alluring and eroticised female personification, a rare depiction of a land or people outside of and beyond the Roman empire's boundaries with whom Rome was not at war (Figure 94). Wearing a large gold necklace and ear-rings, she sits on a rock holding a huge tusk of ivory, and another tusk leans on a rock beside her. To one side of her stands an Indian elephant and on the other side stands a tiger. We are seeing here the wealth of India's raw materials and a highly sexualised image appealing to the Roman elite male gaze.

The Barberini ivory diptych/polyptych (Figure 95) in the *Musée du Louvre* in Paris, probably dating to the sixth century AD, carries a depiction of the emperor Justinian riding on horseback, with what appears to be a captive or subservient Sassanian male walking behind the emperor's horse. In a lower register western and eastern peoples appear paying homage to the emperor. One eastern figure, presumably representing India, carries a huge ivory tusk as a gift, accompanied by a small elephant and a tiger. Why India, outside the Roman empire, would be acting in this subservient manner to Rome is unclear.

Craving for Oblivion

As has been briefly mentioned it would have been interesting to examine the ways in which other non-Roman peoples viewed the Romans themselves, how Romans might have been

Figure 95. Leaf from an ivory diptych (the Barberini Ivory) depicting the emperor Justinian, Constantinople. First half of the sixth century AD. *Musée du Louvre* (Photo: Copyright *Musée du Louvre*).

thought to have been situated in other geographies. Sadly, however, there is little material evidence to allow such an analysis to be undertaken, and nothing in any way comparable to the way that art historians can analyse how white European colonisers and imperialists, traders, and clerics, were commonly portrayed in nineteenth and twentieth century tribal art, particularly in west and central Africa.[21] It might have been thought that some element of self-realisation and humour might have lain in the fact that the Roman mission to conquer and thus civilise some barbarian peoples was at odds with the hyper-masculine Roman behaviour distorted by conquest, and required to achieve that mission: that is violence, aggression, the establishment of hierarchies, and raw displays of power and potency. Did the conquered ever

---

[21] On perceptions of the west from Africa see, for example: Leiris 1934; Menut 2010; Quarcoopome 2010; and Strother 2016.

see through this blatant contradiction? If Roman imperial art was awash with explicit images of violence by the armies of the state did this mean that there was a moral vacuum here at the heart of Rome that others might have been able to identify?

The idea of being able to reveal 'hidden transcripts'[22] which might reveal to us otherwise hidden sentiments expressed through cultural actions like the writing of anti-Roman *graffiti*, the *graffiti* caricaturing of politicians and emperors, or even scatological humour aimed firmly against Rome as an imperial or occupying power, is an appealing one. However, the evidence is not yet there for us to see, though even in Rome itself Plutarch recorded one such subversive protest in the form of the daubing of a statue of Lucius Brutus on the Capitoline Hill in front of the Temple of Jupiter Optimus Maximus in 44 BC with the anti-Caesar slogans 'If only you now lived, Brutus' and 'If only Brutus were alive'. In other incidents discussed by Plutarch he relates the story of Flavius and Marullus who were sent to help restore vandalised statues of Caesar. Were such anti-state transgressions common? Did they occur outside of Rome, particularly at times of outright tension and rupture such as during the Boudiccan uprising in Britain in AD 60-61 or the Jewish revolt of AD 66-70?

There is often a rich semantic ambiguity in some African images of white colonialists which covertly exposes the psychological structures and cultural forces which underpinned cultures of colonialism. Perhaps some of the images produced in Roman Egypt of Roman emperors as Pharaohs, relying on strategies of allegory, silence, and indirection, were not quite as they might now seem to us.

What would sophisticated elite Roman officers have made of the artistic output produced under the rubric of Romano-British art or Gallo-Roman art for instance? Would they have recognised that the figure of Venus on a second or third century AD relief from High Rochester in northern England was meant to be an image of the same goddess as appeared in the form of so many Greco-Roman statues such as the one now-known as the Venus of Capua (dated to AD 117-138)? The anthropological theory of assertive mimesis suggests that in certain cultures at certain times, and in certain contexts, one can identify instances where representation through what might be considered 'bad' copying from an accepted template model can be seen to have been a deliberate act of subversion, a choice, a statement of some kind. The language of artistic expression in conditions of asymmetrical power relationships as existed between Romans and non-Roman provincial peoples might have been a tool in itself. Perhaps we can talk about the non-Roman gaze, a tool in contestations of power perhaps.

The first of two applicable case studies relating to foreigners portraying Romans concerns a number of images of Roman emperors produced in Persia/Iran by the Sassanid dynasty of Persia, the successor power to the Parthians in the east. In the first half of the third century AD the Sassanians, in their attempts to revive the past glories of Darius and Xerxes, pursued a policy of territorial expansion in the region which inevitably brought them into conflict with Rome whose geopolitical interests and influence were now seriously threatened. Initially this proved disastrous for Rome when Shapur I successively defeated three Roman emperors, Gordian III, Philip the Arab, and Valerian, the latter emperor actually being taken prisoner by the Sassanians following the defeat of his forces in AD 260. These great victories were

---

[22] On subversive *graffiti* and 'hidden transcripts' see principally: Morstein-Marx 2012.

commemorated both by a trilingual inscription set up at Naqsh-i Rustam and on five carved relief panels, one at Darabgerd, three at Bishapur, and one at Naqsh-i Rustam.

The most dramatic and striking of these reliefs is carved on a rock face above a pool at Darabgerd, near Shiraz, in Shapur's home province of Fars. Here are depicted Shapur on horseback, laying his hand on the head of the emperor Philip in an act of clemency, while Valerian moves towards him, perhaps also seeking clemency, while Gordian lies on the ground behind Shapur's mount. This particular relief, along with the others apart from one of the Bishapur reliefs, represented a conflated version of the victories of Shapur I and certainly did not reflect real time in its depiction of the three Roman emperors together in a single scene that brought together their three separate traumas.

The exceptional, partially-damaged relief at Bishapur depicts a kneeling figure of a Roman emperor, either Philip or Valerian, between two horses, on which would probably have sat Shapur and his father Ardashir, perhaps in a scene of investiture, one in which the figure of the vanquished Roman emperor seems initially superfluous. Indeed, here he seems to be simply an attribute of the Sassanian victors, in the same way that the figure of the conquered barbarian became an attribute of Roman imperial power. Following this tradition, at Taq-i-Bastan a later rock-cut relief shows the standing figures of Shapur II and his successor Ardashir II, attended by a third figure of a Roman officer lying sprawled on the ground beneath their feet, either dead or caught in an act of prostrate submission. The Roman man here is most likely the emperor Julian, whose ill-fated expedition of AD 362-363 led to his death in battle against the Persians.

It has been suggested that the design of these Persian reliefs, with the exception of the Bishapur investiture scene which is considered to have a local origin, was obviously very heavily influenced by Roman imperial prototypes, but whether such detailed knowledge of the language and structure of Roman imperial triumphal imagery really informed the creation of Shapur I's monuments remains unknowable. Certainly, if viewed as a programmatic sequence it would appear that Shapur I used these images to lay claim to be the successor of Darius and Xerxes and to have power over the Roman usurpers and expansionists. This was a far cry from the victory of Galerius over King Narses of Persia in AD 299 that saw him hailed in some quarters as the new Alexander the Great.

The second case study relating to the portrayal of Romans by non-Roman peoples involves the creation and deployment of equally remarkable images of Roman emperors portrayed as Egyptian Pharaohs in, for example, the Temple of Khnum at Deirek-Haggar, Dahkleh, the Temple of Khnum at Esna, and the Temple of Isis at Shanhur, all in Egypt. In the temple at Deirek-Haggar, Dahkleh can be found images of Titus, Domitian, and Trajan. In the Temple of Khnum at Esna can be seen Domitian, Trajan, Commodus, and Septimius Severus, Julia Domna, Geta, and Caracalla. In the Temple of Isis at Shanhur there has quite recently been found a depiction of Claudius undertaking a ritual to the fertility god Min, again presenting the person of a Roman emperor as a (sexually) potent and powerful individual man.

In Scene 570 at Esna Domitian is depicted smiting enemies before Khnum-Ra. In Scene 520 there Trajan dances before the goddesses Menhyt and Nebetu. In Scene 531 at Esna Commodus in company with Horus and Thoth catches fish and birds in a papyrus thicket using a trap net, in all probability symbolising the killing of his and the gods' enemies. In Scene 496 at Esna Septimius Severus is depicted receiving an *ankh* from the god Khnum-Ra, with his wife and sons in attendance. Roman emperors dressed as Pharaohs were common in Egypt both in wall carvings such as this and in portrait statuary.

Also of interest here is the phenomenon of images of Roman emperors appearing on coin types issued by certain client kingdoms, territories that were outside of the empire but diplomatically tied to Rome and the furtherance of its interests.[23] Sometimes the heads of the client rulers would appear on the coins paired up with the heads of contemporary emperors. Sometimes the client king would seek equity with other members of the imperial house, so as to suggest insider influence and deep relationships. One of the most significant examples of this is on the obverse of a coin of AD 42-43, minted at *Caesarea Maritima*, depicting Agrippa I and Herod of *Judaea* crowning the emperor Claudius with laurel wreaths.

Turning and Turning

In terms of depictions at Rome of non-Roman peoples from within the empire, provincial peoples in fact, in contexts not related to scenes of battle and war, and scenes of enslavement and captivity we can briefly consider depictions of Greeks, of Moesians, and of Jews.

The huge influx of Greek works of art to Rome in a number of waves of acquisition through conquest, following the precedent set by the mass looting of Greek artworks from Syracuse in 212 BC by Marcus Claudius Marcellus and the bringing of all these spoils to Rome for triumphal display, exhibition, and dispersal, was seen by some Roman commentators such as Polybius, Livy, and Plutarch as marking a turning point of some kind and as a kind of corruption by others. These acts together changed Roman art and culture in Rome itself from an Italic phenomenon to a more outward-looking Greco-Roman culture, with Greek art and ideas being accepted and integrated on an unprecedented scale and in a way that was never to be repeated in terms of cultural exchange between Rome and any other contemporary cultures the Romans encountered.

The massed Greek works being brought together in Rome represented an assimilation, almost a capitulation to the aesthetics and values of another, foreign culture. There was no logical reason why Italic art became what we now unquestionably call Greco-Roman art. This phenomenon represented a geography of assimilation, to some at Rome founding a false paradise for the discontented. There, in pursuit of worthless pleasures and luxuries, they discovered their own private hells. Artists and writers were in some way directed to bring the empire alive, opening up new worlds themselves by embracing their clients' Greco-Roman tastes, pointing to the art as much as the *groma* as that which would manoeuvre citizens over the new distances. These images that linked or connected physical and social spaces in Rome, linked Rome with assimilated Greece and not with the mines and quarries of other far away places in the empire: art like this acted as both an enabler and a social integrator.

All of the personifications of foreign countries and peoples so far discussed in this chapter have been images deployed in the service of Roman imperial rhetoric, mostly on imperial

---

[23] On images of client kings see principally: Wilker 2020.

Figure 96. Wall painting with depiction of Macedonia, Villa of P. Fannius Synistor, Boscoreale. First century AD, before AD 79. *Museo Archeologico Nazionale*, Naples. (Photo: Author).

monuments and on coins. Mostly they are identifiable by attributes or costume and appearance, in the case of sculptural images, or by captions, as is the case with most of the coin images. However, there is one particularly interesting exception to this in the form of a wall painting from the Villa of P. Fannius Synistor at Boscoreale on the Bay of Naples. Dating to c. 60-30 BC, and now in the *Museo Archeologico Nazionale* in Naples, it has been suggested that the painting is a version of a Hellenistic original of the second half of the third century BC. Forming part of the decorative scheme in the residential range's main *triclinium* the particular panel in question carries a scene of two women who can both be interpreted as personifications unusually pictured with landscape/geographical elements.

Rather than look directly towards the viewer the women instead rather fixedly stare at each other. The woman in the background of the scene is generally interpreted as the personification of Macedonia in northern Greece (Figure 96). The woman in the foreground is thought to be the personification of Syria or Persia, or of Asia. It is obvious from their relative positioning, their poses, gestures, and their body language that Macedonia is the dominant figure represented here. She is armed, holding a spear with both hands while a large circular shield rests beside her: she wears military headgear but a chiton and mantle rather than body armour. Though larger and in the foreground the subservience of the second female is indicated by her holding her hand up to her chin in a submissive pose. She looks up at Macedonia almost in fear and resignation. She wears generic female dress and is not armed: indeed, we must see her as having been disarmed or defeated.

Macedonia sits on a rocky outcrop or mountain, quite appropriate to the topography of this northern Greek region. The butt end of her spear hovers over what appears to be a patch of blue-grey sea and comes to landfall on a foreign rocky coast. The defeated figure sits against a rocky background. Peter Holliday has suggested that the scene alludes to the famous incident where Alexander the Great of Macedon threw his spear from ship to shore to claim Asia as 'spear-won land'. He also believes that the image was temporally appropriate here in Italy in that it was around this time that Rome was creating and consolidating its own eastern empire.

Why such a scene should be found in a private residence remains a mystery. It could be that the villa's owner-whether the P. Fannius Synistor named on a bronze vessel excavated from the site was its owner or not-was a retired senior officer from the Roman army or an administrator serving in the east at one time. There could be an ethnic link of the owner to Macedonia specifically or to Greece in general.

As well as being a physical city Rome was also to some extent a literary construct, and indeed is much more so today in terms of the reception of Rome by modern academic classicists being mediated by the accumulated texts of its historians, geographers, politicians, jurists, philosophers, rhetoricians, poets, playwrights, letter writers, scientific, medical, and specialist writers, and encyclopedists. In the light of this it seems apposite to ask if the Roman intellectual tradition was in any way affected by the geographical world-views of its rulers and elites? This matter has been investigated by Benjamin Isaac in the broader context of the study of prejudice in the ancient world.[24] Isaac found that there was much more to bind the city-dwelling elites across the empire than there was to separate them. While the flow of wealth, raw materials, and resources from the provinces to the centre helped maintain the power of Rome, is it possible that the flow of ideas from the provinces, particularly the Greek-speaking eastern provinces, represented some kind of geography of intellectual contamination, or was the broadening of Rome's intellectual culture in this way actually the motor for the empire's strength and longevity? In order to answer this question it will be necessary to consider the imported ideas themselves and the provincial cultural traditions from which they emerged. Was the embracing of non-Roman intellectual traditions another form of Roman imperialism?

As will be discussed in Chapter Nine, the emperor Hadrian's philhellenism had a profound impact on Roman culture at the time and through this many ideas about Greek and Hellenic culture and the geography of the eastern empire were presented to the Roman people in a way that had not happened before, perhaps with the exception of Augustus' embrace of a Romanised Egyptian culture. In the case of Hadrian these ideas even impacted on his visual

---

[24] Isaac 2004 and 2011.

identity through his adoption of facial hair and beard, and on the appearance of other elite Romans who then slavishly followed this imperial fashion and on that of many philhellene emperors in the centuries that followed.

It is on the frieze adorning Trajan's Column that as a counterpoint to the numerous scenes of battle against the Dacians we encounter the presentation of images of peaceable provincial citizens, probably in Moesia, images that are more or less otherwise unique in Rome. In Scene XXXVI the emperor appears entering a provincial town with his entourage and army en route to the war. He is met and greeted by the town's citizens-men, women, and children-in an enthusiastic display of homage. This peaceable scene contrasts sharply with the scenes of open warfare and bloodshed elsewhere on the column, setting up for the viewer a striking contrast between civilisation, under Roman rule, and barbarism, on and beyond the frontiers. As Natalie Boymel Kampen has pointed out this scene stresses the importance of provincial families and the guarantee of peace and protection that the Romans could provide for them. The lives of these people had been transformed under Roman rule, their prosperity and safety vouched for by Roman power.

It is interesting to consider examples of the use of images to identify and single out Jews in Roman art, examples involving not the use of images and symbols for self-identification by the Jewish community in Rome or elsewhere in the empire, but rather in official contexts where the identification might have carried some specific message or fulfilled a particular purpose.

The Roman triumph involved a processional parade which might have included in its train a defeated enemy king or general, captives, both military and civilian, spoils of war, exotic animals, and inscriptions of foreign place-names or battle paintings depicting foreign landscapes and topography. Painted bird's-eye views of foreign landscapes allowed viewers to observe a scene from a perspective which would have made them aware of his or her earthbound limitations. There was no event in Rome like the military triumph in which such a diverse number and range of geography products was employed, often in what must have been sometimes bewildering juxtapositions, suggesting contradiction but eliciting positive responses we can assume. Ovid notes that at some triumphs viewers had problems identifying some of the figures of geographical or topographical personifications on parade here, even despite the use of identifying placards and other explanatory schemes.

At Julius Caesar's four triumphs of 46 BC images of the personified *Gallia*, the city of *Massilia*, the marine god *Oceanus*, and the three great rivers the Nile, Rhine, and Rhone appeared. Some triumphal representations may have been painted, others may have taken the form of statue personifications. Actors may have played roles. In the case of representations of cities, small three- dimensional models sometimes may have been carried and displayed: Caesar was said to have ivory models at one of his triumphs while Q. Fabius Maximus had wooden models. Sometimes the city models were much larger, like tower models in a number of stages like a tiered cake. A fragment of a relief from Cherchel, Algeria shows a late Roman triumph in progress, with a large model of a bridge being carried in the procession on a *ferculum*. A caption on a displayed *titulus* names this as *Pons Mulvius*, the Milvian Bridge where Constantine later defeated Maxentius in battle in AD 312.

Figure 97. Inner relief depicting Titus' Judaean triumph, Arch of Titus, Rome. After AD 81. (Photo: Author).

The Judaean triumph of Vespasian and Titus in AD 71 is of particular interest, not only because we have an account of the triumphal procession from the historian Josephus[25] but because parts of this triumph were also represented visually on the carved reliefs adorning the Arch of Titus in the Roman Forum (Figure 97). This coming together of description and visual image remains unique in the context of the analysis of the triumph as an imperial ceremony. This was not just emotional grandstanding, spectacle for its own sake: rather, it had a deeper purpose-to instil at least an idea of geographical knowledge into the minds of the triumph's viewers.

The sacred menorah or candelabrum from the Second Temple in Jerusalem was very clearly depicted as part of the spoils on one of the inner reliefs on the Arch of Titus,[26] and was later displayed as a prize in the Temple of Peace in Rome as one of its treasures of the Roman empire. The Flavian *Judaea Capta* coin types relayed quite stark information to Rome's citizens, including its sizeable Jewish population, about their homeland's subservient relation to the power of Rome. Less blatantly, one particular coin issue of Nerva seems to have been referencing the lifting of Vespasian's Jewish tax, in this case bodying forth *Judaea* not in the

---

[25] Josephus *Bellum Judaicum* 7.132-160.
[26] On Rome and the Jews see, for example: Beard 2003; Fine 2016 and 2017; Grey and Ellison 2022; Millar 2005; Östenberg 2021; Rocca 2017; Shahar 2004; Smith 2017; Yakobson 2021; and Yarden 1991.

form of a captive figure in the form of a female personification, but rather as an image of a palm tree which would have been well understood by its contemporary viewers.

It has been suggested that 'a copy of Jewish Law' was also paraded among the spoils in the Judaean triumph, and it would seem that while clearly focused on the victory in *Judaea* this triumph also strove to celebrate broader aspects of Roman imperial power. The contemporary historian Josephus wrote that 'the magnitude of the Roman empire' was advertised, along with 'the wonderful and precious productions of various nations'.[27]

But the Judaean triumph celebrated in Rome and monumentalised on the Arch of Titus was completely unique, an exception and anomaly, in that it was the only triumph ever held to celebrate the defeat of a provincial population, in this case a section of the province's population in revolt against Rome, rather than the conquering of a foreign enemy or barbarians. For this very reason it is necessary to consider the account of the Judaean triumph in order to gauge whether the exceptional circumstances were in any way reflected in the triumph being markedly different from others or whether the triumph was used to communicate different things to its viewers than communicated to viewers of other triumphs. It also follows that the artworks on the monumental Arch of Titus need re-evaluating from the same standpoint.

Josephus described the triumph: '..nothing in the procession excited so much astonishment as the structure of the moving stages.......many of them being three or four stories high…..The war was shown by numerous representations…affording a very vivid picture of its episodes. Here was to be seen a prosperous country devastated, there whole battalions of the enemy slaughtered; here a party in flight, there others led into captivity, walls of surpassing compass demolished by engines, strong fortresses overpowered, cities with well-manned defences completely mastered and an army pouring within the ramparts, an area all deluged with blood, the hands of those incapable of resistance raised in supplication, temples set on fire, houses pulled down over their owners' heads, and, after general desolation and woe, rivers flowing, not over a cultivated land, nor supplying drink to man and beast, but across a country still on every side in flames'.[28]

The images displayed at the Judaean triumph and the other visual and sensory elements of the spectacle and its after-commemoration comprised what have been called throughout this book geography products, but their value or essence as carriers of geographical and chronological truth was severely compromised by the discourse of reality around the Jewish Revolt and its crushing. That *Judaea* was now being claimed as *capta* implied that the Romans viewed the revolt as so serious that they considered the province altogether lost. This new Year Zero event, the creation of a new Roman *Judaea*, turned the clocks back, whitewashing history. As a geography product the *Judaea* of the triumph and its commemorative coins and monuments was a new place, yet it was still the exact same place that it had been before the revolt. Viewers were being expected to question their previous knowledge of the province and indeed to participate in a realignment of historical and geographical consciousness. In contrast, we can ask why the defeat of the Boudiccan rebellion in Britain was not celebrated by a triumph and a suitably bellicose formal monument either in Britain itself or at Rome?

---

[27] Josephus *Bellum Judaicum* 7.132-137.
[28] Josephus *Bellum Judaicum* 7.139-145.

An inscription[29] on the second victory arch of Titus close by the *Circus Maximus*, now not extant, states that Titus 'subdued the race of the Jews and destroyed the city of Jerusalem, which by all generals, kings, or races previous to himself had either been attacked in vain or not even attempted at all.' A few fragments of reliefs depicting battle scenes might have come from this monument. Rome was itself spatially affected by the building of monuments to the Jewish War, the geography of distant *Judaea* imprinted on the fabric of the city, not only in the form and mass of the two arches but also by the founding of the Temple of Peace discussed elsewhere in this book, and the construction of the great Flavian amphitheatre or Colosseum. The Colosseum is said by an inscription to have been made from *manubia*-'the spoils of war', referring to the putting down of the Jewish Revolt. Josephus tells us that 97,000 Jewish rebels were taken prisoner, and that a significantly large number of these would have been sold, enslaved, and forced to train and fight as gladiators: indeed, he noted that two and a half thousand Jewish gladiators fought and died in the arena at *Caesarea* alone, an extraordinary figure. Thus the Colosseum could itself be said to be a kind of geography product, its dedicatory inscription relating its building to war in a distant land and to spoils from that war, while for some considerable time the playing out of deadly gladiatorial combats involving Jewish enslaved gladiators, until all had died, acted as recurring signifiers of the Jewish War and the rewinning-'reconquest'-of *Judaea*: thus these men themselves unwittingly became active agents in the promulgation of geographic-historical knowledge, reminders of an event and of a distant place.

The destruction of the temple in Jerusalem added a religious element to the victory that allowed Jupiter-the major Roman god and the one most associated with Rome's emperors-to be also presented as a victor of a kind. The image of the menorah on the Arch of Titus represented not a generic trophy object-like a Gallic carynx or trumpet on the Arch of Tiberius at Orange in Gaul for instance-but rather one very specific item whose capture here as an image mirrored its physical capture as a piece of highly-symbolic and very valuable war booty in Jerusalem. In addition, the large gold table for shewbread, the temple veil, and a Torah scroll were also paraded at the triumph. Steven Fine has examined a number of aspects of the reception of the triumphal image on the arch, including the long-held but mistaken idea that the menorah was being carried by Jewish prisoners and not by Roman soldiers, and how the viewing of the image became a focus for Jews travelling to Rome in the later twentieth century.[30] He has also considered whether the accounts of Rabbinic scholars visiting Rome and viewing the Arch of Titus in the second century AD are likely to hold any truth. All of these discussions and digressions show that the image mattered.

In the relatively recently excavated first century AD synagogue at Magdala in Israel there was found a large stone table or table-like object, with decoration carved in low relief, including images of a temple menorah and other religious items, while a large floor mosaic at the fourth century AD synagogue at Hammat Tiberias includes a register of images flanked on either side by a large menorah. Are we seeing here in both cases instances of the symbolic recapture of the Jerusalem Temple menorah that was carried off to Rome, with its reappearance on home-soil in the form of potent images? If this were to have been the case it represents a unique

---

[29] *CIL* 6.9.44.
[30] Fine 2016.

battle of images between centre and periphery, between Rome and the provinces, between the conquerors and the conquered, between the material and the visual.

The Centre Cannot Hold

But it was not only images of non-Roman people that would have been all too familiar to the city's inhabitants. It is of course highly likely that most citizens at Rome would sometimes have viewed or come into contact with foreigners, most commonly merchants and traders and in the form of the city's thousands of enslaved individuals. Some might also have seen foreign ambassadors and their entourages, perhaps hostages (many of whom actually had freedom of movement within the city), exiles and the expelled seeking Roman aid or protection, and elite prisoners.[31] Rome also saw visits from foreign tourists as we might call them. Foreign ambassadors and their entourages, if accepted as formal embassies, came to Rome ahead of February each year in a well-organised system that saw them accepted (or not in some cases), housed, and received. From the second century BC onwards numbers of such embassies increased as Rome's reach and power grew exponentially. It has been suggested that the number of such delegations arriving in Rome was actually quite large in the middle and late Republic, the majority in the crucial years 300-50 BC, with a minimum of c. 1080 such embassies visiting in these years. In 59 BC alone at least 35 such groups were received in the city. As an example, the geographer Strabo wrote about the presence of a number of British tribal leaders in Rome in Augustan times (4.5.3). Why were they there? How did they get there? What other reporting did their presence occasion?

Hostages were certainly paraded in many triumphs but others were forced to remain and live in Rome, sometimes for a short while, sometimes for many years. The Greek historian Polybius is probably the most famous of these elite hostages, along with certain of the young children portrayed on the *Ara Pacis*. Hostages were usually single individuals, though known exceptions included the hundreds of Carthaginian hostages demanded after the Roman victories at the Battle of Zama in 202 BC and in 149 BC at the destruction of the city of Carthage itself.

While one can find many examples of offensive and defensive Roman behaviour towards non-Romans in the city of Rome itself, of prejudice and malice, there is perhaps more evidence for a city that in order to function needed to reflect the diversity and difference of the peoples of its empire rather than necessarily embrace this whole-heartedly. Incorporation and enfranchisement, along with manumission of slaves, became the tools to deal with the commodities that the Romans had made of their subject peoples. Incorporation could also include the introduction of foreign gods and goddesses into the Roman Pantheon, the practice being best illustrated in this case by the bringing of statues and images of foreign deities to Rome, such as those of Juno Regina and Cybele, the *Magna Mater*.

Both positive and negative reports about Rome's cosmopolitan population can be found in the writings of many ancient authors. Athenaeus and Pliny the Elder both wrote about Rome as a model of its own empire. Juvenal wrote of the Syrian Orontes flowing into Rome, as a

---

[31] On visiting foreign delegations in Rome see principally: Westall 2015. On the Roman triumph see, for example: Beard 2001 and 2007; Brilliant 1999; Holliday 1997; Kleiner 1991; Östenberg 1999, 2009a, 2009b, 2014, and 2021; and Östenberg *et al.* 2015.

way to depict the influx, sometimes forced, of outsiders. Ovid in his *Ars Amatoria* wrote about Augustus' *naumachia* or naval games bringing the world into the city.[32]

Tacitus tells us that Claudius made a dramatic speech to the Senate in AD 48 promoting the acceptance of the admission of Gauls into the Roman Senate, while the scabrous historian Suetonius refers to a song about Gauls going from the triumph to the Senate. The snobbish Seneca describes the failed Senate proposal to enforce a compulsory dress code in Rome in order to allow people to distinguish free men from slaves by their attire. Varro provides the advice that it is better to source one's slaves from a number of different places in order to avoid fighting which he suggests will result from having a number of slaves all from the same place. While Cicero may have written that 'denying foreigners access to our city is patently inhumane', there were still incidents of mass expulsions from the city, like the round-up of Jews in 139 BC for being 'not suitable enough to live alongside Romans'.[33]

Rome would not of course have been the only city or town in Italy to have a diverse population, with a multiplicity of languages heard on its streets. While it is incontestable that Latin was the official language in which the Roman empire was governed, many other tongues were spoken, with Greek predominating in the eastern empire. Studies of the *graffiti* of Pompeii suggest that a multicultural community here, permanent residents and those just visiting or passing through, were at least united in the cultural practice of inscribing and writing *graffiti*. Very recently, Kyle Helms has suggested that the eleven examples of *graffiti* on one of the walls in the town's theatre tunnel written in Safaitic script/letters could have been produced by soldiers in the Third Gallic Legion recruited from the nomadic tribes of Harrah in Syria, rather than being inscribed by merchants from the region as has been previously suggested.[34] What is interesting about the Safaitic *graffiti*, whether done by soldiers or traders, is that it was in a number of ways subversive and allows us to identify and isolate a few moments in time when the Roman custom of making a mark using *graffiti* was co-opted by others from a different cultural tradition to literally inscribe their geographic origins on the fabric of the Roman world.

This diversity in the city of Rome was also reflected in the goods and produce from overseas either shipped to Rome or grown there after discovery elsewhere. Statius[35] lists the exotic foods eaten at the *Saturnalia* feast given by Domitian at the Colosseum and a similar listing of abundance is satirically conveyed in the description of Trimalchio's legendary feast in the *Satyricon* by Petronius.

There is a broader issue here relating to geographical provenancing of goods and objects as an economic phenomenon in Roman times[36] which shows that today's western middle class obsessions about which terroir our wines came from and which grower produced our olive oil is not actually a new thing. I have written at length in a previous book-*The Dignity*

---

[32] Athenaeus *Deipnosophists* 1.36; Pliny *Naturalis Historia* 7.2; Juvenal *Saturae* 3.62 ; and Ovid *Ars Amatoria* 1.171-176.
[33] Tacitus *Annales* 11.24; Suetonius *Iulius* 80.2; Seneca *De Clementia* 1.24; Varro *De Re Rustica* 1.17.5; and Cicero *De Officiis* 3.11.47. On the Roman senate and multinationalism see: Eck 2021.
[34] On the Safaitic *graffiti* in Pompeii see: Helms 2021.
[35] Statius *Silvae* 1.6.
[36] On imported foodstuffs see, for example: Brandt 2005; Cobb 2013; Evers 2017; Gowers 1993; McClaughlin 2010; Parker 2002; Rowan 2019; Totelin 2015; and Young 2001. On Roman trade with Han China see, for example: McClaughlin 2010; McClaughlin and Kim 2021; and Young 2001. On the idea of Rome as an empire of plunder see: Loar *et al.* 2017.

of Labour. Image, Work and Identity in the Roman World-[37] about the stamping or paint-labelling (with so called *tituli picti*) of many wine and olive oil amphorae with estate owners' names and sometimes more rarely their actual locations in Roman times. The linking of a product with a place was obviously commonplace at the time, for many such named items to have appeared in the emperor Diocletian's Edict on Maximum Prices of AD 303. A good example is the *birrus Britannicus* or British wool cloak. Provenanced goods and objects could also be geography products.

When away from Rome there were ways in which objects creating a geography of displacement could provide links to home, to the homeland. Standardisation was one such strategy and can be thought to have included the template grid-system for laying out military forts, even down to the regularisation of positions of buildings, a similar grid-template for the planning of new towns and settlements such as veteran *coloniae*, and the attempted standardisation of measurements of distance and weight which represented an attempt to unite across distance rather than divide. Every town had similar types of buildings, such as a forum, civic and monumental arches, public bath complexes and so on. While my experiences only allow me to generalise about the situation with regard to domestic material culture in the north-western provinces, it is interesting to note the incredible similarity between ceramic assemblages from different provinces here, with Gaulish samian pottery, mortaria, cooking pots and amphorae all being remarkably common across a broad geographical area. The same feeling of deja vu must have been evinced when travelling between provinces and encountering more-or-less the same coinage, even if produced at different mints. The same gods occurred across different provinces. The image of the emperor appeared everywhere, leaving aside whether this was a phenomenon driven from above or from below, or more likely a combination of both. Some provinces seem to have been keener on displaying imperial likenesses more than others. In Roman Britain, though we are told that the image of Titus was everywhere, the overall number of imperial images that have come down to us from the province is small: probably a reflection of Romano-British cultural preferences, priorities, and choices. To give one contrasting example, from the small provincial town of Thugga/Dougga in Roman North Africa over forty statue bases for imperial statues have so far been recorded. The Roman poet Ovid wrote from his place of exile at Tomis on the Black Sea that when looking at statuettes of Augustus he 'seems to see Rome itself, because the emperor sustains the face of his homeland'.

In these ways the banal omnipresence of things was enhanced by the power of repetition, of copying, and of attempts at standardisation. Welcome home. Images could travel as well as objects, as can be seen by the trends for copying, duplication, and replication so ingrained in Roman art practice. Such 'travelling' of images around the Roman world is well attested in the case of the *Tellus/Roma* image, while the Aeneas image from the *Ara Pacis Augustae* in Rome can also be seen in a similar version on an altar from Carthage in North Africa which is now in the *Musée du Louvre* in Paris.

The epigrammatist Martial[38] wrote about the vast, ethnically-diverse range of people present at the opening games of the new Flavian amphitheatre. There is no doubt that the Roman arena was a vital venue or *locus* for the managed presentation of certain types of geographical information, most often relating to the origins of the animals used in the *venationes* or

---

[37] Ferris 2021b.
[38] Martial *Liber Spectaculorum* 3.

wild beast hunts there. Again, the presentation of geographical information occasionally occurred in the staging of the fatal charades that constituted mythologically-informed, very theatrically-presented executions. However, there was also the involvement in the arena games of certain stereotypes of gladiatorial fighters, whose origins lay in deep Roman political history and whose involvement in the games was part and parcel of the ideological traditions underpinning the events. There were many individual types of gladiator, each defined by a technical name, such as the *retiarius*, the *secutor*, or the *murmillo*, each of these types of fighter being distinguishable by their costumes, equipment, arms, and style of combat, the latter of course predominantly dictated by the weapons assigned to each type of gladiator. But of great relevance here is the fact that there were also a number of gladiatorial ethnotypes,[39] such as the *thraex* or Thracian, the *samnis* or Samnite, the *gallus* or Gaul, and the *hoplomachus* or Greek hoplite. Each of these gladiatorial ethnotypes had its origins in specific events in Roman history.

But the main point to make here is that though members of the arena audience would all probably have been aware of the names of these four types of gladiator, they might not necessarily have been able to make the geographical and historical links which underpinned the names. Indeed, it is likely that on most occasions the gladiators fighting under these names were in fact of other ethnic origins to the ones suggested by the names and manifested in the distinctive types of costumes. In other words these were costumed fighters, dressed almost like actors, portraying these ethnotypes without necessarily being connected to them in any way. In the case of the continued and regular use of these ethnotypes we are seeing here not only the appearance of each ethnotype as a trigger to memory with regard to geographical origins, but also at the same time as a way to embed certain historical memories in the same ideological programme.

Another entertainment format popular at Rome which lent itself to the situating of what is being called here performative geographies was the *spectaculum* or spectacle, in some instances a harmless quasi-theatrical event and on other occasions an event that mixed spectacle with retributive justice and abject, sadistic cruelty. On a number of occasions it is recorded that the interior arena of the Colosseum was waterproofed, dammed up, and then flooded in order to turn the space temporarily into a *simulcrum* of the Mediterranean Sea on which mock naval battles could be fought for the entertainment of the masses.

In this chapter it has been suggested that images of barbarians on imperial monuments at Rome could have been in many cases deliberate carriers of geographical information, even if largely in a negative form. Images of barbarians were also used on monuments outside of Rome where it must have been the case that such images were deployed in such contexts to put across a different type of geographical information to that given by such images in Rome itself. This was a kind of visual typography, a strategy for giving meaning to images and words, and a method for expressing cultural values, a way to craft and exploit the images as outer shaping-devices that worked upon a viewer to achieve impact and effect. The images offered a particularly poetic instance of the *topos* of ineffability, an attempt to grasp at the infinite, the indescribable, the unutterable, and concretise it into something relatively mundane, where the idea or form became as significant as its content.

---

[39] On gladiatorial ethnotypes see principally: Caldelli 2001; and Janrović 2014.

If something was to have appeared too alien it would have failed to register with its recipients: if it was to be too easily recognised, too easily cognisable, it would never have been more than a reiteration of the already known. Some of these images therefore seem flooded with a grief so old and worn that it must have seemed unearthed, like a fossil of other older, singular, and more potent griefs. The notion of perspective needs to be a recurring motif when examining the images' relationships as they evolved over time, with their vantage points recalibrated by major events and the passing of time. There can be discerned and identified a difference between a mysterious or alien other and an intimately familiar one.

Certainly it would seem that there existed in Rome a visual narrative about non-Roman women that constituted a kind of psycho-sexual geography, as indeed there was also one about non-Roman men. Art such as this was fundamental to the articulation of Roman discourses on ethnic difference. If geography then was a racial-sexual terrain through which women passed, women's bodies would appear to have been considered integral to the production of such spaces: erasure and despatialisation helped question the sense of place. In such instances it is worth considering if bell hooks' idea that an 'oppositional gaze' against the imposition of dominant ways of knowing and looking could be turned on troubling images and be applied to the study of ancient Rome.

If the project of defining Rome's identity and place in the world involved the creation of an archive of geography products around the city we might expect to be able to discern the violent psychic tremors that would have accompanied the editing and maintenance of such a weighted ideological resource, and indeed we perhaps can. The idea of landscapes of destruction or annihilation fits in to this scenario of manipulation: the destruction of Carthage by the Romans and of other towns and villages in whole or in part is well attested; and the deliberate erasure of a whole people or tribe is both attested as historical fact and as a stated Roman war aim or wish. For example, Nero demanded that the *Silures* of Wales be destroyed, referencing the elimination of the *Sugambri* in Gaul. Domitian intended something similar for the *Nasamones*, a Berber people of Libya. The wiping out of a name, its erasure, implied clearing a territory, making it an empty space without the geographic specificity that a resident tribal name gave it. Making it nowhere.

In the next chapter an examination will be made of journeying in the Roman world and in the Roman imagination.

Chapter Nine

Moving Away from the Pulsebeat

The literary, artistic, and even musical theme of journeying is as much about the nature of home as it is about being away from there. The epic journey in Greek and Roman literature was all about the discourse around leaving, travel, and returning home, however that might have been defined. Home may be a fixed spot, a geographically precise spatial location, but it can also be represented by a transportable set of cultural values, making it both a physical journey and an intellectual process. The idea of the journey as both a presentation of geographical reality-moving through space and time-and as a metaphor or, more often, a series of metaphors, was very powerful in Roman times. For instance, the regular presentation of the image of Bacchus/Dionysus as a traveller was often calculated to introduce the viewer to the concept that exposure to the unfamiliar, the foreign, or alien, could bring new knowledge, could be revelatory in an almost mystical sense.

The god of wine and viticulture and of ecstatic celebration, Bacchus/Dionysus was famed for his wanderings which legend had it took him as far as India: in the footsteps of Alexander the Great[1] (Figure 98). The Indian Triumph of Bacchus became a popular subject in Roman art, particularly on sarcophagi, the two best examples of which come from Lyon in Gaul, now in the *Musée Lugdunum* there, and from the *Via Salaria* in Rome, and now in the Walters Art Museum in Baltimore in the United States. Both bear similar scenes of a triumphal procession involving Bacchus and his followers, the Maenads and the Satyrs, and his companion animals. While viewers of these scenes would probably have understood that they were meant to represent the god's Indian triumph simply from the general appearance and composition of the image, the artists in each case had added some specific geographical detail to help the viewer further locate the action, in both cases not necessarily accurately. Images of the labours of the hero-god Hercules were also linked to the broader trends and tropes in Roman geographical art. Certainly the settings for the Twelve Labours were spread over a wide geographical area, including of course the hero-god's excursion into the Underworld.

But journeys within the Roman world were not only made by gods, mythical heroes, and emperors. Journeying within provinces and between provinces can be considered to have been commonplace. But who travelled where and why? And how did the experience of travel translate into geographical knowledge which could be disseminated at an individual level: through writing letters home, through writing travel accounts for private or public consumption, and through orality-that is conversations and travellers' tales?

Many of these journeys would have been made informed by pre-prepared itineraries, so in a way were itineraries made real. Greek geographical writings encompassed four main types of travel narrative, that is: *periploi*-seafaring routes; *itineraria*-linear land routes; *periegesis*-surveys of land, peoples, or topography; *chorographia*-descriptions or details of specific places; or a mixture of these. Some of these writings when originally circulated would have been accompanied by illustrations and probably maps of some kind, and would have been widely used in the Roman world.

---

[1] On Dionysus/Bacchus and geography see, for example: Chuvin 1991; and Jaccottet 2020.

A Map of the Body, a Map of the Mind

Figure 98. Marble sarcophagus depicting the Indian triumph of Bacchus, Rome. AD 260-270. Metropolitan Museum, New York. (Photo: Copyright Metropolitan Museum).

Odyshape

Journeying, in terms of defining origins and thinking about getting from there to here, was the business of Roman historians, emperors, and politicians, creating interlinked and sometimes overlapping narratives that linked Rome and Romans to the Trojans, the Etruscans, and the Sabines in a seamless way that focused on the creation and mutability of the Roman identity, and not on that identity itself. As a result of this bombardment with many geographical messages in images and texts there would have been what we can call an exchange of emotional information. The viewer or reader would have been intended to feel 'that is me' in the world I inhabit, this is what things are like. Individual identities were manipulated in the slipstream of the forging of a national or cultural identity.

Certain mythological journeys or stages on those journeys also featured regularly in Roman art, and in this respect one can think of the journeys of Odysseus, Aeneas, alongside Bacchus/Dionysus and Hercules. In certain contexts the journeys of these figures could be taken to represent concepts as varied as the journey of the Roman people from city state to empire, and of the individual, from birth through life to death. In many ways the geographical information put across in the illustration of these journeys was simply incidental detail and in some cases was actually factually incorrect.

Homer's *Odyssey* is remarkable in that not only are Odysseus and his companions in transit from Troy back home eventually, and certainly not directly or easily, to Ithaca, but en route there, and in the process they encounter many other wanderers of one kind or another, including merchants and traders, adventurer explorers, and criminals in the form of pirates. This is a world of people on the move: this portrayal of intense mobility in the Mediterranean surely reflected the contemporary situation, and we must not think of mobility around the Mediterranean and Aegean seas as being a phenomenon of the Roman period alone.

If I might borrow the word *Odyshape* from the 1981 LP of the same name by the Raincoats this term can usefully be used to invite metaphorical parallels between the travelling (female) body and the non-linear structure of Homer's *Odyssey*. If the travelling epics such as the *Odyssey* and the *Argonautica* represented a genre associated with men in movement, while women either waited at home or were encountered en route, then it is perhaps surprising to read of a number of travelling women in Virgil's *Aeneid*. As pointed out by Alison Keith we encounter here the travelling Dido, Andromache, and other Trojan women.

Basil Dufallo has written at length about how certain Roman Republican poets, principally Horace, Plautus, Terence, Lucretius, and Catullus, reacted to the rapid expansion of the Roman empire in the middle and late Republican period.[2] In the literature of the time, indeed over an extended period of some two hundred years, themes of wandering, of being somehow lost, came to embody broader societal worries about disorientation, and were explored not only through the most obvious tropes of geographical wandering but also less obviously through exploration of the self, the inner spaces of identity, sexual orientation, gender, and change. Disorientation and reorientation seem to have gone hand in hand, and the field of contemporary writing then was not dominated by anxiety. Wandering, roaming, and straying came to signify literary strategies for *finding* something, for seeking reassurance in the fact that getting lost at a time when Rome's power and empire extended into ever-increasing territories, whose strangeness or otherness could now be tamed or changed by conquest or Roman soft power and geopolitical manoeuvrings, was not a negative trope.

In Dufallo's words 'exposure to the attractions of new regions, cultures, and peoples could problematize one's sense of who one was…it could help emphasise the impressive size of Rome's growing empire, the need for Roman power as an organizing force, and the social, cultural, and political potential for expanding into these regions'.[3]

---

[2] Duffalo 2021.
[3] Duffalo 2021: 1.

Latin elegy as a distinct body of literary work, principally the works of Catullus, Propertius, Tibullus, and Ovid, spans the period from the end of the Republic, the territorial acquisitions of Pompey and Caesar, to the extensive land gains for the empire under Octavian/Augustus.[4] This relatively short timespan of seventy or so years saw the Roman mentality change, or be changed, from *having* an empire to *being* an empire, with all that such a change of mindset entailed. In many ways it raised the fundamental question about what it actually meant now to be Roman.

In one of his short Epigrams[5] the poet Martial, who lived AD 40-104 and who was born in Spain, very tellingly provided a virtual catalogue of thirteen notable 'Roman' writers-this should more properly be of writers writing in Latin-none of whom came from Rome, although Martial did not explicitly make this point. Virgil came from Mantua (Mantova), Livy and the writer Stella (none of whose poems survive) from Padua, and Catullus from Verona, all in northern Italy, Ovid came from Sulmona in the Abruzzo region in central southern Italy, Horace (who he calls Flaccus) from southern Italy, and the two Senecas (Younger and Elder), Lucan, Canius, Decianus, and Martial himself (modestly noting 'will not altogether be silent concerning me') all came from Spain, and Apollodorus from Egypt. The epigram was dedicated to Licinianus, also from Martial's home town in Spain, whom Martial hoped would one day be celebrated as a writer alongside these greats. History suggests this never happened. The compilation and presentation of this list marks out a subtle literary geography of a kind, a very deliberate setting out of a very broad definition of what it meant to be Roman at this time. The geography of empire was written through their works and onto them all. Martial's boast about his own assured fame very clearly and deliberately linked him with Bilbilis in Spain, but by placing himself as an equal in a list with, for instance Virgil and Ovid, he has telescoped distance to make the Roman literary world an overlaid plan on the map of empire.

Perhaps the end point of this literary search for soothing lostness was marked by the plot of Virgil's *Aeneid*: a Trojan returning 'home' to the site of Rome through former Greek territory that now fell within Augustus' empire, yet much of which was alien, hostile, and unwelcoming. The irony of all this was surely not lost on its audience of readers. The wandering hero, personified best in the figure of Homer's Odysseus, had been a popular and recurring type in Greek myth and we must not lose sight of the fact that seeking to come to terms with pan-Mediterranean mobility was not something experienced solely by the Romans and in their time alone. Gods wandered too, and the figures of Bacchus/Dionysus and Hercules/Herakles made journeys that allowed them to transcend time, place, and space in the classical imagination. As already noted, Bacchus/Dionysus travelled in India, like Alexander the Great, and often featured in an image of triumph there, even though the country lay outside the Roman empire.

As Jupiter declared in the *Aeneid*: 'For these Romans I have set no bounds in space or time: but have given empire without end',[6] a message tailored to fit well with Virgil's political patron Augustus. Perhaps one of the most famous pieces of Roman geographical writing is

---

[4] On literary journeys and journeying see principally: Duffalo 2021. See also, for example: Biggs and Blum 2019; Blum 2019; Burton 2013; Chinn 2022; Fitzgerald and Spentzou 2018; Fletcher 2014; Green 1977; Hardie 1986; Hinds 2002; Johnston 2019; Keith 2014; Kondratieff 2014; Lindheim 2021; Mayer 1986; Pandey 2018; Rimell 2015; Rimell and Asper 2017; and Skempis and Ziogas 2014.
[5] Martial *Epigrams* 1.61.
[6] Virgil *Aeneid* 1.278-279.

Figure 99. Relief depicting scenes of the Trojan War from the *Iliad* (one of the *Tabulae Iliacae*). First half of first century AD. Metropolitan Museum, New York. (Photo: Copyright Metropolitan Museum).

Virgil's description of the shield of Aeneas in the *Aeneid* which rather than simply being a piece of descriptive writing bodying forth the image of the object is rather a discussion of problematised contemporary geopolitical matters brought to life by *ekphrasis*.[7] Made for him by the smith god Vulcan the shield carried a series of premonitory scenes relating to future events. The images on the shield included: Romulus and Remus with the She-Wolf; the Rape of the Sabine Women; the execution of Mettius Fufetius, king of Alba Longa who had served to betray the Romans; the siege of Rome by Lars Porsena; the attempted sacking of Rome by the Gauls which ended in failure following the discovery of their presence by cackling geese on the Capitoline Hill; dancing and ceremony; the Underworld; the great naval Battle of Actium, won by Virgil's patron, Augustus and his general Marcus Agrippa; and a peaceful and prosperous era for the city of Rome brought about by the triumphs of that self-same benign Augustus whose waging of wars across the Roman world had led to this great peace. The shield was here both like a map and a mirror, a surface for scrying.

Looking for roots and identity in mythological tales of journeying led many Roman emperors and others to actually set out on touristic journeys of discovery themselves. Indeed Hadrian became famed as the most travelled of all Roman emperors. A number of emperors chose to

---

[7] Virgil *Aeneid* 8.

travel to find Troy and explore the mythic roots of Roman culture there in person.[8] The city of *Ilion*-Roman *Ilium*- in Asia Minor, accepted then as the site of legendary Troy, received Roman imperial attention and often direct patronage from the time of Augustus onwards: it became a veritable tourist attraction for elite Romans keen to explore their city's mythic origins, even if they did not altogether buy in to the elements of truth that might have underpinned the fabric of the myth (Figure 99). For their sophisticated tourist market *Ilion* issued souvenir coins that carried images of Trojan heroes and key scenes from the Trojan Wars. In many ways this was the birth of what we would call today literary tourism.

When Augustus visited the city in 20 BC he did so perhaps out of intellectual curiosity: however, we might now see the visit as pivotal in the path towards the formulation of a strategy of claiming Trojan origins for Rome and its ruling class, and to present Troy/*Ilium* as the precursor or antecedent of the city of Rome. The financial sponsorship of major building schemes here revitalised the city, while monuments such as the the Palace of Priam and the Tomb of Hector became vital new landmarks in the reborn city. In the words of Margalit Finkelberg 'for all practical purposes, Troy was reborn.'

It is certainly worth dwelling on the Romans relationship with *Ilium*/Troy and backtracking to the origins of cultural and political tourism at the site. We know from the writings of Plutarch, Herodotus, and Arrian that both the Persian king Xerxes in 480 BC and subsequently Alexander the Great visited the site to see the Tomb of Achilles and the Tomb of Patroclus. Indeed, according to the Greek geographer Strabo, it was Alexander who later provided the funds for the building of the new temple to Athena there in 323 BC which was to become a focus for future visitors, alongside the more ancient monuments. These two powerful figures were not visiting to seek ancestral roots: rather they were in their different ways trying to connect to the heroic past and tales of heroic deeds by Greek heroes in Alexander's case, even to structure their visits around the consideration of the ebb and flow of historical time and circumstances. Roman visitors when they came were often looking for the same things, but with additional motives and feelings. Although by 133 BC Asia had become incorporated as a Roman province, much earlier, in 190 BC, M. Livius Salinator visited the site of Troy, followed later in the year by L. Cornelius Scipio according to Livy.

Roman imperial visits to *Ilium*/Troy continued as a phenomenon long after Augustus' visit of 20 BC. Germanicus came in AD 18, Nero in AD 53, the much-travelled Hadrian in AD 124, Caracalla in AD 214, possibly Constantine in the AD 320s when he was reported to have considered *Ilium* as a candidate for a new capital, and Julian in AD 355. This is actually an extraordinary list of visiting emperors which must have been rather more than simply symbolic, and must have represented some form of memory capture. It seems unlikely that the motives of Augustus, Hadrian, and Julian were identical for instance, and in most cases the emperors had to journey a considerable distance and for a long time to reach the city, though some 'took in' the site as part of broader journeys or regional imperial business.

Even more extraordinary still is Lucan's probably entirely fictional account written in the first century AD of Julius Caesar's visit to *Ilium*/Troy, a visit thought never actually to have taken place, though he just might have detoured here on his way to Pharsalus in 48 BC. Lucan's

---

[8] On tourism to Troy/*Ilion*/*Ilium* see, for example: Erskine 2001; Finkelberg 2021; Minchin 2012; Rose 2013; Rossi 2001; Sage 2000; and Vermeule 1995. Quote is from Finkelberg 2021: 94.

Caesar character is here to look for *his* ancestral Trojan roots and the Trojan roots of the Roman people, and there was much for him to contemplate at the site: *nullum est sine nomine saxum*-'no stone [here] is without an association'.[9] Lucan's creation of this fictional visit many decades in the past evidently sought to cement the mythical origins of the Romans as being at Troy, and in his account he created a fictive geography that dealt not only with spatial dislocation but with temporal dislocation too.

There are also a number of accounts of visits made by certain Roman emperors to the tomb of Alexander the Great, though today we do not know precisely where this was sited. Caesar famously was the first visitor to the tomb in Alexandria, Octavian (later to become Augustus) visited in 30 BC after his great naval victory over Antony and Cleopatra at Actium, Vespasian visited the *Serapeion* in Alexandria in AD 69 to consult the oracle there, and Caligula is recorded as visiting too. Cassius Dio recounted Trajan's visit to the ruins of Babylon in c. AD 116, and records show that Caracalla visited both Troy and Alexander's tomb in Alexandria. Hadrian visited the Tomb of Pompey in Egypt and had the tomb restored, while Septimius Severus also visited there.

Returning to Rome, each of these emperors in their own way transformed the reality of the city by imbuing Rome with memories of their visits and by reaffirming Roman imperial links with past glories and distant places. If we think of Hadrian as the great travelling emperor it is worth also considering the famous Eastern tour undertaken by Germanicus in AD 17-20, as related by Tacitus in the *Annales*. Germanicus, travelling with his wife Agrippina, first travelled to Dalmatia to visit his adoptive brother Drusus, then on to Nicopolis in western Greece, visited the site of the Battle of Actium (or rather viewed it from the coast), stayed on Lesbos long enough for Agrippina to give birth to their ninth child Julia, failed to carry out a planned visit to Samothrace, spent some time in the Black Sea region, visited *Ilium*/Troy and Colophon, spent time in Athens, in Euboea, in Turkey, visited Rhodes, went on to Syria, to Alexandria, and sailed up the Nile. Epigraphic evidence also placed him, Zelig-like, at Olympus, winning a chariot race in which the other participants were likely to have been holding back somewhat. Monuments to Germanicus in the form of arches on the Rhine and at Mount Amanus near Antioch in Syria, a cenotaph at Antioch, and a monumental bier at Epidaphne, the site of his death in AD 19, commemorate his travels.

In many ways the decorated helical frieze around Trajan's Column in Rome represented a record of a series of journeys, and depicted Trajan and his troops in a number of different and distinct geographical settings. This was less so in the case of the frieze around the Column of Marcus Aurelius. The scene in which Trajan and his army enter a Romanised provincial town and are greeted there by a crowd of citizens has been much commented upon, and indeed is unique in that it is one of only a handful of portrayals of non-Roman peoples in imperial art who are not combatants in wars against Rome or the defeated from those wars. Other examples include the crowd of men, women, and children who greet Trajan as he dispenses food alms in a scene on the Arch of Trajan at Benevento and the provincial citizens who greet Septimius Severus and his family on the Arch of Septimius Severus at Lepcis Magna in Libya. The Lepcis and Benevento crowd images would have conveyed a slightly different message to their contemporary viewers, most of whom would presumably have been local to each

---

[9] Lucan *De Bello Civili* 9. 973.

town, than the Trajan's Column crowd scene which was presenting this image to viewers in Rome itself. The fact that in each of the three examples discussed here crowds of people are depicted, indeed mixed groups of men, women, and children in all three cases, seems somehow of significance.

The great journey undertaken by Trajan as depicted on the column frieze would surely have had a certain resonance with those who had come to visit and view the monument when it had become Trajan's tomb, whether it had originally been designed for that purpose from the outset or not. With his cremated remains in a small room inside the base to the column and a gilded bronze statue of the emperor atop the column, placed here as if to allow his spirit to ascend to the heavens in apotheosis, the journey on the decorated shaft clearly linked the two most significant foci of activity at the monument.

Undoubtedly the Roman emperor whose mobility occasioned the most remark was Hadrian.[10] It has been calculated that of his 251 months reign, between 106 and 113 months were spent on the road, mainly, but certainly not exclusively, in the eastern empire, travelling either on horseback or in a mule-drawn vehicle. Three main imperial journeys took place, including the very extended tour of inspection he undertook on his way to take up the purple in Rome after being acclaimed emperor by troops in Syria in AD 117. Later, between AD 121-125 he travelled as far as northern Britain in one direction and the River Euphrates in the other. His third long trip took him to North Africa, the Near East, and the Balkans between AD 128-132. Numerous shorter trips were made, including many within Italy, where inscriptions testify to his imperial munificence in sponsoring and dedicating buildings at many urban centres. Politics and geography in the Hadrianic era were ideologically and conceptually linked. The history of the Roman empire during his reign was written on the body of his lover Antinous: the colonised body of the youth was here also a container of the colonised landscapes of Bithynia and Egypt.

The so-called travel coins of Hadrian (Figures 100-102) represent a remarkable series of images which mark the various stages of the emperor's travels around the empire as symbolic markers of the journeys made and the places visited, capturing *in stasis* the nature of the Roman attitudes towards and relationships with geography, place, and space at the time.[11] These reverse images break down into three categories: firstly a labelled personification with attributes; secondly the figure of the emperor appears sacrificing, accompanied by a personification; and thirdly the emperor is portrayed helping the kneeling personification of a province to get to her feet. Did anyone ever see the whole series of these coins and understand the overarching narrative around travel and discourse about geography and empire?

For the emperor his travels to Egypt and his time spent there with Antinous before the latter's death by drowning in the Nile were monumentalised in the building of the so-called *Canopus* at Hadrian's Villa at Tivoli, in the hills a few miles outside Rome. The *Canopus* at the villa at Tivoli was referenced by this name in the *Historia Augusta*,[12] the name either deriving from that of the Egyptian city of *Canopus* or, more likely, after its famous canal. This narrow, elongated pool

---

[10] On Hadrian and his travels see, for example: Birley 1997 and 2003; Boatwright 1987 and 1999; Destephen 2019; Grey 2016; Kammerer 1972-1973; Longfellow 2009; Speller 2003; and Syme 1988.
[11] On the travel coins of Hadrian see, for example: Kammerer 1972-1973.
[12] On Hadrian's Villa at Tivoli see, for example: Mari 2008.

MOVING AWAY FROM THE PULSEBEAT

Figure 100. Gold *aureus* coin of Hadrian, Rome mint, AD 130-138. Reverse of personification of Africa. One of the travel series of Hadrianic coins. British Museum, London. (Photo: Copyright Trustees of the British Museum).

Figure 101. Gold *aureus* coin of Hadrian, Rome mint, AD 130-138. Reverse of personification of *Aegyptus*/Egypt. One of the travel series of Hadrianic coins. British Museum, London. (Photo: Copyright Trustees of the British Museum).

Figure 102. Silver *denarius* of Hadrian, Rome mint, AD 117-138. Reverse of Hadrian raising the kneeling personification of *Gallia*/Gaul. One of the travel series of Hadrianic coins. British Museum, London. (Photo: Copyright Trustees of the British Museum).

probably also was intended to evoke the waters of the River Nile, and indeed Egyptianising statues of the twinned Nile and Tiber (Figures 103-104) and a crocodile (Figure 105) stood at the pool's edge, while colonised Egypt was further evoked by the caryatids supporting a structure around the pool's sides.

Of course many other journeys were made by Roman emperors or members of the imperial families,[13] journeys made in the service of the empire, for war, to inspect the military, to visit certain provinces, to be seen and to see, even in the exceptional case of the fourteen years old Lucilla, daughter of Marcus Aurelius and Faustina, who was sent to Ephesus in Asia Minor to marry Marcus' co-emperor Lucius Verus who was at war in Syria. The Antonine era has been described as 'a golden age of pilgrimage'. These imperial journeys have been gazetted and discussed together by Helmut Halfmann in his 1986 book *Itinera Principum*,[14] and the maps of each travelling emperor's route, progression, and peregrination provide startling evidence for the high level of mobility connected to imperial power and its link with Roman imperial ideology. These journeys also became in most cases memorialised and thus materialised through their recording and commemoration in written histories, on coins, in the form of honorific statues, of dedicatory inscriptions, and in the commissioning of numerous building projects from roads to fora. Quite a number of emperors died abroad, though some emperors never left Rome and in Late Antiquity some emperors seldom set foot in Rome. The first emperor never to set foot in Rome was Maximinus Thrax, proclaimed emperor by troops in AD 235 after the murder of Severus Alexander.

The experiences of imperial travel superimposed the symbolic upon the realistic, overlaying an imagined map of Rome with a map of its empire, in the same way that James Joyce overlaid contemporary Dublin with the geography of elsewhere in *Ulysses*, the Liffey becoming the Bosphorus through which Jason and the Argonauts had to sail. Rome's Tiber became the Nile, the Tigris, the Euphrates, and the Rhine.

The theme of journeying in Roman art and its links to imperial ideology is perhaps best exemplified and illustrated by the remarkable sculptural groups from an artificial grotto created from a marine cave at Sperlonga, near Terracina in southern Italy[15] (Figures 106-108). The grotto was part of an underground complex of a number of linked rooms associated with an above-ground palatial villa complex that general academic concensus considers probably belonged at one stage to the emperor Tiberius who is usually assigned the role of commissioner of these gigantic sculptures. As an aside, the grotto was enhanced by embellishment with fake stalactites and cave-like encrustations. The statues were found largely broken up and have been painstakingly reconstructed as four main, linked, thematic groups, now on display in the remarkable museum at the site: Odysseus and his companions blinding Polyphemus; Scylla attacking a ship and its crew; Odysseus and Diomedes with the Palladion from Troy; and a warrior cradling a dead companion, quite possibly Odysseus with Achilles. Undoubtedly Hellenistic in style, or harking back to that style, they were probably created by the three Rhodian Greeks whose names appear on a tablet at the site.

---

[13] On other imperial travellers see, for example: Damon and Lalazzolo 2019; Halfmann 1986; Jacobs 2002; and Joska 2022.
[14] For the presently-definitive list of imperial journeys see: Halfmann 1986.
[15] On the Sperlonga sculptures see, for example: Carey 2002; Conticello and Andreae 1974; and Ridgway 1997.

Figure 103. Statue of the personified Nile. AD 133-138. *Museo Villa Adriana*, Tivoli. (Photo: Author).

Figure 104. Statue of the personified Tiber. AD 133-138. *Museo Villa Adriana*, Tivoli. (Photo: Author).

Figure 105. Statue of a Nile crocodile. AD 133-138. *Museo Villa Adriana*, Tivoli. (Photo: Author).

Probably quite confusingly for the Roman people Tiberius's family, the Claudii, claimed descent from Telegonos, son of Odysseus and Circe, making the Sperlonga statue groups coherent in terms of familial mythology. As has been noted elsewhere in this study the geographical nature of Roman elite lineage was a significant factor in defining Roman identity. While Homeric, rather than Virgilian, and not directly stressing Trojan roots for the Roman people as Augustus had been keen to do, this work in its themes of journeying, discovery, sea venturing, and encounters with alien peoples, cultures, and places was very much of its time. If considered alongside contemporary Roman poetic works that also foregrounded journeying the choice of subject matter in the artistic programme makes complete sense.

Death's Echo

Stories about journeys to the Underworld in Roman culture and art were sometimes presented like a filmic space where distance and time had collapsed.[16] More often than not the protagonist set out on an epic journey that was both personal and yet somehow more broadly political, interior and exterior, often taking in confrontation with figures from the past and sometimes vaguely historical events. Impossible scenery implied a purely symbolic geography, resisting through its rigour any definitive or reductive reading.

---

[16] On journeying to the Underworld see, for example: Casagrande-Kim 2012; Felton 2018; and Gee 2020. On the *Cloaca Maxima* see, for instance: Gowers 1995; and Hopkins 2007 and 2012.

MOVING AWAY FROM THE PULSEBEAT

Figure 106. The ship of Odysseus, massive sculpture, Villa and Grotto of Tiberius, Sperlonga. *Museo Archeologico Nazionale di Sperlonga*. (Photo: Author).

As might have been expected in a society where the use, control, and propagation of geographical knowledge equalled power the construction of a 'geography of the Underworld', a kind of map or itinerary for the afterlife, took place in the Greco-Roman imagination. The Greek notion of *katabasis*-literally 'going down'-came to also include trips to the Underworld. In Greek myth Orpheus descended to the Underworld to bring back Eurydice, a tale retold in Ovid's *Metamorphoses*. In the *Odyssey* Odysseus likewise conjured up the Underworld to consult with Tiresias, and there he encountered a veritable host of dead souls, some related, some famous, some quite obscure, all of whom wished to converse with him. In the *Aeneid* Virgil had Aeneas actually descend into the Underworld in order to visit his father. The Romans enthusiastically adopted the myth of Persephone, the Roman Proserpina, once more retold in

A Map of the Body, a Map of the Mind

Figures 107-108. The blinding of Polyphemus by Odysseus, Villa and Grotto of Tiberius, Sperlonga *Museo Archeologico Nazionale di Sperlonga*. (Photo: Author).

Moving Away from the Pulsebeat

Figures 109-110. Sarcophagus decorated with the abduction/rape of Proserpina/Persephone, Rome. Third century AD. *Musei Capitolini*, Rome. (Photo: Author).

Figure 111. End panel of a sarcophagus, with Cupid as Charon rowing across the River Styx in the Underworld, Milan. Third century AD. *Museo Civico Archeologico*, Milan. (Photo: Author).

Figure 112. Detail from a marble sarcophagus, showing Hercules exiting the Underworld with the dog Cerberus, Rome. Third century AD. *Musei Capitolini Centrale Montemartini*. (Photo: Author).

Figure 113. Gold *aureus* of Caracalla, Rome mint, AD 214. Reverse of Serapis with the dog Cerberus seated at his feet to left. British Museum, London. (Photo: Copyright Trustees of the British Museum).

the *Metamorphoses*, along with the tale of Hercules descending to the Underworld to capture the three-headed guard-dog Cerberus. In another popular myth with the Romans Juno descended to the Underworld to recruit The Furies to wreak revenge on her behalf (Figures 109-113).

Through all these stories a complex and detailed topography and geography of the Underworld was created, what Roberta Casagrande-Kim has called 'chthonic landscapes'. All literary accounts agree on the same geography 'facts': that there were entrances, and exits, to the Underworld in the real world, although how many there were and where they were was contested; that there were four principal regions of the Underworld/Hades-Orcus, the Acherusian Plain, Tartarus, and Elysium or the Elysian Fields; that there were four main rivers- Cocytus, Acheron, Styx, and Phlegethon; that there were some primary landmarks linked to individuals like the so-called vestibule, the crossing-point of the Styx where Charon the boatman awaited, the cave of the giant dog Cerberus, the fork in the road between Tartarus and Elysium, and the palace of Hades at the entrance to Elysium; that the landscape around the Styx was marshy, that the landscape of Orcus was heavily wooded; and that character-wise Tartarus was harsh, unforgiving, brutal and punishing, and Elyssium a region of light, pleasure, and reward. There was less concensus about other geographic or topographical 'facts': that there were sub-regions such as *Sedes Beatae*, the *Lugentes Campes*, and *Nitentes Campi*; that there were lesser rivers such as Lethe and Eridanus, and numbers of feeder or branch streams; that different vegetational regimes were identifiable, with Virgil noting elm in one area, myrtles in another, and laurels in Elysium, with ulva seaweed on the shores of the River Styx. All accounts agreed that the Underworld was at the same time inaccessible and remote, reachable and permeable, near and yet far.

The complexity of this geography was deemed to need some kind of real world verification, a site or monument like the *Casa Romuli* discussed in an earlier chapter which served to site Rome's mythic past in its physical and material present, and it would appear that such a

Figure 114. The *Bocca della Verita*, Santa Maria in Cosmedin, Rome. Date uncertain, possibly as early as first century AD. (Photo: Author).

role was played by the identification of an underground cavern or room on the Palatine Hill in Rome as an entrance/exit to the Underworld. Festus, Macrobius, and Plutarch all make mention of this site. The *mundus Cereris*, with its *lapis manalis* or covering stone, was covered throughout most of the year but was uncovered on August 24th, October 5th, and November 8th for a day on each occasion in order to let the spirits of those below come up to mingle with those above on the hills and streets of Rome. Religious rites were carried out on each of the three opening days. Underground Rome, as best exemplified by the great *Cloaca Maxima*, was mapped beneath the city's streets (Figure 114).

If the Roman literary sources created a complex geography of the Underworld, as has been described above, then when attempts were made to visualise the place complexity gave way to a search for clarity, through the use of a reductive imagery that used a very small repertoire of topographical and architectural markers and significant identifiable characters to produce a 'visual synecdoche of Hades'. This is not to say that visual depictions of the Underworld or references and allusions to it were common: far from it. But when they appeared and were viewed in relation to literary accounts they would have conjured forth a completely realised geography of the place to literate viewers.

A generalisation made by Jocelyn Toynbee some years ago still holds true: that in Roman sarcophagus art scenes set in Hades most usually appeared as vignettes on the short ends of

Figure 115. The Velletri Sarcophagus. AD 140-150. *Museo Civico Archeologico Oreste Nardini,* Velletri. (Photo: slide collection of the former School of Continuing Studies, Birmingham University).

sarcophagi and often were images of punishments of individuals such as Sisyphus, Ixion, and Tantalus in Tartarus. Examples of both front and end images include a marble strigillated sarcophagus of c. AD 180 in the *Musei Capitolini Centrale Montemartini* in Rome which has on its front face at the centre a small image of Hercules emerging from a doorway dragging out Cerberus, while on one of the two short ends/sides of the AD 170 Protesilaus and Laodamia sarcophagus in the *Musei Vaticani* can be found on the right side an image of the punishments of Sisyphus, Ixion, and Tantalus.

Only a small number of other artworks relating to the Roman geography of the Underworld will be considered in detail here. Roberta Casagrande-Kim has argued that the well-known Roman funerary image of a pair of doors, though most usually interpreted as a depiction of tomb doors, can sometimes be considered as an image of one of the gates/entrances to or from the Underworld or between the four major regions of the Underworld, particularly if a deity or mythological figure stands by the doors or is entering or leaving through them. Hercules pulling the captured dog Cerberus through the door as just mentioned is a case in point. As noted by Casagrande-Kim, perhaps the best single example of an artwork creating a visual chthonic landscape of the Roman Underworld is the decoration on the Velletri Sarcophagus found near Velletri, some twenty or so miles south of Rome and now in the *Museo Civico Archeologico Oreste Nardini* in Velletri[17] (Figure 115). Dated to AD 140-150, though some date it as late as AD 175, this enormous gable-lidded marble sarcophagus is profusely decorated with

---

[17] On the Velletri Sarcophagus see principally: Marion 1965.

over eighty figures of humans or deities in high relief, many of whom, though not exclusively, have chthonic links. It bears no dedicatory inscription. A great deal of the academic discussion of the sarcophagus has focused on its form and perceived Western Asiatic influences in terms of both form and style: such analyses will not be pursued further here. When opened, the sarcophagus was found to contain the remains of seven adults and two children.

The long sides of the sarcophagus are decorated, suggesting its positioning in the centre of a mausoleum for examination in the round, as are both of the two short sides. The sloping, gabled lid also bears decoration. The decorative scheme will now be briefly summarised, face by face.

In the centre of the main, front long side can be seen the enthroned couple Hades and the abducted Proserpina (Persephone) in the Underworld, with the dog Cerberus sat to their right, flanked by a number of doors through which appear Hercules guiding Alcestis out of the Underworld and Mercury guiding the Greek hero Protesilaus, the first Greek to die in the Trojan War, out to be briefly reunited with his wife Laodamia. On the lower face the abduction and rape of Proserpina can be found, along with a curious image of a cave into which a woman is walking: an inversion of the upper world (in which this scene is set) and the Underworld (represented *above* in the form of the enthroned couple). In total six images of doors appear on the sarcophagus, in this case almost certainly alluding to the geography of the Roman Underworld. Interestingly Hercules enters one of the doors on the short side, then to emerge through a door on the far short end dragging the captured Cerberus with him, pictorially representing his mythic journey into and out of Hades: a journey the length of the body of the sarcophagus. Images of Charon's boat, and most intriguingly, and quite a rarity in Roman art, some scenes of unpleasant activity taking place quite clearly in the Underworld region of Tartarus involving the punishment of Sisyphus, Tantalus, and the Danaids also feature. Given the Hadrianic date usually assigned to this sarcophagus I feel that these scenes foreshadow the fashion for the aestheticisation of images of pain in Antonine art which I have written about elsewhere.

Four regularly-spaced figures of Atlas appear on the front lower long register, holding up ten Caryatids, figures with a symbolic meaning of their own beyond their use in Greco-Roman architectural practice.

On the long back face can be seen the Labours of Hercules, and the hero god also appears on each of the two end faces of the sarcophagus. Very interestingly, on the right end face can be seen two male figures who do not seem in any way connected to the images of Hercules: it may be that they represented the deceased and his heir, something also perhaps suggested by the appearance here of a scene of religious sacrificial procession with bulls.

Rising above the gabled lid at each corner is a *putto*, each holding three-dimensional ribbon garlands of fruit strung out between them, quite standard Roman funerary iconography.

I do not agree with the theory that 'the miniaturisation' of the figures on this particular sarcophagus is a deliberate strategy of ideological intent, rather than simply an artistic strategy and formal strand in narrative portrayal with techniques and presentation informed by external, non-Roman influences as well as Roman tradition. To view the geographical settings here as akin to 'a virtual doll's house' rather negates the sheer presence and materiality of the sarcophagus itself and the subject, seriousness, and electric force of its narrative programme.

Viewed as a whole the scheme of decoration on the Velletri sarcophagus, and indeed this sarcophagus itself, is without doubt unique in Roman art. The sarcophagus itself as a container for the dead here acts as a frame for narratives obviously about belief primarily, but also about moving between two worlds, about the symbolism of journeying, most certainly stressed by the multiple appearances of Hercules, about mobility and geography, about boundaries and their permeability. It also presents to us one of the best geographies of the Underworld in Roman art in terms of its conceptual coherence and complexity. We see entrances into and exits out of the Underworld, a cave, a number of condemned and tortured figures in the region of Tartarus, we see Cerberus, Charon on the shores of the Styx, and we see the enthroned Pluto and Proserpina in his palace at the entrance to Elysium. The possibly very late Hadrianic date of the sarcophagus, its mix of western and eastern style and workmanship, and the Greek Parian marble employed, all speak of its time, when the most travelled emperor up to that time created ideological narratives of journeying in service of the contemporary discourses around Roman imperialism and geography. Even if actually early Antonine, Hadrian's legacy of geographic discourse would still have been current at Rome.

The series of wall paintings from the interior of the Tomb of the Nasonii in Rome also make use of motifs relating to visits to the Underworld and return from there. Discovered on the *Via Flaminia*, this tomb, dating to the second half of the second century AD, was well recorded by an illustrator at the time but then robbed, with little of the painted decorative scheme now surviving. Drawings show that juxtaposed narrative scenes dwelled on the contrast between abduction and return, sometimes mediated by guides 'acting as agents of transition.' Entrance to Hades is thus contrasted with exit from it, without very much illustration of the landscape of the Underworld. Thus Bellerophon and Pegasus are contrasted with Hercules and Alcestis; Hylas and the nymphs are contrasted with Eurydice returned to Orpheus; Mercury leads Alcestis or Laodamia to Pluto and Proserpina, while Adonis is reunited with Venus. Hercules and Cerberus, Europa and the bull, and Pluto and Proserpina appear in small panel scenes.

The narratives found as images on the Velletri Sarcophagus and Nasonii paintings relating to searches for lost or abducted female figures such as Eurydice or Proserpina present journeying as both quest and jeopardy. Freud's theory of the uncanny, of encountering something that is at once familiar and unfamiliar, comes to mind, engendering a feeling of unease at the coterminity of the strange and known. In instances like this the body must have acted or functioned as a site of convergence between the diachronic and the synchronic. Romans had always been curators in the broadest definition of the term: of histories, peoples, and stories. Just as the idea of Rome pivoted on a paradox of doubleness, of knowing and not knowing, so its identity was maintained by recognising the distinction between remembering and the concept of unforgetting.

This eye/land pairing probably formed part of the dismantling capacity of such artworks. Rome was not entirely unique but came to be the primary city in a sea of cities. The Mediterranean and Adriatic seas were landscapes too, facilitating connectivity. Rome came to be not just an isolated microscopic entity composed of stone, brick, and marble but a larger universe comprising land surfaces and surrounding seas, oceans that the Romans could cross to exploit farther lands, the Underworld, and the heavens above.

Postcards from the Edge

A recent fascinating study by Kimberly Cassibry presented an examination of four categories of Roman souvenirs whose contexts of production and use/consumption tell us a great deal about the conceptualising of space in the Roman world and the significance of ideas about the power of place and of memory.[18] These souvenir categories consisted of: what she calls Itinerary Cups, Spectacle Cups, Fort Pans, and Bay of Naples Bottles.

The Itinerary Cups, most commonly called the Vicarello Cups or Beakers and on display in the *Museo Nazionale Romano di Palazzo Massimo* in Rome,[19] were silver vessels produced in Roman Spain, each different but each engraved with an itinerary of places linking Gades (Cádiz), considered the westernmost city in the inhabited world, to Rome. Only four of these vessels are known, all were found at the sacred spring site of *Aquae Apollonis* or Vicarello near Rome and date to as early as the first century BC to the fourth century AD. Did each of the four different inscribed itineraries on the cups represent an actual journey made or a journey to be made? Or were these routes of imaginary journeys, journeys never made? There are so many potential explanations. The fact that the vessels were silver and dedicated as *ex votos* at a religious site dedicated to Apollo adds another layer of complexity to their nature. While they were inscribed with itineraries of successive places they in no way constituted maps of any kind, in that itinerary maps for actual usage would have been set out on parchment, papyrus, wooden or wax tablet, and not on a heavy silver vessel such as these.

As well as bearing different itineraries the cups are of different sizes and vary in form: they are not a set, even if they are being considered as such here. Vessel One has inscribed on it the names of 104 places along the route from Gades to Rome, covering a total of 1840 Roman miles, Vessel Two has 105 place-names over 1842 miles, Vessel Three 107 places over 1840 miles, and Vessel Four 105 places over 1835 miles. These relatively minor but significant differences, taken with differences in layout of names in columns, differences in letter inscribing and in grammatical variations, along with differences between the vessels themselves in terms of size and form, though they bear undoubted similarities, point to the cups representing four distinct instances of dedication (rather than one dedication by an individual or a group of individuals together). Whether each dedication was separated by hours, days, weeks, months, or years cannot be discerned, though knowledge of the first dedication surely informed the subsequent similar acts. It somehow and for some reason set a precedent of some kind that led to the other dedications taking this form at this particular site. They were created almost to be like milestones.

The Spectacle Cups are glass vessels commemorating gladiatorial combat, and thus they could relate to sport in general, amphitheatres and arenas in general, specific named gladiators, and specific named stadia, thus giving them a geographical and locational role. Some of the circus

---

[18] On Roman souvenirs see principally: Breeze 2012; Cassibry 2021; Gianfrotta 2012; Künzl and Koeppel 2002; Ostrow 1979; Painter 1975; and Popkin 2017, 2018, 2022a, and 2022b.
[19] On the Vicarello Milestone Beakers see: Popkin 2022.

Figure 116. The Rudge Cup, schematically depicting Hadrian's Wall and naming some forts along the frontier. Second century AD. Alnwick Castle, Northumberland. (Photo: Tullie House Museum Carlisle and Professor David Breeze).

vessels carry depictions of architectural fixtures very specifically identified as locating them at the *Circus Maximus* in Rome.

The Fort Pans are small enamelled bronze bowls which have been discussed above in Chapter Five, and represent in their decoration schematic representations of the crenellated top of Hadrian's Wall in northern England, accompanied by the inscribed names of a number of military forts along the frontier wall (Figure 116).

The Bay of Naples Bottles are, like the Spectacle Cups, again made of glass (Figure 117). Around a dozen examples are so far recorded. They carry images of buildings, the pier, and sea at the port of Puteoli (seven examples) and at the spa resort of Baiae (two examples), or at both (two examples), and date to the late third to early fourth century AD. The towns are named on the bottles. The findspots include: Italy-Rome itself (one to three possible pieces), Ostia, and Populonia; Spain-Ampurias, Astorga, and Mérida; Portugal-Odemira; and Germany-Cologne. Five of the bottles bear inscribed exhortations to drink and live long or just live long, and one is commemorative of a deceased loved one. A number of fragments from a bottle recently found at Brescia in northern Italy depict a city's buildings in a similar way, but this scene is probably not a depiction of either Puteoli or Baiae. Fragments of gold-glass vessel in the *Landesmuseum* in Bonn possibly depict elements of the cityscape of Trier it is thought. Each vessel bears a uniquely composed image.

The dense scenes of urban architecture and port facilities at Puteoli, with many buildings such as a temple, market hall, baths, and amphitheatre, depicted being labelled with identifying names, are precise and yet impressive and beg the question as to why such detail was being provided here if the vessels were simply souvenirs of some kind. Mount Guaro/Mount Barbaro looms up beyond the city. On the seafront honorific/ dedicatory arches can be seen, along with the harbour and lighthouse. The Baiae scenes include depictions of not only buildings but also the famous oyster farms there-clearly labelled *ostriaria*.

Roman and Italian cityscapes and portscapes appeared in other places and in other media. Portrayals of *Portus* and Ostia have already been mentioned elsewhere in this book. For example, a number of images of fantastic and fantastical port cities appeared on wall paintings from Stabiae and now in the *Museo Archeologico Nazionale* in Naples, a port city appeared on a now-lost fresco from a building on the Esquiline Hill in Rome. A number of oil lamps from sites in North Africa carry decorative depictions of architectural features of a port city, probably Carthage, though they could possibly be of Alexandria or simply have been generic port scenes. The famous lighthouse at Alexandria appeared on an Egyptian glass bottle now in the Regional Museum, Ptuj in Slovenia and possibly is the lighthouse on a mosaic from a private residence in Mérida, Spain, along with images of Oceanus, and the Nile and Euphrates rivers.

Figure 117. The Pilkington Bottle, a souvenir from, and depicting, Puteoli. Third or fourth century AD. Pilkington World of Glass, St. Helens. (Photo: Pilkington Glass Collection. The World of Glass).

It remains difficult to see who these glass vessels were intended for and whether they can strictly be called souvenirs at all in the modern sense of the word. This leading question remains unresolved in Kimberly Cassibry's study as well. It is perhaps surprising that two of the categories of souvenir objects are made of glass and thus would have been more difficult to transport because of their fragility. Pottery souvenir vessels are not known to have been produced. Again, in terms of materials used there would have been a significant cost in producing the enamelled bowls, and more again in manufacturing the silver Itinerary vessels.

Another object which might, or might not, be a souvenir of travel within the empire is an iron writing stylus of the late first century AD from the Bloomberg Site in London, startlingly well-preserved in the former bed of the Walbrook river. Inscribed on the four faces of the shaft is a lengthy inscription in Latin, translated as: 'I have come from the City. I bring you a welcome gift with a sharp point that you may remember me. I ask, if fortune allowed, that I might be able (to give) as generously as the way is long (and) as my purse is empty.' Academic concensus seems to be that the City referred to here was Rome and that this souvenir was then brought back to Roman London. However, an iron stylus would have been a relatively cheap item and, if a souvenir bought as a present, was probably received with a stern countenance and barely-concealed ill grace.

Postcards from the Future

Another topic that needs brief discussion here, although it is one that would probably merit a book on its own, is the use of situational landscape settings in early Christian art in Rome, especially in wall paintings in the city's numerous catacombs and on early Christian sarcophagi. The Christian sarcophagi of Rome have been studied as a group by Jutta Dresken-Weiland who has notarised and quantified the principal decorative themes by chronological periods,[20] a most useful resource for understanding conceptual trends in the imagining of Christianised geographies. No similar catalogue of catacomb fresco images is available, and no attempt has been made here to assemble a detailed catalogue of such images, so the observations that follow should be read as reflecting a general overview rather than a quantified presentation with regard to these. The Roman Christian catacomb paintings and decorated sarcophagi constitute a very different sort of geography product to most of those discussed in this book, in that they were not images deployed in public spaces and places, but here in restricted or private contexts. Three categories of situational landscapes can be identified: landscape or place settings inside buildings which must have been sited in Rome or its environs; landscapes of the Holy Land; and landscapes of the Christian heaven. It would appear that from the second century AD onwards the Christian church was engaged in creating a discourse not only around its own identity but also around the question of how it fitted in amid the spaces and geography of the city of Rome and the Roman empire more broadly. How this self-questioning played out can be seen to some extent in the way that Christian art emerged from Roman art, and how it responded to that art and architecture's repertoire of images, themes, and tropes.

Landscape or place settings inside buildings sited in Rome or its environs was a category which was relatively large in terms of numbers of portrayals of worshippers in house churches and individuals conducting business in home or work premises, and suggests that the Christian community of the city needed to stress here their sense of belonging in and to the city of Rome. The two 'city gate' sarcophagi, from Ancona and Tolentino and dating to the AD 380s, were once thought to be representations of the fortifications of Jerusalem or of heaven but it now seems more likely that these elaborate architectural settings simply represented Italian cities or Rome itself. The large number of appearances of St Peter on the sarcophagi of Rome should not altogether occasion surprise, given his role in the founding of the Christian Church in Rome as a distinct urban religious assembly and sect: his occasional portrayal receiving keys from his jailers neatly suggests that he was here somehow being handed the keys to Rome.

The images of Christian artisans portrayed inside their work premises are very much following a more general trope for framing images of working individuals in Roman art more broadly, as I have written about in a previous study *The Dignity of Labour. Image, Work, and Identity in the Roman World*.[21]

---

[20] Dresken-Wieland 2018.
[21] Ferris 2021b: 199-201.

A number of sarcophagi bore images of Jesus in a detailed landscape setting. On the famous S. Ambrogio sarcophagus of AD 385-390, from Milan,[22] can be seen Jesus standing on a large rock or outcrop, with four springs-perhaps representing the four rivers of Paradise- gushing out of the rock face under him. This could fall under the category of baptismal scenes on sarcophagi which also includes an interesting variant category, represented by a number of examples, in which the imprisoned St Peter strikes his cell walls with a stick to summon forth water. St Peter striking the rock to bring forth water as two Roman guards look on appears on the so-called Arles Sarcophagus II in the *Musée de l'Arles Antique* there. A scene where Peter has been conflated pictorially with Moses striking the rock near Mount Horeb appears in a painting in the Catacomb of Commodilla for instance, though dozens of examples of this scene or variations on it are so far recorded. The imprisonment of Peter in Rome was well documented in Roman Christian art and, as noted above, his incarceration as an enemy of the state represented a brave and subversive image commenting on the very city in which he was held captive.

Landscapes of the Holy Land represent a category seemingly as numerous as the Rome-set interiors. If a message can be discerned from the assemblage of these images it is that in the Roman imagination the landscapes of Latium, Campania, Thessaly, and of Nilotic Egypt no longer held sole sway.

Finally, landscapes of the Christian heaven appear widely in Roman art, probably building upon the earlier tradition of the portrayal of mythical religious landscapes, often called sacro-idyllic landscapes, in which the pagan gods of the Greco-Roman pantheon appeared in scenes of bucolic harmony with Nature.

In the complex known as the *Hypogeum* of Viale Manzoni in Rome are a number of paintings relevant to this discussion. The earliest use of the structure dates to the early third century AD. Two painted scenes represent the return of Ulysses/Odysseus who in one scene is depicted in the disguise of a beggar, and in the other we see a great building, his home palace, outside of which grazes a flock of sheep. Another painting depicts a conqueror on horseback entering a walled city in triumph. A large delegation of citizens waits to greet him at the city gate. Another urban scene consists of a bird's-eye view into a vast courtyard in which a crowd is gathered to listen to a man seated on a raised podium. In both cases we can probably assume that the city is intended to be Jerusalem and that the triumphator and speaker are both images of Jesus.

The term pilgrimage itself is not without baggage and it must be recognised that there is not a consensus among academics as to how or when the word can be applied in relation to pre-Christian and non-Christian contexts, or even if it should be applied at all in describing such situations.[23] Human mobility, journeying to a place or places and back, and seeking or achieving some goal (religious, ritual, spiritual, relating to health or well-being, or personal) at the place journeyed to would all seem to be common elements of most pilgrimages. The act of pilgrimage, the presence and agency of a pilgrim or pilgrims having reached the end target of their journey, and the geo-spatial location, together created 'the place' of pilgrimage. While

---

[22] On the S. Ambroglio Sarcophagus see: Katzenellenbogen 1947.
[23] On pilgrims and pilgrimage see, for example: Elsner and Rutherford 2005; Fear 2005; Graham 2020; Hunt 1982; Kiernan 2012; Kuuliala and Rantala 2020; Markus 1994; Minchin 2012; and Rutherford 2012.

the travels of the Bordeaux Pilgrim, mentioned in the Preface, involved a long and arduous journey, indeed a series of journeys linked as an itinerary, a long distance travelled need not necessarily have been part of the definition of pilgrimage in the ancient world. Relatively localised travel from a town to rural shrine or from countryside to urban temple could still constitute pilgrimage to all intents and purposes.

As Jaś Elsner has acutely pointed out, the Bordeaux Pilgrim,[24] and others like her or him undertaking a Holy Land pilgrimage around the same time, that is over twenty years since Constantine's Edict on Christianity, was also taking a journey into the past. The events detailed in the New Testament of the Bible had taken place hundreds of years earlier and in a country and landscape which would have changed utterly and have been transformed by tectonic historical events. The ancient Jerusalem of the Bible was no longer the same city: not only had the Second Temple and much of the city been destroyed after the Jewish Revolt of AD 66-70 but Hadrian's armies in AD 134 had again sacked the city and it had been refounded as *Aelia Capitolina*. The majority of its most famed Jewish and Christian buildings and sites had been obliterated in the process, although the Burgundian pilgrim in their terse, note-like description of the city in the Itinerary displayed no evident disappointment at this, and indeed briefly alluded to the original natural topography of the place. They cited by name the Mount of Olives, Mount Sion, the valley called Josaphat, and the little hill of Golgotha, along with praise for the church 'of wondrous beauty' built by Constantine at the supposed site of the tomb of Jesus. Such drastic changes as had taken place in the city had not in fact erased its role as a theatre of ritual and the sacred: memory continued to interact here with faith and geography. Travel to and around the Holy Land was by a grid of Roman infrastructure overlying the ancient land and its spaces like the overwriting on a historical manuscript. It would have required a feat of the imagination to travel below and without this enabling grid.

The Christian idea of the defining of buildings and places as 'holy places' evolved throughout the third and fourth centuries AD, and developed from the notion that holy things could occur or had occurred at specific sites which then allowed holiness to be assigned to the places themselves. Such a difference was not purely semantic, but rather it reflected the creation of a geography of Christian spirituality forged out of a need to maintain Christian identity. 'The cult gave place a new significance; it met a felt need to make present in post-Constantin conditions the past of the persecuted church. Christianity could not envisage places as intrinsically holy, only derivatively, as the sites of historical events of sacred significance'.[25] A look at some of the staging and stopping details listed on the *Itinerarium Burdigalense* almost conjures up before our eyes a sacred landscape being created by the joining of dots between one holy site and another. For instance, in the section of the itinerary that listed stops from Caesarea to Jerusalem the pilgrim noted that they were stopping at or passing 'the bath of Cornelius, the centurion who gave many alms'.....'the field in which David slew Goliath'.....'Aser, where was the house of Job'...'Mount Gerizim..[where] Abraham offered sacrifice'....'beyond this, here is a tomb in which Joseph is laid'....'here Jacob saw the vision and the angel wrestled with him....Thence to Jerusalem 12 miles' and so on. This all sounds tremendously exciting as an experience.

---

[24] On the Bordeaux Pilgrim specifically see, for example: Bowman 1998; Douglass 1996; Elsner 2000; and Weingarten 1999.
[25] Markus 1994: 275.

There were many famous pilgrims to the Holy Land, such as Helena, the mother of Constantine, but none left us their itineraries as the Burgundy pilgrim did. St Jerome (died AD 420), originally from Dalmatia, lived for some years in Jerusalem. The middle and later years of his life saw an intensification in the commodification of the region as a holy or sacred landscape: 'They come here from all over the world. The city regurgitates every type of human being; and there is an awful crush of persons of both sexes who in other places you should avoid at least in part but here you have to stomach them to the full.' There was clearly not a warm welcome at Jerome's hillside.

The idea that roads were as quintessentially Roman as aqueducts and bath-houses is ingrained in modern-day thinking, and it is likely that just such an identification was made in Roman times. Eventually roads would be viewed as being tools of the state, and ideological symbols of that state. Building and restoring roads became something that emperors would eagerly lay claim to doing, as evidenced by certain references in Augustus' *Res Gestae* and in the well-attested road building programme of Trajan, particularly in Italy itself. The number of inscriptions recording road-building, most often in the form of inscribed milestones, attests to the need to promulgate the state with the gift of actual mobility, as well as sometimes social and class mobility.

In many instances Roman roads were built along the lines of pre-existing routes. In other words they quite formerly colonised or took over those routes through the Roman road construction methods and new materials used. Ray Laurence has discussed this phenomenon in relation to the Roman *Via Nova* from the boundary of Syria to the Red Sea, not actually an altogether new road conceived under Trajan but rather a Romanisation of a pre-existing caravan route, the Nabatean King's Highway. Many such over-writings of the landscape can be found in both the Republican and imperial periods in Italy and the provinces.[26] Travelling on the roads was already a proscribed activity in most respects: you had to stay on the road and thus have your route dictated to you/for you. Straying off the road and using unpaved tracks would have been unthinkable. Illustrations of travel on coins are numerous, as has already been discussed specifically in relation to the reign of the emperor Hadrian. Under Constantine and some of his successors there emerges a very specific trend towards the favouring of imported marble for milestones along the roads of Italy, perhaps a statement about the costs of large road building and maintenance schemes being funded by monies from across the empire.

Petrified, the Landscape Grows

We must imagine that travel and mobility went hand-in-hand with the winning, creation, and maintenance of empire, though the movement of people must have been for many different reasons at different times in different contexts.[27] Who travelled, why they travelled, and where they travelled makes for fascinating reading in a recent academic study that concentrates on Late Antiquity, from the late third century AD to the seventh century, utilising evidence from

---

[26] On the ideological significance of Roman roads see principally: Laurence 1999 and 2020. On the *Cursus Publicus* see, for example: Kolb 2001; and Lemcke 2016.
[27] On migration and mobility in the Roman world and ancient Mediterranean see, for example: Amiri 2020; Broadhead 2002; Carucci 2017; Clarkson *et al.* 2020; De Ligt and Tacoma 2016; Handley 2011; Isayev 2017 and 2020; Jaccottet 2020; Khellaf 2021; Lampinen and Ferrándiz 2022; Laurence 1999 and 2015; Lo Cascio 2016; Lo Cascio and Tacoma 2016 and 2017; and Woolf 2016.

inscriptions.[28] The study's database consisted of 567 individual inscriptions which recorded 623 foreigners and travellers known in this period. In this study it was found that the reasons for travel recorded in these inscriptions or which can be inferred from a close reading of the inscriptions could be broken down into six main categories. Firstly, there was the retrieval of a deceased's corpse; secondly, religious pilgrimage; thirdly, military service; fourthly, exile or refuge; ecclesiastical or religious business; and sixth and finally immigration. In terms of mobility, it is interesting that the study found that over a third of these recorded movements or journeys were from the eastern parts of the empire, particularly Syria. Handley also looked at the ways in which the naming of geographical origins varied depending on the context of certain categories of inscriptions. For instance, in the case of travelling Gauls moving around within Gaul they were identified or identified themselves most commonly by their *civitas*. However, in the case of Gauls travelling outside of Gaul precise geographical origins were not always stated in inscriptions in a regularised way.

Those who were mobile in the Roman period, that is those who travelled outside of Rome or engaged in cross-provincial travel, were largely either servants of the state-in the army or imperial bureaucracy-or engaged in trade and transportation (that is in service to the economy), or engaged in religious business (that is in service to Roman religion). The elite could travel, as we have seen they did to Greece for instance, but most others did not. There has quite rightly been a great deal of writing about diversity in Rome and across the Roman empire in the last few years, but once the main groups of travellers have been discounted from discussions of origins, it is inevitable that most of the rest of the instances of diversity at particular locations are linked to forcible relocation and to the system of slavery. Slavery did not altogether otherwise remove the identity of individual slaves in all situations and contexts, and if a slave were to become a freedman or freedwoman their personal situation changed considerably, though the informal patronage system between freed slaves and former master or mistress could still apparently be an onerous obligation and difficult relationship. As an illustration of how diverse a household might have been in Rome in terms of the origins of the slaves there an inscription of c. 48-46 BC[29] relating to the purchase of a burial plot and a tomb built there is highly informative. But we must take care to realise that though this household could be presented as representative of the idea of diversity that would be altogether mistaken, given the differentials in power and control within the house.

It is a curious phenomenon that the Romans often wondered conceptually about their own city and its place in the world, both geographically and spatially and historically. Discourses around location, uniqueness, origins, and future trajectory were vital to the definition of the city's sense of self, as has argued throughout this study. However, bold certainty was not always the driving narrative of some of these discourses. Doubt and alienation were ever present too. *Ilion/Ilium* was described by Caesar in Lucan[30] as *Romana Pergama*. Anxiety about the possibility of a founding of a 'new Rome' was quite evident and real for many years: Constantinople became 'the new Rome' or 'the second Rome' or 'the Rome in the east' in AD 324[31] (Figure 118),

---

[28] Handley 2011.
[29] *CIL* 1.2965a.
[30] Lucan *De Bello Civili* 9. 998-999.
[31] On Constantinople as 'the new Rome' see, for example: Grig and Kelly 2012; Isaac 2021; Kaldellis 2020; and Melville-Jones 2014.

Figure 118. Sardonyx cameo of the Tyche of Constantinople crowning the emperor Constantine with a laurel wreath: known as the *Gemma Constantiniana*. Probably AD 315. *Rijksmuseum van Oudheden*, Leiden. (Photo: Author).

and Aulus Gellius called *coloniae* 'little Romes'.[32] With the administrative division of the Roman empire into a western and eastern empire, so a number of 'little Rome's' emerged as new imperial centres, Trier, Milan and Aquiliea in the west and Spalato, Sirmium, Thessalonica, and Serdica in the Balkans, and Nicomedia and Antioch in the east. Herodian summed up a prescient view that heralded the emergence of a new attitude to spatial thinking when he declared[33] that 'Rome is wherever the emperor is'.

This all very much ties in with the concept of presented literal geographic information being intended to be used by its viewers and readers in 'world-writing', creating and shaping views of the world, in other words making their own mental maps. In this way some Romans made a world of their own, but it was not within everyone's power to do the same. They were not attempting a completely neutral rendering of the real world, certainly not at times of rapid expansionism-in trade, geographic knowledge, and minds. Ultimately this probably led to the creation of a patchy but strangely unified picture that flattened out the military and political conflicts that underwrote this drive towards expansionism, and indeed difference in general. Passing through circuits of meaning, these geographic images contained not just information

---
[32] Gellius *Noctes Atticae* 16.13.8-9.
[33] Herodian *Hadrianus* 1.6.5.

but also emotions, abstract concepts, and ideas in a covert spatial form. The Romans were not only writing about the world but also writing upon it. The colonial geography associated with Senecan drama was as much a cultural thing as it was ideological, one aspect of this being the positing of territory as a woman waiting for exploration and ownership. Symbols of alliances, conquests, wealth, and land itself invaded and defined Roman culture.

As images Roman elite women could become immersed in imaginative geographies of their own bodies, the images representing and demonstrating the renown of ancestors (familial ties and lineage, themselves a kind of social and political geography). If Roman elite marriage was an alliance, then women in such relationships and positions contained symbolically within their bodies joined wealth and property, joined estates and lands, that is shared places, and enhanced political power. They became focal points in Roman society, maps of familial and political ties. The non-elite adapted this strategy and discourse simply to record and display family history, something quite similar but less deep. The Roman state and individual emperors could even create fantastical and mythological familial and geographical links, as with Romulus and Remus, Aeneas, and certain Greco-Roman or even foreign deities.

In some ways it could be said that Roman elite women in their Coan silk and using Syrian myrrh were mapping the geography of the Roman empire on their bodies. Most remarkable in this respect was the fashion at Rome for wigs made of hair from women in Germany. Ovid in the *Ars Amatoria* very specifically referred to this hair as being from women of the *Sugambri*, 'a gift of a race that Romans have triumphed over'.[34] This raises all sorts of difficult questions and debates which can be informed by the quite recent discourses around women's hair. It is unlikely that this hair was freely given, that it was in Ovid's word 'a gift'. Was it purchased and traded once a market for such a 'product' came into being? Was it forcibly cut off captured women's heads, or off those of enslaved women, in both cases the women presumably having no alternative course of action other than to acquiesce to this act. Historically it is recorded that following the quelling of unrest among the *Sugambri* across the Rhine the tribe were forced to return Roman standards and to provide hostages to Augustus: one wonders if the cutting off of hair from female hostages took place at this time. This perhaps showed that for Roman women the way to embrace the empire was to physically *become* it, as far as that was possible. These maps of desire have left no traces in the archaeological record.

The cutting of hair and the shaving of heads are highly symbolic acts in both mythologies and in many historical circumstances: they are most usually disempowering or emasculating acts, often with a gendered power subtext, and sometimes with a racial one.[35] We have only to think of the well-documented history of the hair trade in the nineteenth and twentieth centuries, involving everything from long luxurious tresses to balls of comb waste, to understand how such a curious but highly lucrative trade was built upon and relied upon disparities in wealth, values, cultural mores, and opportunities between those parting with their own hair for money and those dealing in it, and most particularly those consumers buying the wigs, hairpieces, and extensions that formed the main end product of the trade. Accounts of hair markets in certain rural regions of nineteenth century France describe dealers often conducting public haircuttings themselves, until the practice came to be seen as unseemly and degrading, and local bans were put in place in some areas. The enforced shaving of the heads of those in

---
[34] Ovid *Ars Amatoria* 1.14.45-46.
[35] See Chapter Eight Note 16.

British prisons, workhouses, and certain hospital facilities, before its banning in the 1850s, also now appears to us as cruel, alarming, and degrading. The shaving of hair from arrivals or inmates at Nazi concentration camps and labour camps and the forced head-shaving of alleged female collaborators after the liberation of Paris in August 1944 are highly-relevant too in this context.

Ovid in the *Fasti* wrote: 'To other nations territory is granted with fixed boundaries: the extent of the city of Rome and of the world is one and the same'[36] Put into such a broader context the map of Agrippa can be seen as just one of a series of marks left by Augustus on the city of Rome, on its form, fabric and appearance, on Italy, on the known world as then was, and on the bodies and minds of Romans and non-Roman visitors to the city. The Portico of Nations built by Augustus took its inspiration from that first attempt by Pompey to visually demonstrate Rome's wider location. Agrippa's world map did not mark the start of something but rather surely marked the end point of something, creating a connection with every place and thus with no place at all in the end. The effect of such a series of geography products needs to be appreciated. Perhaps, to paraphrase from Baudelaire's *Correspondances*, the effect was like prolonged echoes that merged far away in a dark and profound oneness as vast as night, as vast as light, they answered to each other.

At times spectacle could be said to have overwhelmed narrative in these quite formal images of empire. They used a combination of historical material and generic conventions and archetypes in order to represent a narrative of the Roman past, present, and future; a set of overlapping narrative ideologies relating to the themes of nationhood, empire, class, masculinity, race, and state militarism. Such art helped Romans to understand the workings of memory in colonial spaces where topographies of remembering would have coexisted with topographies of suffering. It is hard to tell from Roman Nilotic scenes that Egypt was also a place in which great violence took place and was then both concealed and memorialised at Rome. The concealment of things only magnified their absence here.

Rome was designed to be full of spaces that could so easily be filled with ideological content. Most of the empire was way beyond that present space, and was characterised by a view towards it, rarely from it. The conception of a centre and periphery was false, as the spaces of empire were not voids, but territories of knowledge and sources of raw materials and goods. Transforming a once boundless world into provinces and territories, military districts, and client kingdoms helped make sense of the pace of Rome's expansion and ambition. Chronological distinctiveness was not always apparent in some cases though. It was now but it was always. It was both on the surface and penetrated deep within landscape and self. An invisible umbilical chord from centre to peripheries enabled the real and imaginary to cohabit. This allowed entities, time, and space to collapse together in an interconnectedness, the benefits and possibilities of exchange being demonstrated by the embeddedness of flows of knowledge both to and from Rome, and flows of people on journeys.

I have argued that many different things, objects, images, and texts, constituted geography products in the Roman world. They became a series through accretion rather than design, but they would appear now to be interpretable as a single, total work of art -a *gesamtkunstwerk*.

---

[36] Ovid *Fasti* 2.683-684.

Their examination today allows the viewer-who can see/read them all-to travel forward to the past and backwards to the future, a hauntological reminder that the past is now and the present in the future. The idea of the Roman empire as a kind of 'third zone', disconnected from the landscape and the knowledge of its history, is attractive but ultimately flawed as a concept. For the Romans the landscape, whether at home or overseas, was a vessel of both culture and possession, also a container of dispossession, sadness, and loss.

The idea that journeying around the city and the wider world was sometimes akin to moving through a labyrinth, in which knowledge, intelligence, preparedness, and often luck came together to plot a navigational course is perhaps reflected in the popularity of images in Roman art and literature relating to the Perseus story. The most dramatic and aesthetically-pleasing labyrinth mosaic was found in a villa in Orbe-Boscéaz in Switzerland. At the centre of the labyrinth can be seen Theseus and the Minotaur, and at the four cardinal points around its outer perimeter are crenellated gateways, with towers at each corner. At its simplest symbolic level the encounter between Theseus, representing civilisation, and the Minotaur, representing the unknown, the monstrous, possibly also the alien or barbarian, reflected the standard Greco-Roman discourse around identity and difference.[37] The maze can be seen as being unknown territory, primal forest or marshland even, beyond which lay sanctuary and safety in the form of the city, as represented by the gates and towers. Another good example comes from Loigersfelder in Austria. From a Republican era house beneath *Piazza San Giovanni in Laterano* in Rome and now in the *Centrale Montemartini* outpost of the *Musei Capitolini* comes another pavement dating to 100-80 BC in which the four part labyrinth is framed by crenellated city walls and their lookout towers and gate.

The body, in the form of both male and female statues, emerged at Rome as a space of unfathomable warmth, depth, and opacity, its presence serving as a reminder of the origin of an idea, place being privileged in this way, allowing for the creation of random, aleatory meanings. In a sense this evoked the impossibility of woman herself, since the female image here appeared only as an object of exchange: at best she served as and marked a potential space to be journeyed to, conquered and exploited. Hélène Cixious in *The Laugh of the Medusa* wrote about 'writing through the female body'-'*l'écriture féminine*'. The body could be understood as a site of natural desire and its satisfaction was here replaced by the culturally and politically-constructed body. The body in ideological bondage was a trope open to many sophisticated explorations of gender dynamics and sexual politics; image risked turning everything and everyone into a commodity, with exchange value usurping human dignity. Perversely the images did not look outwards but turned in on themselves: often fragmented, elliptical, and impressionistic they did not so much convey reality so much as the complexity of language, with all its slippage, play, and ambiguity, creating the possibility for the viewer to be part of the process. The open-endedness must have been appealing, and instead of getting bogged down with expectations of meaning, message, and closure they contained together a strange idioglossia or internal logic. Acceleration was on the cultural agenda. Like lyrics these pieces mapped unique subjective landscapes. The city's streets, its paths to possibility, might have offered no solace, but the world, the real, was not an object. Rather it was a process hedged with fascinating ambiguities and contingencies.

---

[37] See Chapter Seven Note 4.

What was to all intents a spatial turn in Roman thinking and ideology was an art not here to mirror the world but to create a new one, in a tour de force of distance and remove. Ultimately, the production, use, and display of geography products at Rome was concerned with testing limits, setting boundaries: between here and there, between self and other, between the sacred and the profane, between male and female, queer and straight, between the poetic and the demotic, and between the living and the dead. Mythological geographies and imagined interior spaces related to Ovid's creation of the *locus amoenus* (literally the pleasant place) and the *locus horridus* (the unpleasant space) as articulated in his *Metamorphoses*.[38] This would appear to be not simply a literary construct but rather a definition and display of spatial complexity and of dissolving boundaries, and a proposition regarding the construction of personal geographies. For example the *locus amoenus*, by its very definition a good place one would have thought, is often in the book the setting for horrendous and unpleasant events, often a realm of dark, umbral characters. Equally, the *locus horridus* is presented as a series of places where not everything is indeed negative. Ultimately, Ovid's images, these visions, constituted another kind of alternative geography. They vouched that the earth not only wept and roared but that it also hosted, enveloped, and transported. If the flaneur poem from *Ars Amatoria*, with all its trappings of male predatory sexual behaviour, was about a specific day and what might be done on that day, then the *locus* elements in the *Metamorphoses* extended the timeframe and broadened the context to encompass endless empire.

The viewing of Ruth Ewan's beautiful and contemplative artwork *Back to the Fields*, first displayed at Camden Arts Centre, London in 2015/2016, requires the viewer to engage in a leap of faith, to find this display of Nature a synecdoche for the terror and bloodshed of the French Revolution, representing as it does the bodying forth of the imagery associated with the French Republican Calendar of 1793-1805. This symbolic return of the land to the people aptly demonstrates how beauty can emerge and even thrive in any temporal *locus horridus* and suggest to us how in the beauty of so much Roman art can be found the lees of violence, bitterness, and exploitation. There would appear to have been a need to embrace doubt and paradox, though the binding thread was the fact that though all was revealed, everything was still a secret. The Romans did not fragment the world: they just happened to notice that it was fragmented and exploited the fact.

Of course, the issue of migration in the Roman world has been addressed both in terms of migration into Rome and migration as a phenomenon within the empire at certain times. Boundaries, frontiers, and permeability were important concepts. Mobility would appear to have been both a structural and structuring element of most of the pan-Mediterranean societies that historically preceded Rome and then of the Roman empire itself. Voluntary travel though would seem to have generally been short distance, local, or inter-regional, rather than extra-provincial. Exceptions most obviously were travelling high status individuals: emperors and their officials and courts, ambassadors, governors, and intellectuals including writers. Epigraphic records suggest that those long-distance travellers outside of this elite bubble were principally merchants and craftsmen, as indeed might have been expected. The travelling Bordeaux Pilgrim and the travelling Egyptian landowner Theophanes going from Egypt to Antioch quite tellingly were travelling in the fourth century AD when long-distance

---

[38] On the idea of the *Locus Amoenus* and *Locus Horridus* see, for example: Bernstein 2011; and Fröhlich 2023.

mobility and travel obviously was still not common but which had filtered down the social and professional hierarchies of earlier times.

Migration into Rome would have been on a number of levels: forced migration of subjugated and enslaved persons, that is involuntary as opposed to voluntary migration of merchants and traders. Some migration might have been permanent, some temporary, some seasonal. Some might have migrated in reaction to political changes and so on. The existence of certain regular and well-established migration streams can be assumed. Estimates of the number of migrants in Rome as a percentage of the city's overall population at the time of Augustus suggest 20-30 per cent, or even as high as 33 per cent, in a population of c. 650,000. An earlier influx of migrants into Rome in the final two centuries of the Republic is estimated to have increased the city's population by tripling or even quadrupling it. Analysis of inscriptions produces a minimum figure of 5 per cent of Rome's population being foreigners from outside Italy. Migration within and across the empire was linked more to occupation, status, and mobility. There was a noticeable and noteworthy trend in Roman historical writing for the interrupting digression into the origins of certain peoples and their migrations. This trend can be found fully developed not only in the works of late Republican and Augustan writers such as Livy and Sallust but continued into late Roman histories such as that of Ammianus Marcellinus. Migration into the Roman empire from beyond its frontiers is another issue altogether.

A funerary monument from Rome[39] dedicated by Iulius Secundus to his late wife Cornelia Tyche and his late daughter Iulia Secunda, both of whom would appear to have died in a shipwreck en route to or from Spain, is not only a record of a journey made but not completed, but also a detailed exposition of geographical information. 'You two were exhausted by the violence of the sea on the Phocaean coast where Tagus starts and the famous river Hiberus ends and both flow the one into the Ocean and the other into the Tyrrhenian.'

Ultimately tales or images of voyages, of journeys made, will point the reader or viewer home. Images that were constantly changing, never the same, never remaining. A self-contained series of fragments such as this, each would have left sharp memories and fully-realised environments in their wake. As the viewer sought direction the image was already gone. Partly carved, partly mapped, they carried an abstract concept linked to geography and place overlain on human form: they were almost somewhere else, and as time passed they became the past themselves, caught between stillness-being of a place-and motion-travel there.

In this chapter it has been suggested that the depiction of both real and mythological journeys and travel formed part of an overall narrative of geographical messaging that in some cases acted as a metaphorical framework for the consideration and discussion of issues that were more mystical and philosophical than quotidian. If the emperor Hadrian best represented the travelling figure whose peregrinations almost took the form of a beating of the bounds of the Roman empire it can at least be suggested that at many places along his routes he left some form of marker in terms of confirming his presence there. Certainly we talk about the nature of these markers in terms of the impact that they made in each individual province where

---

[39] *CIL* 3.3107.

such construction or infrastructure projects took place: the frontier system of Hadrian's Wall in northern England was one such marker.

The observations made by Katja Pilhuj in her book *Women and Geography on the Early Modern English Stage*, particularly with regard to the differences between geography and cosmography, well illustrate how world-views can be created, manipulated, and maintained. Roman geography products such as literary and historical texts and writings, actual maps, estate and cadastral plans, chorographies (descriptions of smaller, local areas which can also include paintings), genealogies, and other cartographic products also aided world-building. The effect of reading or viewing texts and images as geography products would have varied from individual to individual but would have included using the geographic information provided: to locate past events, such as battles fought, an extremely important thing in Roman life; to view a place and the regions adjoining it; to calculate, estimate, or simply appreciate the distance of a place from Rome (from us); to view and appreciate the size and extent of the Roman dominion, and perhaps maybe to also consider that of other empires past and present (the swell and flow of history); to plan journeys to distant lands and places or to understand other's travels in some sort of broader context.

Roman ideologies of scientific conquest are often credited with propounding much later interventions and events. Roman geographic images must therefore be evaluated by us in terms of the processes of their conception, creation, and reception instead of as the end product representing a static artefact, murmurations, with only a visual, immediate, and unchanging value.

In the next and final chapter an attempt will be made to present an overview of the main argument in this book, that the Romans gained geographical knowledge through the contemplation of a wide range of what have been called here geography products and how these images of empire can be read today.

Chapter Ten

Slouching Towards Empire

As will have become apparent through reading this study the Roman use of geography products to explain the world changed over time. Things were not the same in the reign of the emperor Constantine as they were in the age of Augustus. But the centreing of Rome in a wider world had not altogether been replaced by the founding of a number of new Romes, and imaginative geographies of Hades had not been usurped by geographies of a Christian Heaven and Hell.

The nature of these geography products varied considerably, from an elegy or poem to an ethnographic account, from an itinerary recorded on a souvenir silver vessel to a sketch map on a piece of leather, from a portable sundial held in the hand to a massive decorated sarcophagus, from a display map of the city of Rome to the smell of exotic foodstuffs cooking in the city's streets.

Their presence in Rome or in another Romanised context acted to add knowledge, to make the world larger, and to make another dimension smaller by placing an emphasis on the lack of visible horizons. One would have got a glimpse of something both contemporary and very ancient, mythic even, seeming to point back to long before Romulus, especially when the images resounded with chilly authority in the central spaces of the city. There were numerous visual presentations about the wider world, from region to province to territories outside the empire, each nested inside the other *ad infinitum* like Russian *matryoshka* dolls. The assertion of Pythagoras that he already had lived four lives, his body and mind together exploring the transmigration of the soul via reincarnation, traversing an ageing shepherd, a young goat, a fir tree, and a store of charcoal, provides a model for Rome's constant rebirthing and transmigrating. Inevitably it is human nature to map out, measure, and quantify, because this provides a sense of security from which to perceive the world: to contain, in order to come to terms with vast space. It might seem to us that they were suffocating in a sarcophagus of data.

As Gore Vidal says in Fellini's film *Roma* of 1972, 'what better place than Rome to wait for the end?' Ancient Rome must have seemed quite claustrophobic if you stood in one place: but if you moved around it, like Ovid, it must have opened up to the world beyond. Cicero wrote[1] 'Plans have been formed in this state, o judges, for destroying the city, for massacring the citizens, for extinguishing the Roman name'. This sentence carried within it a key to understanding the conceptualisation of Roman identity, of Romanness, of *Romanitas*: his very careful, precise, deliberate, and legalistic use of the words and terms *civitas* (state), *urbs* (city), *cives* (citizens), and *nomen Romanum* (the Roman name) was essential, as all these words together designated essential aspects of Rome and being Roman.

Shadowplay

It has quite rightly been observed that there were few advances in the creation and dissemination of written ethnographic knowledge after the earlier principate. However, as

---

[1] Cicero *Pro Murena* 80.6.

this study has demonstrated there were constant attempts to visualise certain aspects of geographical knowledge and to create new geographical products for doing so. The cultural history of Roman geography can be written in ways other than simply presenting and analysing literary sources. As to contemporary historical sources, in many instances the impression given to the reader would have been that most foreign landscapes described were landscapes of war, in terms of the written presentation of accounts of wars in Parthia, Dacia, *Germania*, or Britain for instance, a feeling probably backed up by the visual presentation of conflict in these lands on numerous imperial monuments in Rome and elsewhere.

It has not been argued in this study that there was any kind of ongoing, extended programme of the presentation of geographical information to the Roman public by the Senate in Republican times or by Rome's emperors: the accumulation of isolated, single presentations ultimately constituted a series, even if never actually planned or intended to become so. If there was a motive behind the presentation to viewers of artworks that conveyed geographical information it was on many occasions an attempt to domesticate the distant, and for the people of Rome to know the previously unknowable. These and other geography products all contributed together at different scales to make and sometimes unmake Roman social memory. The focus on individual pieces as works of art rather than conveyors of information unfortunately downplays the connections between the material and the invisible, an important dynamic to be sure. Ovid's poem *Ars Amatoria* perhaps represents the only material trace of such an encounter and response which as an emotional geography refuses easy interpretation. It refuses closure and linearity in favour of fragmentation and indeterminacy, being a journey not around the world, mediated by roads and itineraries, but of self-discovery.

This study has been all about the ways in which geographical and spatial information was sometimes conceptualised and visualised in the Roman world and how such information might have been understood by its viewers, particularly when mediated through the control and often the manipulation of the processes and contexts of presentation. It has also been about landscapes and the sense of place in the Roman imagination. The idea that such visualisation was part of a larger phenomenon, of what one ancient historian has called 'illiterate geography',[2] is interesting, but under that pejorative title is not ultimately appealing. The social dynamics of Roman art and its cultural value did not rest on the proposition that much Roman art was simply images for those who could not read. In the cases of many of the images and artworks discussed in this book in certain contexts certain types of images probably would have acted almost as mnemonic devices, setting up and triggering chains of thought and associations in the viewer's mind, sometimes deliberately so it would appear, and possibly accidentally in other situations.

It was not just illiterate versus literate, but something to do with the dual character of violence and history at the time. A geography of obscenity, ecstatic and anti-tragic it discovered in the contemporary world of Rome visible signs, apprehensible epiphanies of a universal link between the organic and the inorganic, dynamically identifying different processes in emblematic correspondences between flesh and earth, cloud and breath, blood and water, rock and body. Land, sea, underworld, and heavens converged. The sea, rivers, and roads connected all these disparate places, like death in a sense, as it highlighted their memories,

---

[2] On 'illiterate geography' see: Dueck 2021.

Figure 119. Example of a Roman 'geography product'. The silver Parabiago plate, bearing an image of Cybele and Attis in a cosmic setting. Mythological figures present include river deities and *Tellus*. Fourth to fifth century AD. *Museo Civico Archeologico*, Milan. (Photo: Author).

both individual and collective. We should not fail to recognise that it was the Mediterranean and Adriatic seas themselves which constituted the most potent images here, a metaphorical matrix of all Rome's potentiality, natural and cultural, engulfing and transforming the constructions of humans.

This was a complex story of geographic and temporal entanglements that questioned standard narratives of slavery, manumission, citizenship, and mobility, mediated by a dominant art which developed alternative and hybrid artistic techniques and narrative forms which broke with the former aesthetic and formal hegemony of earlier Roman art. This helped to open up a rift between what the viewer desired and expected of the world and what they felt when viewing: it was not necessarily important what they thought, but that they just thought. This would have destabilised supposed boundaries and rendered differences more subtly. To us, viewing them from a distance mediated by time and intrusive knowledge they feel almost like miniature exercises in ethnographic parataxis, that is they perform an uncanny ethnography of the abstraction and precarity of the present moment, circling around notions of memory, regret, resilience, compassion, loyalty, hate, and war. They were a time capsule of historical moments and of futures to come.

Eventually there was created an umbrella-like superstructure or rather superstructures, consisting of what can be called a technoscape, a mediascape, and financescape network, with culturally-shared assumptions for the Romans, underneath which was a large number of subcultures, each with unified codes and roles. The interculture was a cross-cutting system where whole sections of Roman society acted out the role of subcultures, with the interplay of consumer and state, immigrant culture and mother culture, and audience and artwork. An industrial interculture, a diasporic interculture, and an affinity interculture can potentially be identified. Transformations such as these are best portrayed in the style of a thick description, to be viewed from a variety of angles. The twentieth century European art movement Fluxus was interesting in that it regarded every object as an event and every event as having an object-like quality, acting as a conduit through which ideas, possibilities, and personalities flowed, bringing about a multi-layered phenomenon. In ancient Rome this was as much about local (Roman) identity, the energy of the city of Rome, the importance of independence, the removal of rules, the depths of art, and the otherness and drama of discovery. Geographical artworks of the kind discussed in this study could make things happen just by existing, and thus by helping to make connections in the viewers' minds. Everything was linked, especially in those contexts and situations where disorder created connections and a glorious resonance. It made people see the present, something that is often harder to do than seeing the future. Ivan Chtcheglov, an inspiration to the Situationist International, noted that in a growing city three steps cannot be taken without encountering ghosts, bearing all the prestige of their legends: such encounters exploiting tense, sense, and absence.

The scrutiny which images of non-Roman bodies were under in the city was almost a kind of surveillance. The Romans understood that the infrastructure project of empire, the lines and networks of roads, involved a complex intervention in the spatial structure of the entire known world as it was then, which went far beyond the traditional concerns of the surveyors or *agrimensores*. Monuments in Rome became part of an architecture of longing, aimed at the development of observation, of differentiation, a general passport to perception. Disarming and exhilarating in their beauty, they promoted recognition of insight coming from experience and evaluation, itself resulting from comparison. As a series they balanced relativity, instability, and interaction: variants versus variety.

Rome was an importer of goods and ideas and not really an exporter. What made it truly unique was the fulgurating character of the ideas it brought in and contained. It can be argued that this concealment of these issues was the foundational gesture of the state art, and from it was born intermediality of a kind which lifted the veil that obscured the viewing in many instances. Many of the artworks discussed in this study offered replication and profusion, seriality and singularity, a magnificent continuum, one large multiple, endlessly and subtly developing. But they seem to have had their own future obscurity programmed into their relationship with the world from the very beginning.

The policy of removal of foreign antiquities and parts of monuments to Rome had a casuality all of its own, becoming part of the wider process of mentally locating Rome as the centre. They represented a chronicle of the Romans, wedded not simply to territory or bloodlines, but to certain ideals and ways of thinking. Perhaps more than anything else it was this, along with the Latin language which defined Romanness. What Rome increasingly lacked as time moved towards Late Antiquity was filigree and shadow, ambiguity and ache, ghosts and spectres. It is

Figure 120. Example of a Roman 'geography product'. Small sarcophagus decorated with images of personified river deities, probably from Rome. Second or third century AD. (Photo: Duke's Auctions).

praxis that maintains contact with the totality: in the perspective of praxis every fragment is or becomes the totality.

When the Romans turned to geographical images of the human body and its experiences to search for a new identity, it would seem that traditional notions of beauty and containment no longer related to the expansion of the boundaries of their world. This sense of boundaries marked was never final: they had their pressures, their faultlines, and their porousnesses, but they were nevertheless elastic. Alliances and diplomatic collusions were the ultimate proof of this. But there were also distinctions, limits to these collusions, a power differential which needed to be maintained.

Unknown Pleasures

A recurring theme throughout this study has been the identification of the myriad ways in which cultural memory, cultural traditions, and cultural encounters fed into and fed off Roman ideology and art. For all its technical verve and power Roman art's palette was always somewhat limited, rarely admitting nuanced moods, anything of drift, reflection, or loss. Roman geographical images almost dared their viewers to measure themselves and their world against them, with sharp overtones of siege and fear, loss and nostalgic regret. Like the 'man of the crowd' in the writings of Baudelaire and Walter Benjamin they were in the crowd but not of it. The language and metaphor of foreignness permeated Roman life, not just as a device to deliver a message, but more as a demonstration of the omnipresence of the foreigner in the Roman mind.

The sociological typology of these images was turned into something more strange and satisfying through the visionary dimension of the scene-setting and by context. The images described distant locations in an exuberant way that foregrounded knowledge and its transmission over experience. They also highlighted and encapsulated rather than denied the ambivalence many people would have felt about their city being part of a larger whole that they themselves had no direct experience of. They were viewing something that aimed to replace their own personal experience with a mediated narrative, hovering between projection and self-recognition, with events happening off stage and in their peripheral vision. This allowed the ordinary citizen with no experience of foreign travel to imagine the power plays behind conquest, violence, and the creation of empire, and to explore those relationships from a place of relative calm and ease. Certain tropes were deployed to a knowing audience, beating familiar pathways, both subverting and not diminishing expectations, creating reciprocity between the viewer and the presenter of the image. Each knew what these signifiers denoted: the process could be recuperative. These circumstantial geographies represented substance, not truth, and tapped in to contemporary desires.

It was proposed to viewers that the world beyond Rome was something to behold, not to buy not to be sold, it was something that they could potentially hold. This cultural obsession with the idea of place permeated not only Rome's political and ideological life but its religious life too, certainly up to and including the role of place in the mythology of Christianity, and had its origins in the Roman elite's over-riding interest in ancestral connections to specific locales.

Rome and its citizens had slipped inexorably from dreams and expectations into an age of greatness, conjuring vivid, concrete images from the quotidian dullness of city life. The complexity and constraint of formal presentation nevertheless was a powerful manifestation of Roman state power, constantly confronting notions of difference and belonging in both a literal and a figurative set of journeys mediated by images. The vast majority of cases discussed during the course of this study have consisted of situations in the city of Rome itself, though many other contexts away from the centre of empire have also been considered. Again and again, thoughts return to the sculptures from the *Sebasteion* building in Aphrodisias,[3] as discussed at length in Chapter Eight. Here, women in particular could not express their feelings, their own being, through art or see it in much of the art: rather, they had it expressed for them by men in an equivalent to what Susan Sontag has called 'sentimental fantasies' of suffering. Images of anger and violence were counterpoints to passivity and powerlessness. There is no way in which the violence that underpinned Roman imperialism can be explained away as a side effect of economic expansion and necessity. That this violence dictated how the Roman state negotiated its place in the wider world is undeniable. That its impact on certain foreign lands and foreign peoples was catastrophic and entirely detrimental is beyond doubt.

Non-Roman women were routinely hypersexualised and objectified in Roman culture. War in the Roman masculine imagination was somehow deeply erotic. Even Roman mythology cast violence as regenerative, viewed against the degenerative violence of others. This all took place in an increasingly connected world that offered myriad opportunities both licit and illicit for exploiting women in the name of Rome. Rome was made of the earth, flesh, ocean, blood, and bone of all the places it conquered and all the people it encountered. In a way it was comparable to the fictional *Aleph* created by Jorge Luis Borges in the short story of the same name, a very small iridescent sphere which contains every other place on earth- the teeming seas, daybreak and nightfall, multitudes of peoples, urban labyrinths, unending watching eyes, all the mirrors of the world, and yet none of them truly reflecting Rome itself. It was suggested in an earlier chapter that if there was one place in Rome which played the role of a Roman Aleph, it was the *Templum Pacis* or Temple of Peace, with its extraordinary collection of

---

[3] See Chapter Eight Note 11.

Figure 121. Example of a Roman 'geography product'. Small bronze figure of the Tyche of Antioch. First century AD. Metropolitan Museum, New York. (Photo: Copyright Metropolitan Museum).

treasures, each a repository of unique cultural value both on an individual level and in terms of each item being part of a larger collection with its own meaning and emotional value, its botanical garden, and its one-time housing of the great Severan marble map of the city of Rome.

The monumental *porticus* or portico, and portico-temple had a long history 'as a type of public, politicised museum'[4] in the words of Elizabeth Macaulay-Lewis, that would lead to the creation of the *Porticus Pompei*. The *Porticus Metelli,* later replaced and renamed as the *Porticus Octaviae,* was a centre for the display of art collections, the make-up of which can be partially reconstructed by the accounts given of these buildings in Pliny's *Naturalis Historia*. M. Caecilius

---
[4] Macaulay-Lewis 2009:

Metullus had his portico complex, which included the Temple of Juno Regina and the newly-built Temple of Jupiter Stator, built in 146-143 BC to display artworks to the Roman public brought back as loot from his military campaigns in Macedonia, and other Roman works such a statue of Cornelia, wife of Scipio Africanus and mother of the Gracchi. While these were not 'geography products' as such, as defined earlier in the book, they nevertheless would not only have testified to Metullus' victories in Greece but also to the dramatic geo-political shift in relationships between Rome and Greece, with Macedonia now a province of the emerging Roman empire. Macedonia in particular as the home of Alexander the Great would have been highly evocative as a touchstone to many Romans. Alexander was quite literally evoked in the Portico by the display there of the twenty five looted equestrian statues of his *heteri* or comrades killed in the Battle of the River Granicus that made up the so-called Granicus Monument by the famous Greek sculptor Lysippus. The Granicus statues were displayed again in the *Porticus Octaviae*, constructed 33-27 BC, as well as the Cornelia statue and many Greek statues, and a painting of Alexander the Great. The painting and the Granicus figures probably helped cast Octavian/Augustus here as the new Alexander, expanding Rome's empire, prestige, and reach through territorial expansion.

Pompey's Theatre-Portico-Temple complex dedicated to *Venus Victrix* in 55 BC has often been called Rome's first museum, most convincingly by Ann Kuttner who particularly stressed the garden aspect of the complex.[5] In his Asian triumph of 61 BC Pompey had displayed, along with other more standard prizes and booty, a number of living trees, probably including balsam from Judaea, Ethiopian palm trees, and possibly Asiatic planes. The defeated Mithridates' library of plant books had also been taken as war booty, and as Pliny tells us in the *Naturalis Historia*,[6] that their translation became 'a victory that was as much a profit to life itself as it was to the Republic.'

It would seem that we can also talk about 'triumphal trees', indeed about botanical imperialism and a botany of desire. Studies of the ancient plant trade also highlight how imported exotica were used to create mini, domestic, colonial landscapes in both public and private gardens in Rome and Italy: this form of cultural exchange through horticultural design, technology, and plants was not altogether a brand new phenomenon.

The *Templum Pacis* was used to display artworks from two principle sources, the Second Temple in Jerusalem, that is war booty such as the temple menorah, shewbread table, and trumpets, and items from Nero's *Domus Aurea* in Rome. The Jerusalem artefacts represented not only the suppression of the First Jewish Revolt but also acted as a signifier for Judaea itself. A statue of the flooding of the Nile displayed there could have referenced Egypt's location as a base for the launching of Rome's response to the Jewish revolt or to Vespasian's supposed favouring by the eastern deity Serapis. The *Domus Aurea* material was extensive and varied, with no underlying ideological message linking the works, other than their being on public display rather than in the private collection of a now-despised emperor in the case of Nero. It could have included the famous statues of dying 'Gauls', Roman copies of Pergamene originals. Yet as Josephus observed, one was seeing here items that previously one would have had to travel to multiple locations to view-'into that temple were collected and contained everything which men previously travelled all over the world to see, eager to look at them

---
[5] Kuttner 1999.
[6] Pliny *Naturalis Historia* 25.5-8.

individually when they were in different lands'.[7] Peace and violence, conquering and culture came together here.

It is uncertain whether the gardens of the complex were used to grow plants and trees brought from abroad as cuttings and saplings. The suggestion that the gardens associated with the temple complex could have been used as a kind of imperial botanical gardens, nurturing and growing plant species brought here from overseas, might well be tentative but is not altogether without interest. Just as I was writing this section of the book London's Kew Gardens, one of the world's largest botanic gardens, announced in its next ten-year plan that they intend to examine the collections' historical origins and present information to the visiting public on the British imperial and colonial origins of much of the plant and seed collection here in Asia and Africa.[8] The document produced by Kew pledged to 'decolonise' the collection through addressing the 'exploitative and racist legacies' of its gathering and creation. The Director of Kew is quoted in the newspaper article as stating that: 'For more than 260 years, scientists from Kew have explored every corner of the world, documenting the rich diversity of plants and fungi. We were beacons of discovery and science, but also beacons of privilege and exploitation.'

The creation of Roman gardens and the propagation and growing of trees and plants at that time was work undertaken by slaves or members of the plebeian class. The use and enjoyment of the gardens and plantings was more-or-less the exclusive preserve of the Roman elite, though some public gardens were created in Rome. In Augustan times the greening of the city in this way went hand in hand with the transformation of Rome from a city of brick to a city of marble. It was no coincidence that plant iconography was significant in Roman art at the same time, as can be seen in the case of the vegetal motifs on the reliefs of the *Ara Pacis Augustae* and in the remarkable wall paintings from the garden room of the Villa of Livia at Prima Porta. To what extent the creation of plant, specimen, and seed collections was ever formalised as part of the Roman project of imperialism must remain uncertain, but there is no doubt that from the first century BC onwards there was a huge interest among the Roman elite in reading about plants and in plant encyclopaedias. Pliny the Elder's writing on plants was possibly influenced by his access to looted gardening and natural history texts brought back to Rome as war booty. The very fact that Pliny also mentions the presence of trees brought from overseas as being 'captives' in triumphs after the time of Pompey is hugely significant in this respect.[9]

Another recent phenomenon of interest and relevance here has been the interest shown in the collecting of trees for his garden by the former Georgian Prime Minister Bidzina Ivanishvili, initially triggered by the online dissemination of an almost surreal photograph of a large tulip tree being transported across the Black Sea. The arduous and expensive business of locating, buying, uprooting, curating, shipping, and replanting specimen trees for Ivanishvili's arboretum formed the subject of a documentary film by Salomé Jashi released in 2021. The film quite clearly sets out the sheer scale of such an undertaking and even without any explicit historical contextualisation provides insights into how plant and tree collecting

---

[7] Josephus *Bellum Judaicum* 7.162.
[8] *Guardian* 19 March 2021.
[9] On Pliny the Elder and his *Naturalis Historia* see principally: Beagon 1992 and 1996; Carey 2003; Evans 2005; Foss 2022; French 1994; French and Greenaway 1986; Laehn 2013; Murphy 2004; Pollard 2009; and Roller 2022.

Figure 122. Example of a Roman 'geography product'. Black and white mosaic of ships approaching the lighthouse at *Portus*, outside Tomb 43 *Necropoli di Porto*, *Isola Sacra*, Ostia. Second to third century AD. (Photo: Author).

might be motivated, and how such interventions can disrupt not only home ecosystems but also channels of memory. Some local people interviewed on screen appear to suggest that uprooting and removal of 'their trees' causes some kind of psychic rupture, and that without these trees to anchor communal memories those seemingly embedded memories will themselves start to fade and eventually disappear. This rupture moves the trees out of the diegetic world of human drama into another and totally fantastic realm centred on money, occasionally scientific and botanical research, and sometimes mania.

It would therefore not seem inappropriate for a botanical collection to have been housed at the Temple of Peace in Rome, but evidence for this is however purely circumstantial. There is no doubt though that the Roman elite interest in the management of country estates and gardens was part and parcel of the way in which elite identity was forged and maintained, and that those who read the works of writers such as Pliny were themselves buying in to the Roman imperial project through their interest in imported plant species. Imported plants which subsequently came to be cultivated in Italy included the pomegranate, certain strains of lemon, peach, and walnut.

It is perhaps the extraordinary breadth of the Roman engagement with geography which magnified its definition to include not only alien places and peoples, but landscape and topography, hydrology and the nature of the oceans, quasi-geology, journeying, the heavens and the Underworld, and even the winds and the weather. A great deal of writing about the relationships between war, landscapes, and weather can be found in many descriptions of battles in the histories of Livy. It has been suggested that bad weather was among a range of factors, including unknown or inhospitable terrain, used by Roman writers to explain certain historic Roman military defeats. In such accounts the weather and landscape topography, and Nature, in other words the environment in its totality, became enemy foes in themselves. Geography became flesh.

Of course, sometimes in both Greek and Roman literary accounts weather events were used as markers of omen, as harbingers of (disastrous) events to come. As the vertical sheets of torrential rain lashed down to the ground, interrupting battle, the water would soak into the ground, percolating down and feeding out in rivulets, natural runnels, and streams that would themselves feed into rivers. Surely not coincidentally, the scene directly below the so-called Miracle of the Rains on the Column of Marcus Aurelius in Rome is of a great river, high up to its banks, as if enhanced and nourished by the tributary infillings resulting from the deluge above. The viewer surely must have felt the same awe that the soldiers would have experienced seeing serpentine rivers flowing through seemingly limitless forests, sheer-sided rock escarpments vanishing into sky, and apocalyptic clouds looming over wispy treetops. This was both a real and a metaphorical journey, an adventure story, and a quasi-history of the world delivered in the artwork of a single monument. In the works of the Brazilian documentary photographer Sebastião Salgado miners crawl like termites up the muddy sides of giant pits, refugees cling to life in dusty wastelands, and villages burn, all images eerily reminiscent of the images of struggle and destruction on Marcus' column. But the Roman empire, for all its rapacity, only destroyed a little bit of the periphery of the wider world: there were untouched lands and peoples beyond. These deep spaces seemed to recede into infinity, containing the ages.

Much of the discussion in this study has centred on the presentation of geographical information to people in Rome, or how the very fabric of the city or of specific spots in the city could potentially inform about other cultures, other peoples, and other lands. However, it must not be forgotten that geographical knowledge could also be acquired by travel, again as we have seen, and indeed by an early form of tourism which included the opportunity to acquire not only knowledge but also souvenirs which then took on the role of mnemonic devices.[10] While we cannot talk of mass tourism and destination holidays, it would seem that Greece in general and Athens in particular exerted a strong pull on the cultural imagination of the Roman elite who travelled there in quite significant numbers. It is curious that ancient historians do not write about a 'Grecomania' in the same way that they seek to highlight 'Egyptomania', given that the cultural intertwining of Rome and Greece and influences running both ways at different times was far deeper and altogether more complex than the relationship between Rome and Egypt.

Travel looked back to earlier myth, and yet at the same time forward to historical journey as jeopardy, as in the case of journeys to Hades. As we saw in Chapter Nine an emphasis on journeys and journeying helped to acclimatise people to the idea of change, of progress, and

---

[10] On mnemonic systems-'memory palaces'- in Greco-Roman thought, and in later periods see principally: Spence 1985.

of surprise. The journeys of Bacchus/Dionysus came almost to be symbolic of other journeys of conquest, first those of Alexander the Great and then those of Augustus. Journeying in the footsteps of Bacchus became like some kind of eternal return, both a discovery and a homecoming at the same time.

The ancient mapping of the afterlife constituted a kind of spiritual geography: the afterlife journey harmonised the soul with the contemporary idea of what constituted the universe. Again, in relation to journeying there is the creation of the composite image which might best be called 'the winnowing oar', in relation to the prophecy said to have been made to Odysseus that he must wander the earth until he found a people who would mistake a rowing oar for a winnowing fan or paddle.

That much geographical information in the Roman world either took the form of itineraries-routes from A to G, via B, C, D, E, and F, with distances between each set of places given-or was structured in a similar form to an itinerary cannot be disputed. To read a number of quite recent studies of Roman geographical knowledge and knowledge-exchange is to come away with the idea that such an ordering of geographic information ranks a poor second to accurate mapping and spatial accuracy, when in fact itineraries were perfect tools for their time, aiding in the management of both short and long-distance travel, and quantifying the potential length and duration of journeys. Modern motorway maps, with sequentially-numbered junctions, railway maps, tube or metro maps, and indeed Satnav routes are all schematised itineraries of sorts, co-existing with broader systems for accurately depicting spatial complexity from Ordnance Survey maps to online Google Maps. A.J.P. Taylor's thesis that railway timetables inadvertently dictated the quickening pace of mobilisation of national armies on the eve of the First World War might equally be a framework to have applied to Rome's rapid territorial expansion linked to its road building and road network creation.

Intuitive geography was employed by sailors on sea journeys, but this eventually fed into or shadowed scholarly geography and fully-reasoned geographies. What has been called 'illiterate geography' or 'common sense' geography is not a descriptive phrase I fully empathise with in terms of its pejorative connotations and the suggestion of opposed elite levels of knowledge. Geographical knowledge was not simply about maps and spatial understanding. Ancient critical theory was a broad field encompassing both the rational and the irrational, the literate and the non-literate. Geographical knowledge was sometimes the product of the Roman cultural and political systems, teasing out its political relevance in explorations of family roles, desire, and subjectivity. It unsettled conventional ways of understanding the world, by allowing viewers to think beyond the limited categories previous orthodoxy took for granted. It pointed towards what cultures find difficult to name and know.

The term 'illiterate geography' has recently been used, occasionally interchangeably with 'common sense geography', by Daniela Dueck in an excellent monograph study of the phenomenon in classical Athens and Republican to early imperial Rome.[11] While the word 'illiterate' is used here to contrast such a geography with ancient literate or written geographical studies or descriptions, it is not a term that I particularly like-it seems at once both negative and pejorative-and indeed 'illiterate' could have been replaced with 'non-

---

[11] Dueck 2021.

Figure 123. Example of a Roman 'geography product'. Part of frieze depicting captured, bound barbarians, Trier. First century AD. *Rheinisch Landesmuseum*, Trier. (Photo: Author).

literate' without loss of meaning. The elite male audience for literary geographies, that is written texts, would always have been relatively small.

Dueck's project included not only the consideration of visual evidence, as in this present study, but also information that could be gleaned from the texts of contemporary public speeches delivered in Athens and Rome, from the texts of ancient plays, and from the analysis of proverbs and idioms in ancient Greek and Latin. It is certainly worth briefly summarising her conclusions on the analysis of these written sources in the Roman period, particularly her quantification of the data revealed by analysis, as this can help throw light on the visual evidence and provide a broader context for certain aspects of the visual presentation of geographical and topographical information in the Roman world.

The surviving texts of contemporary public speeches delivered in Rome are few compared to those from ancient Athens, and indeed most of the Roman speeches that have come down to us were all given by Cicero and preserved in probably polished versions in his copious writings. Analysis of Cicero's speeches when read alongside his letters and other writings provides a fascinating snapshot of the geographical scope and scale of his world and the spatial ambit of his life. We know where he lived, travelled, and worked and his other geographical connections as manifested by family ties and friendships revealed in the letters. Dueck counted 66 references to places in Italy, including Sicily, in Cicero's 52 surviving public speeches, delivered over a thirty years period, 18 in mainland and island Greece, 20 elsewhere in Europe, 26 in Asia Minor, four in Asia, and seven in Africa.[12]

Although Dueck quantifies geographical references in the plays of a number of Greek writers (Aeschylus, Sophocles, Euripides, Aristophanes, and Menander) I will not discuss these further here, even though some of these by-then old and classic plays would have been performed and read in ancient Rome, with the resulting opportunities for the transfer of geographical information to contemporary Romans from the works of these long-dead writers almost in a hauntological way.

The numbers of geographical references in the texts of ancient Roman comedies by Plautus and of tragedies by Seneca are almost equal, though there is a difference in emphasis on location bias. Both refer a great deal and more-or-less equally to Greece and its islands, Plautus more to Italy than Seneca, Seneca more to the rest of Europe and to Asia than Plautus, with Asia Minor being referenced more by Plautus, and Africa just slightly more. As to the comedies of Terence, geographical references and settings are very few in number, 17 in total, of which none was a reference to Italy and in which Greece figured heavily, and obviously were otiose to the requirements of the working, plotting, and understanding of the plays.[13] But what does all of this potentially mean? It would seem that the Roman theatre-goer was exposed to a great deal of geographical referencing, some clear and correct, some obscure or even obtuse, and some incorrect or even misleading. None of these playwrights was a Roman state stooge, placing and embedding geographical information at the behest of the Roman state for benevolent or nefarious purposes, and we must therefore ask why so much information of this kind was presented in the plays of theirs that have come down to us? As has been noted by Dueck, with regard to Seneca, 'remote peoples are typically associated with riches and unique exotic products'.[14]

Taken together as a group, Roman poets and playwrights 'delivered to their audiences a specific understanding of the size of the world, of routes within it, of places and peoples'.[15] However, not all of this information was correct or strictly true. For instance, it has been noted that certain elements of Seneca's geography were either inaccurate or vague and not properly defined. In many ways this is not altogether surprising, but it is interesting to consider the meaning and impact of this when compared with the relative absence of topographical detail or specificity in relation to civic monuments in Livy's writings on the city of Rome. Livy's methodology contrasted considerably with the quite specific contextualisation of monuments in Rome's cityscape and the topography of the city in the writings of Dionysius of Halicarnassus for example. Were these different works tailored to their reception by different audiences or could there have been some other explanation for this curious contrast in presentation and contextualisation?

Equally interesting is Dueck's analysis of Latin proverbs and idioms such as 'Africa always brings something new' and 'as much grain as Africa reaps'.[16] On certain Republican coins a

---

[12] Dueck 2021: 58 Table 2.14.
[13] Dueck 2021: Plautus 98 Table 3.8; Terence 102 Table 3.9; and Seneca 104 Table 3.10.
[14] Dueck 2021: 106.
[15] Dueck 2021: 107.
[16] Dueck 2021: 116.

Figure 124. Example of a Roman 'geography product'. Relief from sarcophagus depicting a banquet scene and a wind god, possibly Rome. Third to fourth century AD. *Rijksmuseum van Oudheden*, Leiden. (Photo: Author).

female personification with sheaves of wheat and ears of corn often appears with African animals, making much the same point as the latter proverb cited here. Many geographical idioms were based on ideas centred around ethnicity and difference and, more often than not, on prejudices of one kind or another. 'The more the Parthians drink, the more they become thirsty' one went. Gauls were associated with credulity, Greeks as frivolous or idle, and so on.

Such proverbs and idioms with geographical subject matter or links occur in most languages and cultures, and often have a national or regional circulation. For instance, in English the proverb 'it's like taking coals to Newcastle' is nationally known. To explain to an international reader it means that Newcastle, a once-famed shipping port for British coal, was a place that needed no more coal taking there. Again, 'you could ride bareback to York on that', referring to a completely blunt knife, is a phrase I've only ever heard my Yorkshire mother utter. Such geographical idioms could also relate to humour and jokes, and while quite rightly in Britain the genre of jokes known as 'Irish jokes', with Irishmen the butt of ridicule based upon a misconceived lack of intelligence or nous are now frowned upon, in Ireland such jokes are told about the supposedly stupid or naive 'Kerry man'.

Transmission

There is a vast difference between the level and detail of geographical knowledge required to inform political and strategic intelligence decisions and simple background knowledge intended merely to inform those who need to be able to cope with the certainty of uncertainty in a shifting, unpredictable world. The latter was part of a broader suite of strategies required to acclimatise Rome's citizens and other inhabitants there to the ever-changing nature of what it was to be Roman when the mother city was but a centre rather than simply its own enclosed world. The difference between *orbs* and *urbs*. Throughout this book it will have become apparent that Rome's success was very much reliant on its ability to integrate and absorb exogenous elements into its host body, be they territories, peoples, cults and religions, and indeed cultural practices. This assimilation was, of course, not without its limits and exceptions, and more often than not achieved by force of arms rather than just by force of will. The spilling of blood could be regenerative for the Roman state in the same way that religious sacrifice underwrote individual piety. Violence was such a deeply entrenched threat in this world that it permeated and infected all human interactions. Trauma was not just an individual event: it was a psychosomatic experience on a social scale. Comparing a society or country to a body made an authoritarian or reactionary ordering of the world seem inevitable, somehow immutable, but the Romans could also liken the body to society, the body to a temple, to a city, to a fortress and so on.

One of the most noticeable omissions in the most recent studies of exchanges of geographical knowledge in Roman times has been discussion of oral communication between friends and acquaintances, either based on first-hand knowledge on the part of one of the parties or based on second- or even third-hand information. Knowledge exchanged in this way need not necessarily have been accurate or precise, though of course it could well have been, and it might indeed have been muddled or sketchy, even inaccurate or exaggerated for effect. However, it is likely that sharing in an experience in this way, through travellers' tales or accounts given by soldiers or bureaucrats who had served in the provinces, perhaps had a more profound effect on those hearing such stories than their gleaning of information through official channels in the form of imperial propaganda.

The Romans were able to use art and culture more broadly to become a pathway for knowledge: they synthesised a great deal of disparate information in a way that transcended summation or homage to become instead an act of artistic assimilation that acquired its own generative power. Desire was laid bare by an astute, unflinching elegance. Part of what bothers me is that formalising such an idiosyncratic and unwieldy thing as desire is practically impossible, or at the very least fraught with difficulties. This study has had to circle and fold back on itself as present analysis and ancient Roman practices give meaning to each other. There is a kind of mimetic evocation of minds at work in the way that Montaigne talked about '*la peinture de la pensee*', that is the painting of thought. If the Romans were always seeking to imagine themselves anew by the presentation of geographical exemplars, and by other strategies, then these openly-expressed mediations on identity led towards some kind of reckoning with the great betrayal of the Roman ideal by the need to constantly redefine Romanness in relation to others and not simply on its own terms.

SLOUCHING TOWARDS EMPIRE

Figure 125. Example of a Roman 'geography product'. Cupids carrying a crocodile, Oxyrhynchus. Fourth to fifth century AD. *Rijksmuseum van Oudheden*, Leiden. (Photo: Author).

A version of what has been termed in other contexts *nostalgie de la boue*, literally a nostalgia for mud or dirt, for lowlife, for simpler things and times, resonated throughout the Roman period. There seems to have been a fear that the Romans were not who they claimed they were, a state addressed by national rituals of expiation and the moments of moral reckoning that inevitably came. That this could be made to seem part of the ordinary horizon of expectation constituted what was up till then an unparalleled violence which was being done to the contemporary sense of reality, to the citizens' humanity. This was part of a process of moving towards a world in which nothing important was regional, local, limited: a world in which everything that could circulate did so, and in which every problem was, or was destined to become, worldwide: escaping to a place of familiarity could have meant updating old stories for a world they were not created to anticipate. Looking at Roman geographical art and its deployment suggests that the aesthetic and conceptual motivations behind this art were part of an active process, rather than the fixed singularity of an artwork being consummate. They presented their mysteries without recourse to the expositional, they succeeded in their embodiment of ideas in contrast to an illustrative description of ideas. The geographical art was never declamative or stable in its meaning, but engaged with an infectious energy redolent of an unfathomable interconnectivity, enacting the mobility of how meaning was made over the stasis of what was meant. Part of the making of such works was just saying it, over and over. The ability to project events with some accuracy into the future enlarged what power consisted of at Rome, because it was a vast new source of instructions about how to deal with the present, with waning, decline, entropy, as opposed to uncontrollable growth for all time.

Throughout this study it has been made clear that geographical information in the Roman world could be presented in a myriad of different forms, and that its reception was entirely dependent on context. The idea of the *flaneur* or the psychogeographer 'floating' around the streets of Rome and almost subconsciously imbibing information about Rome's place in the world as he or she walked is an attractive one. But equally a walker could have been totally indifferent to the messages we believe the city was there to deliver to its citizens. Images and inscriptions, architecture and artworks are only informative if their viewer can be or is prepared to be responsive. How much of this geographical information might have been taken in, and the intellectual reaction to it, again would have depended on the identity of the viewer. Were they a citizen of Rome? Were they of the elite, were they senators or military men? Were they elite women? Were they freedmen or slaves? Were they merchants or traders, artisans or labourers? Were they visitors to the city?

As was discussed in Chapter Eight it is altogether possible to argue that the presence in Rome of hundreds if not thousands of images of captured or subservient male and female barbarians could be accounted for in a number of ways, not directly linked to geographical information exchange. In relation to images of female barbarians their presence in some cases could have been in some ways similar to the early modern poetic invocation of (female) bodies as representing new, undiscovered, or exoticised lands. If then further applied to images of male barbarians this poetic sexualisation might have taken on homo-erotic overtones and alluded to issues focused around ideas of domination and of subservience.

Non-Romans and non-Italians otherwise very rarely appeared as subjects in images, though the Romanised Moesian provincials greeting the emperor Trajan as he enters their city en route to war in Dacia, as appear on the frieze around Trajan's Column in Rome, represent a particularly interesting exception to this rule. Another exception is marked by the occurrence of a number of images of Egyptian priests in certain contexts in Roman Italy. Visually very identifiable by reason of their distinct clothing in the case of priestesses and by the shaved heads and diagnostic linen clothing of male priests, these were officiates in the cults of Isis and her consort Serapis, the former of these cults being particularly popular in Italy from the first century AD onwards, peaking under the Severan dynasty. Temples to Isis-*Isea*- in Italy are known to have existed in Rome, Ostia, Puteoli, Pompeii, Herculaneum, and Benevento, to name but a few places. It is difficult to know which images of exotic priests such as these related to cultic membership or adherence and which were simply part of the overall Roman discourse around the exoticisation of all things Egyptian.[17] In Pompeii, for example, nineteen images of Isis have so far been recorded and twenty six of Isis-Fortuna. Of some significance is the role that the goddess Isis was deemed by the Romans to have played in relation to the protection of shipping and mariners and, to a lesser extent, the same attribute attached to Serapis. For a society whose raison d'etre was principally war and the power of military might inextricably linked to maritime trade and commerce, seeking protection from them in these roles would have been logical and sensible. The place of non-Roman gods such as these in the

---

[17] On the cult of Isis in Greece and Italy see principally: Mazurek 2022.

state pantheon, and their depiction as images in the form of cult statues or portable figurines for display in home shrines, is also of great interest.

If nothing else it has been demonstrated in this book that the dissemination of geographical and topographical knowledge at Rome and more broadly in other locales was a carefully-calculated and well-calibrated exercise, involving the presentation and revelation of information on a number of different levels, from the specific to the general, and from factual to allusive. What was presented or revealed was part of a cumulative process that revolved around the idea that Roman self-image was principally defined by the peoples and places outside of Italy and within its empire. In other words it helped clarify the fundamental question as to what it meant to be Roman. The presentation of geographical and spatial information about the city of Rome itself to its inhabitants was less intensive, but equally pervasive and ideologically mediated. In many ways the city of Rome acted as a synecdoche, metonymy or even epitome for the broader contemporary world and its representation. Its use of imagery in imparting geographical knowledge to its citizens spoke clearly of presence, mutability, and possibility in many forms. The presentation to the people of Rome of much of the geographical information discussed in this study was undoubtedly ideologically driven. However, a great deal of it was also incidental in terms of its form and the methods of its dissemination. The ideological information formed part of a much broader didactic programme, while the more informal passing on of knowledge was simply the result of the regular mass consumption of images. This mass of information, this plethora of geographical detail, allowed its recipients and viewers the opportunity to consider Rome's place in a broader world and thus to locate the individual Roman themselves. All the senses often could be involved.

But this was not some kind of nostalgia for what could not be, a journey on a road to nowhere. Rather, this self-enclosed aesthetic system elided geography with anthropology, this unity in being of the personal and the social at its peak made sensate in the form of didactic images. But the viewing audience was not a blank canvas: the hyper-politicised urban plebs of Republican and early imperial Rome were anything but. Like the voice in Verlaine's *Langueur* they perhaps could have proclaimed 'I am the Empire in its decadence', even if these encounters with presented geographic images were intermittent and evanescent. Such images as an assemblage were united more by the myths that they intrinsically connoted and the environment and contexts within which they were framed than by the aesthetic or formal qualities they outwardly displayed. Taken together, as in this present study, it can be suggested that they did not have the specificity and tautness of a true series from which certain aesthetic conclusions or imperatives could be drawn. Rather, they served as a useful coded expression for a whole array of different messages and practices that existed as an extended and multifarious group with partial resemblances and differences, but not sharing any core aesthetic messages that made them one. The series, if considered as a single artistic project, approximated a kind of rapture by setting up a fluid interplay of visual and textual refrains, framing and contextualising existing archival images elsewhere in the city, creating a continuous frame of action and reactivation across the group. These social preoccupations and aesthetics must have resonated with the practices of information presentation and exchange, oscillating between the negative and the affirmative, though obviously framed as constructive endeavours.

Reception depended on a pool of shared knowledge, common capabilities and interests, leading to the expansion of consciousness to a shared space beyond the framing/contextualisation of the images. These images of foreign bodies in space and time were richly layered, but their very being, their materiality, was underpinned by violence, exploitation, and death. This stunning ambivalence allowed political action to be presented as a creative act, one that relied on the ability of the viewer to imagine a future that encompassed consideration of this bold insistence on negative spaces. While being able to consider multiple images together allows us a much better grasp of how they functioned as less a series of isolated compositions rather than as stages in an ongoing attempt to uncover the precise nature of *Romanitas*. Repetition led to often messy, obsessional, and overlapping messages, undermining the singularity and sometimes great power of each individual work. Such a juxtaposition of the single image and the series was both fresh and reactionary, in that it was either a rejection or a response, negative or positive, to all that had preceded it in pre-imperial times.

The power of any illusion such as this is that it draws its significance from what it is not: a common reference point, a coming together of time and space mediated by the central authority, deprived of all reference points and all coordinates. This reflected an understanding of social space in political terms and would appear to have been denying the innocence of a single monument. Rather, the Roman state metaphorically dug up all the unpaid debts of Roman history, like museum curators, and at every moment old images were ripe for effective rediscovery, their confrontation with particular events then giving them a largely new meaning.

The unity of Roman imperial power and messaging was threefold: coercion, seduction, and mediation. Understanding of its messages marked the spontaneous coherence fusing attempts to enrich the everyday lived experience. People were being given the tools to aid their own subjectivity, out of which they had to build a world that everyone else would recognise as their native land. Some though would have found nothing in other people apart from their own absence. Geographic images reinvested appearance in reality: a bridge between imagination and reality was being built. Uptake could never have been uniform-that is the unique power of every individual viewing experience-and people must have always been at different stages of consciousness and determination. The real power in providing this information must have lain in the degree of collective unity it attained without losing its variety. It strengthened the wealth of individual possibilities in the unity of federated subjectivities. Those who could not see themselves in images of other lands and peoples were perhaps condemned for ever to be strangers to themselves and to the proposed universal project of empire which ultimately functioned as part of an order of appropriation and possessions. This was a project requiring participation.

A citizen of Rome in the first century AD would have felt very differently to a citizen of the fifth century in terms of thinking about their place in the wider world. To begin with, the founding, or rather rededication, of the new Rome at Byzantium-Constantinople in AD 330 under the emperor suddenly moved the 'centre of the Roman world' to the east. Of course, there had been harbingers of this need to decentralise affairs away from the city of Rome, and the division of the Roman world into a western empire and an eastern empire represented just such an event. At the same time Rome was transforming itself from a pagan city to a Christian one. Roman Christians would also be considering whether the centre of their world

Figure 126. Mosaic panel bearing a personification of the Tyche of Antioch merged with a portrait of a Hellenistic ruler (Arsinoe II perhaps), Thmouis, Egypt. As early as 200 BC. Graeco-Roman Museum, Alexandria. (Photo: Copyright *Centre d'Études Alexandrines*).

was to be found in Italy or in the Holy Land. If this was quite literally an existential crisis, it needs to be considered whether the last great imperial monuments at Rome-including the *Arcus Novus* and the Arch of Constantine-carried any geographical information which helped Roman viewers of these monuments to understand more about themselves. The building of Christian churches was literally writing on the surface of the city of Rome, altering its geography and its religious topography, and shifting the political focus of the city away from the forum and associated buildings, including the state's pagan temples, at its centre towards the peripheries, creating a number of new foci of influence.

Insight

In late 2006 to early 2007 the author Toni Morrison acted as a guest curator at the *Musée du Louvre* in Paris on an exhibition project entitled L'*Étranger Chez Soi* or *The Foreigner's Home*, involving the creation of a number of guided visitor routes through the rooms housing the museum's collections of ancient art and the contextual labelling of individual artworks or artefacts along the routes to explain their inclusion and significance in the grand overarching theme of home/exile and longing/belonging. I am particularly interested in the theoretical underpinning and rationale for the project and this has been articulated in the exhibition's catalogue, in a film and filmed interview with Morrison based on the project, and in some of her other writings collected in the volume '*Mouth Full of Blood....*'. It was argued in the exhibition that the present day should not be seen as a unique period in terms of the mass movement of people, and such movements in the ancient world linked to colonisation and empire-building were equally linked to trade, political intervention, and to the resulting violence, persecution, exile, and in some cases poverty. Borders, frontiers, porous places, vulnerable points, customs barriers and posts, hidden routes all exist now or existed previously somewhere. Soldiers, bureaucrats, merchants, traders, intellectuals went one way and slaves the other. The journey of the colonised to the seat of the colonisers could be both involuntary, as in the case of slaves, or voluntary, as in the case of foreign merchants. But it was still a world which enfolded within itself all there was in its orbit. Mania, preciousness, and vanity. A duty to constantly recall the gods from banishment through the medium of politics.

If today ideas about race, identity, and foreignness inform the discourse around politics and art it was rather different in Roman times. Again, if family, language group, country, gender, religion, and race are the principal factors used to differentiate between 'them' and 'us', then difference in Roman times was complex but subtly different. There was not however a disregard of the margins and the marginal peoples who lived there: if anything there was at times an over-concern with the frontiers of the Roman empire. Engulfing, accelerating erasure, the flattening out of difference and of specificity created or took place at borders: real, metaphorical, and psychological.

There can be no disputing that blood lay at the roots of Roman civilization and Roman religion. That rape, abduction, murder, blood sacrifice, and revenge occurred so commonly in Greco-Roman mythology, that these essential allegories of who the Romans were helped usher in a world of unfixed meanings and constant dread, particularly away from the centre.

There was certainly very little negative rhetoric among Roman writers about the acquisition of foreign goods and materials by the Roman state, even if there was quite considerable criticism of what was seen as the dangerous indulgence in luxury by individuals, with simplicity and abstinence being projected as good Roman qualities. Overindulgence and a propensity towards luxury was most certainly decried in some quarters as being decidedly un-Roman. Mining, logging, trawling, scorching, levelling, and even poisoning or polluting lay behind this. Rare elements and materials were clawed from the earth. The modern idea of so-called 'conflict materials' which links the excessive western consumption of metals such as tin and tantalum with violence, war, and human suffering, and environmental degradation is a useful lens through which to view certain aspects of Roman materials exploitation.[18] The circular economy, in which the life-cycle of materials becomes a closed loop rather than a linear model of consumption whereby a line with a beginning and an end existed, was not one that made economic sense in the ancient world. The late antique practices of the recarving of statuary, the recutting of gemstones and cameos, the reuse of architectural stonework may have been both the result of shortages of raw materials at this time, or the collapse and disruption of supply lines, or a new cultural choice.

This was all part of a kind of post-political politics, with Republican ideas and ideals replaced by imperial diktat and certainty. It was reflected in the growth of public and semi-public identity presentation by the emerging freedmen and freedwomen class, and by an emerging, though not overarching, ideology that married pragmatic realisation with the needs, demands, and

---

[18] On Roman asset-stripping see, for example: Loar *et al.* 2017.

Figure 127. Example of a Roman 'geography product'. Nilotic-themed wall painting, Pompeii. First century AD, before AD 79. *Museo Archeologico Nazionale*, Naples. (Photo: Author).

desires of the Roman people as a whole. Desire here was not necessarily physical or erotic, but rather, as in French critical theory of the 1970s as promulgated in *Semiotexte* or *Tel Quel*, a force that refuses to recognise geographical or political boundaries and which was increasingly manifesting itself in Roman society by the flows of capital, goods, and culture from outside Italy. The exaltation of desire was expressed in the escalating and intensifying presentation of geographical images to the Roman public as signifiers of this change. In those sections of Roman elite society with a puritanical streak, where luxury was a dirty concept linked to a disdainful suspicion of vulgar materialism, this must have seemed flagrantly heretical and running against the grain of Roman culture. This did not mean that the images' viewers saw utopian yearnings for purity, perfection, and the absolute in them or indeed encoded in the new consumerism linked to empire and expansion. It might be thought that these constant visual references to geographical information were simply anodyne, but it can conversely be argued that on closer inspection they turn out to be images of contradictions, with an aporia-a void in the fabric of their meaning-lurking at their heart and in their seriality. This apparent surface emptiness was also a fullness, discourse as a lexical maze. The series became a chain along which desire traversed endlessly, looking to heal some lack at the heart of Roman being. It was both a project of undoing and unsettling, and of integration and fulfilment.

If it was possible to map all known ancient trade-routes and all attested journeys made in ancient Roman times the spatial data presented is likely to cast a new light on the way in which the regularity of contact between citizens or dwellers in Rome and those of distant provinces and lands beyond fed in to the dissemination of geographical knowledge at Rome as a whole, even if only at anecdotal level.

Opaque Manifesto

The ancient city of Rome, like many cities since, became a dwelling place of diasporas, a set design of the spaces between cultures, a locus of transience, and a place of cross-cultural encounters, some highly eroticised or sexualised as we have seen.

If the art of sculpture can be accepted as freezing the body, indeed as otherwise only death can do, then it can be seen as being transformed in the process into a *nature morte*. The overlap between anatomy and the mapping of the body was best represented in Rome by the display of the Hanging Marsyas statue, but the colours of Rome were themselves the colours of human hair and flesh. In a similar way Vitruvian Man had the buildings and cityscape of Rome written on his body. Women in Roman art were often simply collapsed with the terrain in a type of representation that feminised the very notion of land and space itself: the lands were not simply feminised but rather subjected to a gendered act of mapping. If the landscape became a body, the body conversely became a site of mapping, and yearning for a place might have turned into a bodily desire, a physical matter. People became places, living maps, as lined faces so often detail the human life-course and ultimately become the maps of our deaths. Mapping like this defied decolonising: they were not just in the service of domination, and partial mapping might be thought to have resisted a univocal and totalising vision.

Time and discourse at Rome were not only understood spatially but were mobilised in imaginative ways. As mental mappings allegories both portrayed and promoted the presence of women in the geographical field, though not all such figures were women, if we consider the Titans and Atlas, for example, in this way. The issue at stake was not simply one of iconography. Viewers would have been caught up in the very flux of psychogeography, interiorising the layout of the city, practiced in each of its pivots and sites of junction, digesting every single point of entry and exit. Such navigations would have connected distant moments and far-apart places by absorbing and connecting visual spaces. Narratives would have risen, built, unravelled, and dissipated, revealing potential sites of opening, an elsewhere that was nowhere. An obsession with searching and finding hints of distant lands and peoples would have both revealed and covered a fear of being lost. Presentation of a series might have aimed to corrode the opposition between mobility-immobility, inside-outside, private-public, and dwelling-travel, with architecture in Rome being a map of both dwelling and travelling.

Geographic images had a mirroring effect, the views of the city of Rome and vistas of foreign lands offered back to the urban audience for viewing. The effect of this must have been cumulative. Whichever path was followed there were different points at which fragments came together narratively. By making tours and detours, turns and returns, the viewer opened up on different vistas of the production of space. If the art of memory was an architectonics of recollection, then so were these a stimuli individually and as a series, making a vast haptic terrain, ultimately constituting a movement from the purely optic to the haptic. Spatio-visual arts created a bond between architecture, travel culture in all its forms, the history of visual art and its well-established tropes, and memory and map-making in its broadest, non-literal, meaning.

SLOUCHING TOWARDS EMPIRE

Figure 128. Example of a Roman 'geography product'. Ceramic oil lamp decorated with image of eroticised woman (caricature of Cleopatra?) on the back of a crocodile. Provenance uncertain. AD 40-80. British Museum, London. (Photo: Copyright Trustees of the British Museum).

What was produced was a mixture of utopias, centred on imperial harmony, and dystopias, relating to conquest and enslavement. The drive to possession and domination created an erotics of knowledge, a spatial curiosity. The geographies of space and the body were combined in this age of exploration and empire. The reproduction of the corporeal landscape turned the body into a site of exhibition, and in this respect one has only to think of the elite's ancestral *imagines*. Statue images acted to build a landscape linked with the opening up of new worlds of knowledge, as well as geographic new worlds, powerfully creating the feeling of simulated travel.

An image in the city was not a singular image but rather a fragment of display. It had to be seen in conjunction with a crowd of other images and to be read in the course of spectatorial motion. The geographical images of concern throughout this book overcame the form of the outer world-principally space, time, and casuality-to create and debate with the inner world of attention, memory, imagination, and emotion, a decomposition of history as it blurred reference to a specific time and space. The play of distance and proximity possessed a force that lay not in their aesthetic value but in their emotional power: taken singularly most were banal or routine, a collection transformed into narratives by way of emotive mobilisation. New images were added and incorporated, changing the form and territory of an ever-growing atlas: cartographic thinking like this need not always have been coloured by the impulse to conquer or by the language and ideology of power and its tendency to unify, to stifle the exploration of difference.

The use of geographical tags in many contexts allowed the Romans to highlight issues around the discourse centred on connectivity and interconnectedness, facilitated by mobility. If the use of geographical epithets could add value and a whiff of exoticism to a product, such as the (Egyptian) bean or (Ethiopian) cumin, then that product had assigned to it a symbolic value as well as a botanical and monetary one. The mantra-like repetition and listing of ethonyms, tribunyms, and lists of submissive barbarian kings, in Augustus' *Res Gestae*, on his *Tropaeum Alpium* at La Turbie, on the *Arcus Claudii* in Rome, and on the Mausoleum of the Plautii at Ponte Lucano served a similar purpose.

It was a series as a montage accentuating visual similarity as well as distinction. Contiguous, interlinked, organically connected to the inside and the outside. Somehow the images also portrayed absence, in fluid intersection, passing from space to space, in joined trends of public consumption. The images became an environment, an architecture. Rome was not just 'here': it also aspired and strove to be 'there', making it somehow in-between, but not in Homi K. Bhabha's sense of a zone of cultural intersections and marked hybridity. In Dutch art of the seventeenth century, where picturing and mapping coincided, a geography of transits was created, this movement, this geography of passage presupposing access, interface, border, and threshold.

Pliny's project to encyclopedise the entire world and its contents, and present it as Rome, was both prophetic and timely, at once both up to date and instantly obsolete. This built-in obsolescence was a product of the times themselves: catching *in stasis* a constantly changing world was a project doomed to magnificent failure. He proposed a way for Rome not just *to have* an empire, but rather how *to be* an empire, but the moment of his proposition passed almost instantaneously. Towers of Babel turned to bridges. Memories composed now a single memory, assuming the changing form of a shadow of moments past. It represented a time when poetry, plays, literature, images, and objects, paintings, sculpture, mosaics, words, voices, raiment, hair, bodies, food, drink, minerals, pearls, and animals could become geography products. The babble, Babel, of foreign voices as the city awoke brought the birth of a whole world with it, like Luigi Russolo's *Risveglio di una Città* (*Awakening of a City*) or Bob Graeffinger's jazz composition *City of Glass*' final movement *Reflections,* it created an almost musical vantage point from which the city could be viewed as a whole, entering from a morning world. Plurality became both Rome's strength and its weakness. As has been shown in this study topographical, celestial, marine, and wind charting looked at together seem to have expressed a desire of the Romans to also map the unmappable, just as both the map and the garden were themselves kinds of imaginary topographies.

Empire was here presented as the conquest of the discontinuity between individuals, with the city reconstituting itself as an oneiric location. Geography products formed a palimpsest, each recontextualising the other in a cancelled future. The art of Roman freedmen also sometimes learned to engage with geography if only to firmly locate the commemorated deceased within a specific place in the city of Rome.

Figure 129. Example of a Roman 'geography product'. Mosaic depicting a Nilotic scene, Pompeii. First century AD, before AD 79. *Museo Archeologico Nazionale*, Naples. (Photo: Author).

The display of geographical artworks at Rome did not represent the choice of one specific materialisation of the possible at the expense of all or some others. In terms of monumental display in purely architectural terms its characteristic feature was its insistence on popular consent, or rather individual integration in the world of ideological conditioning. Success lay in both the actual physical organisation of the city and in the setting up of this stable information system: people were required to identify with their own environment and that of others mediated by the presentation of selective detail and information. As the Roman empire expanded so too did the need for the minds of its citizens to be expanded exponentially too in every direction, ceaselessly creating new possibilities and eschewing old frontiers. The possibility of ambiguity, misunderstanding, compromise, or misuse was built in to geographical presentation underwritten by ideological motives, as was conscious choice and gamble.

While it is true to say that the Roman empire at its peak was a multicultural empire it was in many respects actually monocultural, in that while the mainstream Roman/Italian dominant culture interacted with cultural aspects from outside the Greco-Roman core it was not changed ideologically or politically as a result of this. Chronology, politics, economics, environmental change, and expediency were probably more telling factors driving change. With the very notable exception of Greek and then Egyptian art and material culture it is not possible to suggest that the art and material culture of Roman North Africa, of Syria and the Roman east, of Roman Iberia, of the Danube provinces, and of the north-western provinces had any identifiable impact on cultural life and art at the centre. Contacts and trade with other great powers like India and China again seem not to have shaken the core culture and inspired influences we can detect. What did the owner of the Indian ivory statuette or figure of Lakshmi from the *Casa dei Quattro Stili* in Pompeii see when he or she looked at this artwork? What did visitors to the house make of it when viewing it? We will never know. Yet in Gandhāran art can be found a profound reaction to contact with Greco-Roman art and culture.

Quiet Mapped Waters

Of course, I have only been able to suggest ways in which the presentation and reception of geographical information might have corresponded or differed according to context and circumstances. An image might have triggered no response or even a different one to that intended or expected. If the Latin word *hostis* could mean guest, or stranger, or enemy then ambivalence in response to images might also have been equally subtle and multilayered. To welcome in the strange or alien could be enriching, but it could also be dangerous or at least be thought to be so, even in the face of evidence to the contrary. Exposure to the desire of others, to their violence, to their neediness, to their gaze, created a shared destiny, a new identity, through no agency of the viewer's own, as they were confronted with their own prejudices and privileges.

There is no doubt that many people viewed identity as being situationally constructed, but a number of important questions might go unasked and unanswered if this situation is taken at face value or questions will simply be given schematic or provisional answers. The recurrence of these intercalations between political ideology and images was not always simply a repetition of the same thing. There were undoubtedly significant shifts and developments in the way that these encounters were negotiated and whether a meaningful historical trajectory can be discerned in the multiplicity of meanings presented and entangled in the process.

Especially dramatic and portentous was how empowerment with geographic knowledge, when it took place, found expression in new aesthetic practices. Cultural power here was in many ways analogous to economic power, building on the ideas of Pierre Bourdieu. This was not to deny the inherent tension here, that the full pursuit of economic capital was usually incompatible with the full pursuit of cultural capital. Indeed, the process became altogether saturated with aesthetic discourses at the expense of other interpretations. A genealogy of such geographical images, in Foucaultian terms, was not seeking to provide a continuous history, a seamless narrative, but rather to focus on certain eruptions, breaks, and displacements of the cultural field. It stressed heterogeneities and specificities. Genealogies focused on struggle and competition and were interested less in the narrative of events than in patterns and structures.

Everything in the geographical field must have acted like a citation, embedded discourse, mention rather than use. The false symmetries of good sense reflected techniques of compression and collage, myths of abjection and omnipotence. Homesickness was presented here as a pathology and became a conduit through which a world could emerge. Any attempt to constitute the images as settled carriers of meaning ran aground on their incompleteness

Figure 130. Example of a Roman 'geography product'. Sarcophagus carrying image of cupids/*putti* operating ships in a busy harbour, with buildings behind forming an urban backdrop, Rome. Third century AD. *Terme di Diocleziano Museo Nazionale Romano*, Rome. (Photo: Author).

and inconsistency, as against the latent power of symbols and sigils, and the potential ecstasy of communication.

Everything that had directly lived under the Roman system moved into a representation, caught between everywhere and nowhere, geographical images embodying a very particular sensibility and representing a site or *locus* of deluded aspirations, and yet at the same time probably occasionally freeing the imagination and shackled ambition. All of this emerged in tandem with the defining motifs: consumption, art, the scrambling of linearity, the debasing of perceived truths or grand narratives, the collapse of representation into reality: tensions born of such developments were engaged with and played out in various ways. The ideology lay in its practice, in its ability to pleasure, surprise, transgress, inspire, question and imagine, a mode of critique rather than a definite answer or solution. It confronted, challenged, and gave vent to a message that was both resonant and ambiguous, and it provided a space or spaces to reflect upon or explore the social tensions contained within it. It helped collapse the distinction between here and there and, in the words of Jean Dubuffet, sought instinct, passion, caprice, violence, [and] madness. Such images demonstrated a conflict between feelings of being rooted and rootless, belonging and not belonging, place and displacement.

Compassionate scrutiny probably would have revealed the moral complexity of many geographical images, a desire to be fully part of the world mediated by images of oscillation and unsettledness and shadows. Revelation would have helped shatter the sense of continual change, initiating the reuse of images to allow the past to exist in the perennial present. No longer were lost or marginal cultural resources recovered and reimagined: they existed concurrently to be collated, imitated, decontextualised, and disarmed. The differences that had always been contested merged or conflated, and point and purpose hardened as cultural forms evolved away from their initial stimulus. Cultural changes enabled competing sites of attention to operate contemporaneously and probably sometimes competitively too, continuing to provide a means of agency and a platform, even as its cultural traces were archived, appropriated, and often historicised.

A process of critical engagement was required though, as these images must sometimes have appeared to be contradictory and formative, implicit and explicit, liberatory and reactionary. Meanings were projected but also cultivated from within, shaping the dialogues that ensued as cultural spaces were opened up. The exploration of sexual, psychological, and delinquent extremes was surely a bi-product of the viewing of such images by certain viewers, as has been argued in a number of places in this study.

Taken together, the collective assemblage of Roman geography products and geographical images were like a series of return journeys which could not be judged on output alone. There was so much more hidden, suggested, left to the imagination. The mystery was left caged. It remained an exception to the rule. Moments of viewing must have helped provide insight into life as a series of potentially random journeys that one might take or might imagine oneself taking, showing some viewers some of the possibilities.

But can we really talk of a form of globalisation existing in the Roman era? Certainly connectivity and the establishment of networks can be thought of as prerequisites for this, but the slave economy was such that its very nature argues against anything other than relations between many peoples in the Roman world and Rome the centre as being like a bridge. As Martin Heidegger wrote: 'the bridge is a site. It locates a space in which the earth and the heavens, the divine ones and the mortals gain a joint presence.'

The Romans had many ways to manage geographical information, not only by preparing texts of one sort or another and by literally inscribing themselves on the land of others, and had many scientific ways of handling geographical data. The surveyors' *groma*, the cadastral plot, the drawn or inscribed ground-plan of a building, the city plan, the travel itinerary. These were things to help order, understand, and control space and place. But in order to conceptualise geographical knowledge and information art needed to be employed to spark the imagination, to suggest connections, to scare, to bewilder, or even to reassure.

To paraphrase the poet Adrienne Rich, the images are purposes, the images are maps. Indeed, a map is not the space mapped. The great power of illusion, the often unspoken dynamics of society or community, and the excitement and thrill of pursuing empire and expansion all helped shape the people of Rome. The actual and emotional geography of the place in which they lived were intertwined, presented in a way that focused not so much on specific events but rather on subtexts, atmospheres, and perceptions. Though argued here not to have been direct, the message was nevertheless clear and present in the architecture of every image discussed in this book, and in the sounds, the gestures, and the representations on which they were asked to turn their attention.

Through this study there has run the seemingly simple image of a river, but the complex interplay of geology, topography, hydrology, and climate/weather that creates each river and makes it unique from moment to moment means that like the river this argument has run, turned, erased and replaced itself multiple times along its meandering course. As the French philosopher Henri Bergson observed, art is often a powerful way to overcome our limits: limits in both time and space, limits of the mind itself, and limits of perception and capability.

Appendix

I Remember

I remember the past and sense its pastness
I remember being flooded with a grief so old
I remember the butcher's shop on the Viminal Hill
I remember how even the cars in Rome looked ancient
I remember how all that was solid melted into air
I remember the maidens of Caryae
I remember *Er Buchetto*, *porchetta* rolls and very young wine
I remember being in Rome and not doing as the Romans did
I remember the mountain, the sherds beneath my feet
I remember the Planetarium, always closed
I remember the Ship of Aeneas
I remember the *Basilica S. Maria degli Angeli e dei Martiori*, its metal gates of doom
I remember the umbrella pines: they were all dying, like the past
I remember he went to Rome and all I got was this stylus
I remember the *Automobile Club D'Italia*: now there are no cars
I remember the gladiators outside the Colosseum, selling cheap souvenirs rather than death
I remember casting these images into a future yet to come
I remember the shoemaker at the *Porta Fontinalis*, stripped down and bare-chested
I remember that it was all ultimately unexplainable
I remember the *lapis manalis*, entrance to the other life
I remember that Rome is real and not an object
I remember the *Casa Romuli*
I remember that Rome was where the emperor was
I remember the flies at the *Ara Maxima*
I remember arriving at Termini station: the Aurelian walls and sleepers on cardboard boxes
I remember the Temple of Peace, the statue of a cow
I remember keeping company with the past
I remember the *ficus ruminalis*: it used to be there but now it's here
I remember Rome as the ghost of all our lives
I remember the raising of the obelisk
I remember the city awakening, voices babble, Babel
I remember how the colours of Rome are the colours of human hair and flesh
I remember the Coan silks and the Syrian myrrh
I remember a crisis of confidence, a creasing of certainty
I remember the sundial in his hand and how inaccurate it was
I remember the face of the emperor, the face of the homeland
I remember the marble map of the city

I remember the Rubicon swelling up
I remember the wigs of foreign hair
I remember 'a new Rome', and then another
I remember surveying the whole world from the seven hills
I remember Eurysaces, the most famous baker in twenty first century Rome
I remember promises of empire without end

Bibliography

Adams, C.P. and R.M. Laurence (eds) 2001 *Travel and Geography in the Roman Empire*. London: Routledge.

Albano, C. 2001 Visible Bodies: Cartography and Anatomy. In A. Gordon and B. Klein (eds) 2001 *Literature, Mapping, and the Politics of Space in Early Modern Britain*. Cambridge: Cambridge University Press: 89-106.

Albu, E. 2005 Imperial Geography and the Medieval Peutinger Map. *Imago Mundis* 57: 136-148.

Albu, E. 2008 Rethinking the Peutinger Map. In R.J.A. Talbert and R. Unger (eds) 2008: 111-119.

Aldrete, G. 2006 *Floods of the Tiber in Ancient Rome*. Baltimore: Johns Hopkins University Press.

Alexianu, M. 2019 Ovid: the Double Face of the Danube. In L. Mihailescu-Bîrliba (ed) 2019 *Limes, Economy and Society in the Lower Danube Provinces*. Colloquia Antiqua 25, Leuven: Peeters: 1-9.

Alston, R. 2018 The Utopian City in Tacitus' Agricola. In W. Fitzgerald and E. Spentzou (eds) 2018: 235-260.

Alvar, J. 2007 *Romanising Oriental Gods. Myth, Salvation, and Ethics in the Cults of Cybele, Isis, Serapis, and Mithras*. Leiden: Brill.

Angius, A. 2020 Places of Political Interaction and Representation in the City of Rome. In M.L. Caldelli and C. Ricci (eds) 2020: 27-68.

Armstrong, R. 2013 Journeys and Nostalgia in Catullus. *Classical World* 109(1): 43-71.

Armstrong, R. 2019 *Vergil's Green Thoughts: Plants, Humans, and the Divine*. Oxford: Oxford University Press.

Arnaud, P. 1984 L'Image du Globe dans le Monde Romain: Science, Iconographie, Symbolique. *Mélanges d'École Française de Rome* 96(1): 53-116.

Arnaud, P. 2008 Texte et Carte de Marcus Agrippa: Historiographie et Données Textuelles. *Geographia Antiqua* 16-17: 73-126.

Arnaud, P. 2016 Marcus Vipsania Agrippa and His Geographical Work. In S. Bianchetti, M.R. Cataudella, and H.-J. Gehrke (eds) 2016 *Brill's Companion to Ancient Geography. The Inhabited World in the Greek and Roman Tradition*. Leiden: Brill: 205-222.

Assenmaker, P. 2021 Neptune dans le Panthéon d'Imperator Caesar: de l'Art de Récupérer un Dieu Hostile. In Y. Berthelet and F. Van Haeperen (eds) 2001 *Dieux de Rome et du Monde Romain en Réseaux*. Bordeaux: Ausonius Éditions: 181-209.

Audley-Miller, L. and B. Dignas (eds) 2018 *Wandering Myths: Transcultural Uses of Myth in the Ancient World*. Berlin: De Gruyter.

Augoustakis, A. and R.J. Littlewood (eds) 2019 *Campania in the Flavian Poetic Imagination*. Oxford: Oxford University Press.

Austin, N.J.E. and N.B. Rankov 1995 *Exploratio. Military and Political Intelligence in the Roman World from the Second Punic War to the Battle of Adrianople*. London: Routledge.

Avi-Yonah, M. 1954 *The Madaba Mosaic Map with Introduction and Commentary*. Jerusalem: Israel Exploration Society.

Awan, H.T. 2003 *Dominus Aquarum: Nilotic Scenes in Roman and Early Byzantine Art*. College Park: University of Maryland Press.

Bach, S. 2020 *Espace et Structure dans les Métamorphoses d'Ovide*. Bordeaux: Ausonius Éditions.

Bakker, E.J. 2019 In and Out of the Golden Age: a Hesiodic Reading of the Odyssey. In T. Biggs and J. Blum (eds) 2019: 11-30.

Balland, A. 1984 La Casa Romuli au Palatin et au Capitole. *Revue des Études Latines* 62: 57-80.

Balsdon, J.P.V.D. 1979 *Romans and Aliens*. London: Duckworth.
Barber, P. 2004 Was Elizabeth I Interested in Maps-and Did it Matter? *Transactions of the Royal Historical Society* 14: 185-198.
Barber, P. and T. Harper 2010 *Magnificent Maps: Power, Propaganda, and Art*. London: the British Library.
Baronowski, D.W. 2011 *Polybius and Roman Imperialism*. London: Bristol Classical Press.
Barrett, C.E. 2013 Nilotic Scenes, Egyptian Religion, and Roman Perceptions. *Journal of Ancient Egyptian Interconnections* 5-4: 3-5.
Barrett, C.E. 2019 *Domesticating Empire: Egyptian Landscapes in Pompeian Gardens*. Oxford: Oxford University Press.
Barrett, C.E. 2017 Recontextualising Nilotic Scenes: Interactive Landscapes in the Garden of the *Casa dell'Efebo*, Pompei. *American Journal of Archaeology* 121.2: 293-332.
Barrett, C.E. 2018a Battle Between Pygmies, Crocodiles, and Hippopotami. In J. Spier, T. Potts, and S.E. Cole (eds) 2018: 252-253.
Barrett, C.E. 2018b Nilotic Scenes in Roman Art. In J. Spier, T. Potts, and S.E. Cole (eds) 2018: 250.
Barry, F. 2007 Walking on Water: Cosmic Floors in Antiquity and the Middle Ages. *Art Bulletin* 89(4): 627-656.
Bartman, E. 1991 Sculptural Collecting and Display in the Private Realm. In E.K. Gazda (ed) 1991: 71-88.
Bassani, M. 2019 Shrines and Healing Waters in Ancient Italy. Buildings, Cults, Deities. In M. Bassani, M. Bolder-Boos, and U. Fusco (eds) 2019: 9-20.
Bassani, M., M. Bolder-Boos, and U. Fusco (eds) 2019 *Rethinking the Concept of 'Healing Settlements': Water, Cults, Constructions and Contexts in the Ancient World*. Oxford: Archaeopress.
Batty, R. 2000 Mela's Phoenician Geography. *Journal of Roman Studies* 90: 70-94.
Baxa, P. 2010 *Road and Ruins: the Symbolic Landscape of Fascist Rome*. Toronto: University of Toronto Press.
Beagon, M. 1992 *Roman Nature. The Thought of Pliny the Elder*. Oxford: Clarendon Press.
Beagon, M. 1996 Nature and Views of Her Landscapes in Pliny the Elder. In G. Shipley and J. Salmon (eds) 1996 *Human Landscapes in Classical Antiquity: Environment and Culture*. London: Routledge: 284-329.
Beard, M. 2003 The Triumph of Josephus. In A.J. Boyle and W.J. Dominik (eds) 2003 *Flavian Culture: Culture, Image, Text*. Leiden: Brill: 543-558.
Beard, M. 2007 *The Roman Triumph*. Cambridge: Harvard University Press.
Beaulieu, M.-C. 2016 *The Sea in the Greek Imagination*. Philadelphia: University of Pennsylvania Press.
Bedon, R. 1988 Les Remparts Urbains dans l'Iconographie Gallo-Romaine. In Actes du Colloque a l'École Normale Supérieur de Sèvres 1988 *Le Monde des Images en Gaule et dans les Provinces Voisines*. Paris: Éditions Errance: 47-62.
Bejaoui, F. 1999 Découverte dans l'Antique Haidra: la Méditerranée sur une Mosaique. *Archéologie* 357: 16-23.
Bejaoui, F. 1999/2000 Iles et Villes de la Méditerranée sur une Mosaique d'Ammaedara. *Comptes Rendus de l'Académie des Inscriptions et Belles-Lettres* 1997: 827-860.
Bekker-Nielsen, T. 1988 *Terra Incognita*: the Subjective Geography of the Roman Empire. In A. Dansgaard-Madsen, E. Christiansen and E. Hallger (eds) 1988 *Studies in Ancient History and Numismatics Presented to Rudi Thomsen*. Åarhus: Aarhus University Press: 148-161.
Belayche, N. 2003 Tychè dans les Cites de la Palestine Romaine. *Syria* 80: 111-138.

Bell, S.W. 2022 Images and Interpretation of Africans in Roman Art and Social Practice. In L.K. Cline and N. Elkins (eds) 2022: 425-463.

Bellori, G. 2021 *Forma Urbis Romae. I Frammenti Farnesiani della Pianta Marmorea Severiana*. Rome: Arbor Sapientiae Editori.

Bellucci, N.D. 2021 *I Reperti e I Motivi Egizi ed Egittzzanti a Pompei: Indagine Preliminare Per Una Loro Contestualizzazione*. Oxford: Archaeopress.

Bergmann, B. 1991 Painted Perspectives of a Villa Visit: Landscape As Status and Metaphor. In E.K. Gazda (ed) 1991: 49-70.

Bergmann, B. 1992 Exploring the Grove: Pastoral Space on Roman Walls. In J.D. Hunt (ed) 1992 *The Pastoral Landscape*. Studies in the History of Art 36, Washington: University Press of New England: 21-46.

Bergmann, B. 2002 Art and Nature in the Villa at Oplontis. In C.Stein and J.H. Humphrey (eds) 2002 *Pompeian Brothels, Pompeii's Ancient History, Mirrors and Mysteries: Art and Nature at Oplontis, and the Herculaneum 'Basilica'*. Journal of Roman Archaeology Supplementary Volume 47: 115-118.

Bergmann, B. 2008 Staging the Supernatural: Interior Gardens of Pompeian Houses. In C. Mattusch (ed) 2008 *Pompeii and the Roman Villa: Art and Culture Around the Bay of Naples*. London: Thames and Hudson: 62-64.

Bergmann, B. 2018 Frescoes in Roman Gardens. In W.F. Jashemski, K.L. Gleason, K.J. Hartswick, and A.-A. Malek (eds) 2018: 278-316.

Bergmann, B. 2023 *The Roman Art of Landscape*. In J. Powers (ed) 2023: 29-42.

Bernstein, N.W. 2011 *Locus Amoenus* and *Locus Horridus* in Ovid's *Metamorphoses*. *Wenshaw Review of Literature and Culture* 5.1: 67-98.

Berthelot, K. and J.J. Price (eds) 2019 *In the Crucible of Empire: the Impact of Roman Citizenship Upon Greeks, Jews and Christians*. Leuven: Peeters.

Betancourt, R. 2020 *Byzantine Intersectionality: Sexuality, Gender and Race in the Middle Ages*. Princeton: Princeton University Press.

Bevan, L. 2006 *Worshippers and Warriors. Reconstructing Gender and Gender Relations in the Prehistoric Rock Art of Naquane National Park, Valcamonica, Brescia, Northern Italy*. British Archaeological Reports International Series 1485. Oxford: Archaeopress.

Bianchetti, S., M.R. Cataudella, and H.-J. Gehrke (eds) 2016 *Brill's Companion to Ancient Geography: the Inhabited World in Greek and Roman Tradition*. Leiden: Brill.

Bielfeldt, R. 2018 Candelabrus and Trimalchio: Embodied Histories of Roman Lampstands and Their Slaves. *Art History* 41.3: 421-443.

Biggs, T. and J. Blum (eds) 2019 *The Epic Journey in Greek and Roman Literature*. Cambridge: Cambridge University Press.

Birley, A.R. 1997 *Hadrian. The Restless Emperor*. London: Routledge.

Birley, A.R. 2003 Hadrian's Travels. In L. De Blois, P. Erdkamp, O. Hekster, G. De Kleijn, and S. Mols (eds) 2003 *The Representation and Perception of Roman Imperial Power*. Amsterdam: J. C. Gieben: 425-441.

Bishop, C. 2019 *Magnum Opus*: Atticus, Cicero and Eratosthenes' Geography. *Rheinisches Museum* 162: 265-291.

Bispham, E. and D. Miano (eds) 2019 *Gods and Goddesses in Ancient Italy*. London: Routledge.

Black, J. 1997 *Maps and Politics*. London: Reaktion Books.

Blake, S. 2012 Now You See Them: Slaves and Other Objects as Elements of the Roman Master. *Helios* 39: 193-211.

Blondel, J. and J. Aronson 1999 *Biology and Wildlife of the Mediterranean Region*. Oxford: Oxford University Press.

Blouin, K. 2014 *Triangular Landscapes: Environment, Society, and the State in the Nile Delta Under Roman Rule*. Oxford: Oxford University Press.

Blum, J. 2019 "What Country, Friends, Is This?": Geography and Exemplarity in Valerius Flaccus' *Argonautica*. In T. Biggs and J. Blum (eds) 2019: 59-88.

Blunt, A. and G. Rose (eds) 1994 *Writing Women and Space: Colonial and Post-Colonial Geographies*. London: Routledge.

Boatwright, M.T. 1987 *Hadrian and the City of Rome*. Princeton: Princeton University Press.

Boatwright, M.T. 1998 Luxuriant Gardens and Extravagant Women: the *Horti* of Rome Between Republic and Empire. In M. Cima and E. La Rocca (eds) 1998 *Horti Romani: Ideologia e Autorappresentazione*. Rome: L'Erma di Bretschneider: 71-82.

Boatwright, M.T. 1999 *Hadrian and the Cities of the Roman Empire*. Princeton: Princeton University Press.

Boatwright, M.T. 2014 Visualizing Empire in Imperial Rome. In L.L. Brice and D. Slootje (eds) 2014 *Aspects of Ancient Institutions and Geography: Studies in Honor of R.J.A. Talbert*. Leiden: Brill: 235-259.

Bolder-Boos, M. 2019 Hercules and Healing. In M. Bassani, M. Bolder-Boos, and U. Fusco (eds) 2019: 133-140.

Bosak-Schroeder, C. 2020 *Other Natures: Environmental Encounters with Ancient Greek Ethnography*. Oakland: University of California Press.

Bosio, L. 1983 *La Tabula Peutingeriana. Una Descizione Pittorica del Mondo Antico*. Rimini: Maggioli Editore.

Bouet, A. 1998 La Mosaique de la Via Marsala à Rome (Regio V): le Plan des Thermes d'une Association d'Athlètes? *Mélanges de l'École Francaise de Rome* 110-112: 849-892.

Bowerstock, G.W. 2005 Foreign Elites in Rome. In J. Edmondson, S. Mason and J.B. Rives (eds) 2005: 53-62.

Bowman, A.K. and G. Woolf (eds) 1994 *Literacy and Power in the Ancient World*. Cambridge: Cambridge University Press.

Bowman, G. 1998 Mapping History's Redemption: Eschatology and Topography in the *Itinerarium Burdigalense*. In L.I. Levine (ed) 1998 *Jerusalem: Its Sanctity and Centrality to Judaism, Christianity and Islam*. New York: Continuum Press: 163-187.

Boyle, A.J. and W.J. Dominik (eds) 2003 *Flavian Rome: Culture, Image, Text*. Leiden: Brill.

Bradley, G. and J.-P. Wilson (eds) 2006 *Greek and Roman Colonization: Origins, Ideologies and Interactions*. Swansea: Classical Press of Wales.

Brandt, J.R. 2005 "The Warehouse of the World": a Comment on Rome's Supply Chain During the Empire. *Orizzonti* 6: 25-47.

Braund, D. 1996 River Frontiers in the Environmental Psychology of the Roman World. In D.L. Kennedy (ed) 1996 The Roman Army in the East. Journal of Roman Archaeology Supplementary Series 18: 43-47.

Breeze, D. 2013 *Roman Frontiers in Their Landscape Setting*. Charles Parish Lecture 2011. Newcastle upon Tyne: Literary and Philosophical Society of Newcastle Upon Tyne.

Breeze, D. (ed) 2012 *The First Souvenirs: Enamelled Vessels from Hadrian's Wall*. Kendal: Cumberland and Westmorland Antiquarian and Archaeological Society..

Brélaz, C. 2021 Claiming Roman Origins: Greek Cities and the Roman Colonial Pattern. In J.J. Price, M. Finkelberg and Y. Shahar (eds) 2021: 100-115.

Bricault, L., M.J. Versluys and P.G.P. Meyboom (eds) 2007 *Nile Into Tiber: Egypt in the Roman World: Proceedings of the IIIrd International Conference of Isis Studies, Leiden May 11-14, 2005,* Leiden: Brill.

Brilliant, R. 1967 *The Arch of Septimius Severus in the Roman Forum.* Memoirs of the American Academy in Rome 29.

Brilliant, R. 1999 "Let the Trumpets Roar!". The Roman Triumph. In B. Bergmann and C. Kondoleon (eds) 1999 *The Art of Ancient Spectacle.* New Haven: Yale University Press: 221-230.

Broadhead, W.M. 2002 Rome's Migration Policy and the So-Called *Ius Migrandi. Cahiers du Centre Gustave Glotz* 12: 69-89.

Brock, A.L., L. Motta and N. Terrenato 2021 On the Banks of the Tiber: Opportunity and Transformation in Early Rome. *Journal of Roman Studies* 111: 1-30.

Brockliss, W. 2019 *Homeric Imagery and the Natural Environment.* Cambridge: Harvard University Press.

Brodersen, K. 1995 *Terra Cognita.* Studien zur Römischen Raumerfassung. Hildesheim: Georg Olms Verlag.

Brodersen, K. 2001 The Presentation of Geographical Knowledge for Travel and Transport in the Roman World: *Itineraria non tantum adnotata sed etiam picta*. In C. Adams and R. Laurence (eds) 2002 *Travel and Geography in the Roman Empire.* Routledge: London: 7-21.

Brodersen, K. 2004 Mapping in the Ancient World. *Journal of Roman Studies* 94: 183-190.

Brody, L.R. and G.L. Hoffman (eds) 2014 *Roman in the Provinces: Art on the Periphery of Empire.* Chicago: Chicago University Press.

Brown, N.G. 2020 The Living and the Monumental on the *Anaglypha Traiani. American Journal of Archaeology* 124 (4): 607-630.

Bruneau, P. 1991 Topographie et Histoire Religeuse. *Bulletin de Correspondance Hellénique* 115: 379-386.

Buchardt, J. 2023 *Lifescapes: the Experience of Landscape in Britain, 1870-1960.* Cambridge: Cambridge University Press.

Burton, P.J. 1996 The Summoning of the *Magna Mater* to Rome (205 BC). *Historia* 45: 36-63.

Burton, P. 2013 Enter the Muse: Literary Responses to Roman Imperialism 240-100 BC. In D. Hoyos (ed) 2013 *A Companion to Roman Imperialism.* Leiden: Brill: 99-112

Buxton, R. 1992 Imaginary Greek Mountains. *Journal of Hellenic Studies* 112: 1-15.

Buxton, R. 2016 Mount Etna in the Greco-Roman Immaginaire: Culture and Liquid Fire. In J. McInery and I. Sluiter (eds) 2016: 25-45.

Caldelli, M.L. 2001 Gladiatori con Armaturae Etniche: Il Samnes. *Archeologia Classica* 52: 279-295.

Caldelli, M.L. and C. Ricci (eds) 2020 *City of Encounters: Public Spaces and Social Interaction in Ancient Rome.* Rome: Edizioni Quasar.

Cameron, A. 2015 City Personifications and Consular Diptychs. *Journal of Roman Studies* 105: 250-287.

Campbell, B. 1996 Shaping the Rural Environment: Surveyors in Ancient Rome. *Journal of Roman Studies* 86: 74-99.

Campbell, B. 2000 *The Writings of the Roman Land Surveyors; Introduction, Text, Translation, and Commentary.* London: Society for the Promotion of Roman Studies.

Campbell, B. 2009 River Definitions in Roman Technical Literature. *International Journal of Landscape Archaeology* 6:188-193.

Campbell, B. 2010 Managing Disruptive Rivers. In E. Hermon (ed) 2010 *Riparia dans l'Empire Romain*. British Archaeological Reports International Series 2066, Oxford: Archaeopress: 317-328.

Campbell, B. 2012 *Rivers and the Power of Ancient Rome*. Chapel Hill: University of North Carolina Press.

Campbell, B. 2015 Watery Perspectives: a Roman View on Rivers. In T.V. Franconi (ed) 2015 *Fluvial Landscapes in the Roman World*. Journal of Roman Archaeology Supplementary Series 104. Portsmouth, RI: Journal of Roman Archaeology: 23-32.

Carettoni, G., A. Colini, L. Cozza, and G. Gatti (eds) 1960 *La Pianta Marmorea di Roma Antica. Forum Urbis Roma*. Rome: Danesi.

Carey, S. 2002 A Tradition of Adventures in the Imperial Grotto. *Greece and Rome* 49: 44-61.

Carey, S. 2003 *Pliny's Catalogue of Culture: Art and Empire in the Natural History*. Oxford: Oxford University Press.

Carroll, M. 2018 Temple Gardens and Sacred Groves. In W.F. Jashemski, K.L. Gleason, K.J. Hartswick, and A.-A. Malek (eds) 2018: 152-164.

Carucci, M. 2017 The Dangers of Female Mobility in Roman Imperial Times. In E. Lo Cascio and L.E. Tacoma (eds) 2017 *The Impact of Mobility and Migration in the Roman Empire*. Leiden: Brill: 173-190.

Casagrande-Kim, R. 2012 *The Journey to the Underworld: Topography, Landscape, and Divine Inhabitants of the Roman Hades*. PhD Thesis, Columbia University.

Cassibry, K. 2021 *Destinations in Mind: Portraying Places on the Roman Empire's Souvenirs*. Oxford: Oxford University Press.

Casson, L. 1994 *Travel in the Ancient World*. Baltimore: Johns Hopkins University Press.

Castagnoli, F. 1987 *Oceani Solium, Diebus Solia*. Rome: L'Erma di Bretschneider.

Castriota, D. 1995 *The Ara Pacis Augustae and the Imagery of Abundance in Later Greek and Early Roman Imperial Art*. Princeton: Princeton University Press.

Castro-Páez, E. 2023 *Geografía y Etnografía en la Literatura Greco-Latina*. Barcelona: Bellaterra.

Chaniotis, A. 2018 *Age of Conquests: the Greek World from Alexander to Hadrian*. Harvard: Harvard University Press.

Chevallier, R. 1998 *Voyages et Déplacements dans l'Empire Romain*. Paris: Colin.

Chinn, C.M. 2022 Empire and Italian Landscape in Statius: *Silvae* 4.3 and 4.5. In M. Horster and N. Hächler (eds) 2021: 353-371.

Chuvin, P. 1991 *Mythologie et Géographie Dionysiques: Recherches sur l'Oeuvre de Nonnus de Panopolis*. Clermont Ferrand: Adosa.

Ciarallo, A. 2001 *Gardens of Pompeii*. Rome: L'Erma di Bretschneider.

Cifani, G. 2018 Visibility Matters: Notes on Archaic Monuments and Collective Memory in Mid-Republican Rome. In K. Sandberg and C. Smith (eds) 2018 *Omnium Annalium Monumenta: Historical Writing and Historical Evidence in Republican Rome*. Leiden: Brill: 390-403.

Cima, M. and E. La Rocca 1998 *Horti Romani: Atti del Convegno, Roma 1995*, Bullettino della Commissione Archeologica Communale di Roma, Rome: L'Erma di Bretschneider.

Cima, M. and E. Talamo 2008 *Gli Horti di Roma Antica*, Rome: Quaderni Capitolini. Electa.

Clarke, J.R. 1996 Landscape Paintings in the Villa of Oplontis, *Journal of Roman Archaeology* 9: 81-107.

Clarke, J.R. 2003 *Art in the Lives of Ordinary Romans. Visual Representation and Non-Elite Viewers in Italy, 100 B.C.-A.D. 315,* Berkeley: University of California Press.

Clarke, J.R. 2007 *Looking at Laughter: Humor, Power, and Transgression in Roman Visual Culture, 100 BC-AD 250*. Los Angeles: University of California Press.

Clarke, K. 1999 *Between Geography and History. Hellenistic Constructions of the Roman World*. Oxford: Oxford University Press.
Clarke, K. 2008 Text and Image: Mapping the Roman World. In F.-H. Mutschler and A, Mittag (eds) 2008 *Conceiving the Empire: China and Rome Compared*. Oxford: Oxford University Press: 195-214.
Clarkson, J., P. James, K. McDonald, L. Tagliapietra, and N. Zair (eds) 2020 *Migration, Mobility and Language Contact In and Around the Ancient Mediterranean*. Cambridge: Cambridge University Press.
Cline, L.K. and Elkins, N. (eds) 2022 *The Oxford Handbook of Roman Imagery and Iconography*. Oxford: Oxford University Press.
Coarelli, F. 1991 Le Plan de Via Anicia: un Nouveau Fragment de la Forma Marmorea de Rome. In F. Hinard and M. Royo (eds) 1991 *Rome: l'Espace Urbain et ses Répresentations*. Paris: Presses de l'Université de Paris-Sorbonne: 65-81.
Cobb, M.A. 2013 The Reception and Consumption of Eastern Goods in Roman Society. *Greece and Rome* 60(1): 136-152.
Cody, J.M. 2003 Conqueror and Conquered on Flavian Coins. In A.J. Boyle and W.J. Dominik (eds) 2003: 103-123.
Coleman, K. and P. Derron 2014 *Le Jardin dans l'Antiquité: Introduction et Huit Exposés Suivis de Discussions*. Entretiens sur l'Antiquité Classique 60, Vandoeuvres: Fondation Hardt.
Colpo, I. 2010 'Ruinae..et putres robere trunci.' *Paesaggi di Rovine nel Paesaggio nella Pittura Romana (I Secolo A.C.-I Secolo D.C.)*. Rome: Antenor Quaderni.
Comment, B. 1999 *The Panorama*. London: Reaktion Books.
Conan, M. 1986 Nature Into Art: Gardens and Landscape in the Everyday Life of Ancient Rome. *Journal of Garden History* 6: 348-356.
Conticello, B. and B. Andreae 1974 *Die Skulpturen von Sperlonga*. Antike Plastik 14. Berlin: Gebr. Mann.
Coombe, P. 2022 Aspects of the Iconography of River Gods in Roman Britain. In M. Henig and J. Lundock (eds) 2022 *Water in the Roman World: Engineering, Trade, Religion and Daily Life*. Oxford: Archaeopress: 105-126.
Coppola, D. 2010 *Anemoi: Morfologia dei Venti nell'Imaginario della Grecia Arcaica*. Università degli Studi di Napoli Federico II. Pubblicazioni del Dipartimento di Discipline Storiche 24. Naples: Liguori Editori.
Cornwell, H. and G. Woolf (eds) 2023 *Gendering Roman Imperialism*. Leiden: Brill.
Coulston, J.N. 2001 Transport and Travel on the Column of Trajan. In C. Adams and R. Laurence (eds) 2001: 106-137.
Crawford-Brown, S. 2022 Down from the Roof: Reframing Plants in Augustan Art. *Journal of Roman Archaeology* 35.1: 33-63.
Crofton-Sleigh, L. 2016 'The Mythical Landscapers of Augustan Rome', in J. McInerny and I. Sluiter (eds) 2016: 383-407.
Croisille, J.-M. 1965 *Les Natures Mortes Campaniennes*. Brussels: Collection Latomus 76.
Croisille, J.-M. 2010 *Paysages dans la Peinture Romaine: Aux Origines du Genre Pictural*. Paris: Picard.
Croisille, J.-M. 2015 *Natures Mortes dans la Rome Antique. Naissance d'un Genre Artistique*. Paris: Picard.
Damon, C. and E. Palazzolo 2019 Defining Home, Defining Rome: Germanicus' Eastern Tour. In T. Biggs and J. Blum (eds) 2019: 194-210.
D'Ancona, M.L. 1950 An Indian Statuette from Pompeii. *Artibus Asiae* 13(3): 166-180.

Daniels, S. and D. Cosgrove 1993 Spectacle and Text: Landscape Metaphors in Cultural Geography. In J. Duncan and D. Ley (eds) 1993 *Place/Culture/Representation*. London: Routledge: 57-77.
Dang, T.K. 2021 Decolonizing Landscape. *Landscape Research* 46(7): 1004-1016.
Davenport, C. 2020 Roman Emperors, Conquest, and Violence: Images from the Eastern Provinces. In A. Russell and M. Hellström (eds) 2020: 100-127.
Davies, M.I.J. and F.N. M'Mbogori (eds) 2013 *Humans and the Environment: New Archaeological Perspectives for the Twenty-First Century*. Oxford: Oxford University Press.
Davies, P.J.E. 2011 *Aegyptiaca* in Rome: *Adventus* and *Romanitas*. In E.S. Gruen (ed) 2011: 354-372.
Davies, S.H. 2020 *Rome, Global Dreams, and the International Origins of an Empire*. Leiden: Brill.
Dawson, C.M. 1944 *Romano-Campanian Mythological Landscape Painting*. New Haven: Yale University Press.
De Angelis, F. (ed) 2013 *Regionalism and Globalism in Antiquity: Exploring Their Limits*. Leuven: Peeters.
De Bellefonds, P.L. 2011 Pictorial Foundation Myths in Roman Asia Minor. In E.S. Gruen (ed) 2011: 26-46.
De Caro, S. 2001 *La Natura Morta nelle Pitture e nei Mosaici delle Città Vesuviane,* Naples: Museo Archeologico Nazionale di Napoli and Electa.
De Certeau, M. 1984 *The Practice of Everyday Life*. Berkeley: University of California Press.
De Grummond, N.T. and B.S. Ridgway (eds) 1997 *From Pergamon to Sperlonga: Sculpture and Context*. Berkeley: University of California Press.
Dekker, E. 2009 Featuring the First Greek Celestial Globe. *Globe Studies* 55/56: 133-152.
Delano Smith, C. 1982 The Emergence of 'Maps' in European Rock Art: a Prehistoric Preoccupation With Place. *Imago Mundi* 34: 9-25.
Delano Smith, C. 1990 Place or Prayer? Maps in Italian Rock Art. *Accordia Research Papers* 1: 5-18.
De Ligt, L. and L.E. Tacoma (eds) 2016 *Migration and Mobility in the Early Roman Empire*. Leiden: Brill.
De Los Úbeda, A. (ed) 2011 *Roma. Naturaleza e Ideal (Paisajes 1600-1650)*. Madrid: Prado Museum.
Dench, E. 1995 *From Barbarians to New Men. Greek, Roman and Modern Perceptions of Peoples of the Central Apennines*. Oxford: Oxford University Press.
Dench, E. 2005 *Romulus' Asylum: Roman Identities from the Age of Alexander to the Age of Hadrian*. Oxford: Oxford University Press.
Denson, R. 2022 Order Among Disorder: Poseidon's Underwater Kingdom and Utopic Marine Environments. In H. Williams and R. Clare (eds) 2022: 147-164.
DeRose Evans, J. 2009 Prostitutes in the Portico of Pompey? A Reconsideration. *Transactions of the American Philological Association* 139: 123-145.
Desbiens, C. 2003 Colonialism and Landscape: Postcolonial Theory and Applications. *Annals of the Association of American Geographers* 93(2): 915-917.
Destephen, S. 2019 The Time Travelling Emperor: Hadrian's Mobility as Mirrored in Ancient and Medieval Historiography. *Scripta Classica Israelica* 38: 59-82.
De Vos, M. 1980 *L'Egittomania in Pitture e Mosaici Romano-Campani della Prima Età Imperiale,* Leiden: Brill.
Diederich, S. 2021/2022 Empire and Landscape in the *Tabula Peutingeriana*. In M. Horster and N. Hächler (eds) 2021/2022: 372-397.

Dietrich, N. 2017 Pictorial Space as a Media Phenomenon: the Case of 'Landscape' in Romano-Campanian Wall-Painting. In F. Lissarrague, E. Valette, and S. Wyler (eds) 2017 Cahiers des Mondes Anciens 9: 1-27.

Dilke, O.A.W. 1962 The Roman Surveyors. *Greece and Rome* 9(2): 170-180.

Dilke, O.A.W. 1971 *The Roman Land Surveyors. An Introduction to the Agrimensores.* Newton Abbot: David and Charles.

Dilke, O.A.W. 1985 *Greek and Roman Maps.* Ithaca: Cornell University Press.

Dilke, O.A.W. 1987a Maps in the Service of the State: Roman Cartography to the End of the Augustan Era. In J.B. Harley and D. Woodward (eds) 1987: 201-211.

Dilke, O.A.W. 1987b Roman Large-Scale Mapping in the Early Empire. In J.B. Harley and D. Woodward (eds) 1987: 212-223.

Dorcey, P. 1992 *The Cult of Silvanus: a Study in Roman Folk Religion.* Leiden: Brill.

Douglass, L. 1996 A New Look at the *Itinerarium Burdigalense. Journal of Early Christian Studies* 4: 313-333.

Dresken-Weiland, J. 2018 Christian Sarcophagi from Rome. In R.M. Jensen and M.D. Ellison (eds) 2018 *The Routledge Companion to Early Christian Art.* London: Routledge: 39-55.

Dueck, D. 2000 *Strabo of Amasia: a Greek Man of Letters in Augustan Rome.* London: Routledge.

Dueck, D. 2012 *Geography in Classical Antiquity.* Cambridge: Cambridge University Press.

Dueck, D. 2021 *Illiterate Geography in Classical Athens and Rome.* London: Routledge.

Dueck, D., H. Lindsay and S. Pothecary (eds) 2005 *Strabo's Cultural Geography. The Making of a Kolossourgia.* Cambridge: Cambridge University Press.

Dufallo, B. 2021 *Disorienting Empire: Republican Latin Poetry's Wanderers.* Oxford: Oxford University Press.

Duffy, C. 2021 'Famous From All Antiquity': Etna in Classical Myth and Romantic Poetry. In D. Hollis and J.P. König (eds) 2021: 37-54.

Dwyer, O.J. and D.H. Alderman 2008 Memorial Landscapes: Analytic Questions and Metaphors. *GeoJournal* 73: 165-178.

Eck, W. 2021 The Imperial Senate: Center of a Multinational *Imperium*. In J.J. Price, M. Finkelberg and Y. Shahar (eds) 2021: 29-41.

Edlund-Berry, I. and J.M. Turfa 2019 *Lacus* and *Lucus*: Lakes and Groves as Markers of Healing Cults in Central Italy. In M. Bassani, M. Bolder-Boos, and U. Fusco (eds) 2019: 141-156.

Edwards, C. 2003 Incorporating the Alien; the Art of Conquest. In C. Edwards and G. Woolf (eds) 2003 *Rome the Cosmopolis.* Cambridge: Cambridge University Press: 44-70.

Egerton, F. 2012 *Roots of Ecology. Antiquity to Haeckel,* Berkeley: University of California Press.

Elkins, N.T. 2015 *Monuments in Miniature: Architecture on Roman Coins.* New York: American Numismatic Society.

Elsner, J. 2000 The *Itinerarium Burdigalense*: Politics and Salvation in the Geography of Constantine's Empire. *Journal of Roman Studies* 90: 181-195.

Elsner, J. and I. Rutherford (eds) 2005 *Pilgrimage in Graeco-Roman and Early Christian Antiquity: Seeing the Gods.* Oxford: Oxford University Press.

Erdkamp, P., K. Verboven, and A. Zuiderhoek (eds) 2015 *Ownership and Exploitation of Land and Natural Resources in the Roman World.* Oxford: Oxford University Press.

Erskine, A. 2001 *Troy Between Greece and Rome: Local Tradition and Imperial Power.* Oxford: Oxford University Press.

Erskine, A. 2012 Polybius Among the Romans: Life in the Cyclops' Cave. In C. Smith and L.M. Yarrow (eds) 2012 *Imperialism, Cultural Politics, and Polybius.* Oxford: Oxford University Press: 17-32.

Evans, H.B. 1997 *Water Distribution in Ancient Rome: the Evidence of Frontinus*. Ann Arbor: University of Michigan Press.

Evans, J.D. 1991 The Sacred Figs in Rome. *Latomus* 50.4: 798-808.

Evans, R. 2005 Geography Without People: Mapping in Pliny's *Historia Naturalis* Books 3-6. Ramus. *Critical Studies in Greek and Latin Literature* 34(1): 47-74.

Evers, K.G. 2017 *Worlds Apart, Trading Together: the Organisation of Long Distance Trade Between Rome and India in Antiquity*. Oxford: Archaeopress.

Fabbri, L. 2019 *Mater Florum. Flora e il Suo Culto a Roma*. Florence: Leo S. Olschki.

Fabre-Serris, J., A. Keith and F. Klein (eds) 2021 *Identities, Ethnicities and Gender in Antiquity*. Berlin: De Gruyter.

Fabrizi, V. 2021 War, Weather and Landscape in Livy's *Ab Urbe Condita*. In B. Reitz-Joosse, M. W. Makins and C.J. Mackie (eds) 2021: 38-61.

Fantham, E. 2009 *Latin Poets and Italian Gods*. Toronto: University of Toronto Press.

Fantham, E. 2012 Images of the City: Propertius' New-Old Rome. In E. Greene and T. Welch (eds) 2012 *Propertius*. Oxford: Oxford University Press: 302-319.

Farney, G.D. 2007 *Ethnic Identity and Aristocratic Competition in Republican Rome*. Cambridge: Cambridge University Press.

Favro, D. 1996 *The Urban Image of Augustan Rome*. Cambridge: Cambridge University Press.

Favro, D. 2005 Making Rome a World City. In K. Galinsky (ed) 2005 *Age of Augustus*. Cambridge: Cambridge University Press: 234-263.

Favro, D. 2006 The Iconicity of Ancient Rome. *Urban History* 33(1): 20-38.

Fear, A. 2005 A Journey to the End of the World. In J. Elsner and I. Rutherford (eds) 2005 *Pilgrimage in Graeco-Roman and Early Christian Antiquity: Seeing the Gods*. Oxford: Oxford University Press: 319-331.

Feeney, D. 1988 *Literature and Religion at Rome: Cultures, Contexts, and Beliefs*. Cambridge: Cambridge University Press.

Felton, D. (ed) 2018 *Landscapes of Dread in Classical Antiquity: Negative Emotion in Natural and Constructed Spaces*. London: Routledge.

Fernández-Götz, M., D. Maschek, and N. Roymans 2020 The Dark Side of the Empire: Roman Expansionism Between Object Agency and Predatory Regime. *Antiquity* 94: 1630-1639.

Ferris, I.M. 2000 *Enemies of Rome. Barbarians Through Roman Eyes*. Stroud: Sutton.

Ferris, I.M. 2009 *Hate and War. The Column of Marcus Aurelius*. Stroud: the History Press.

Ferris, I.M. 2013 *The Arch of Constantine. Inspired by the Divine*. Stroud: Amberley Publishing.

Ferris, I.M. 2018 *Cave Canem. Animals and Roman Society*. Stroud: Amberley Publishing.

Ferris, I.M. 2021a *Visions of the Roman North. Art and Identity in Northern Roman Britain*. Oxford: Archaeopress.

Ferris, I.M. 2021b *The Dignity of Labour. Image, Work and Identity in the Roman World*. Stroud: Amberley.

Fine, S. 2016 Who Is Carrying the Temple Menorah? A Jewish Counter-Narrative of the Arch of Titus Spolia Panel. *Images: a Journal of Jewish Art and Visual Culture* 9(1): 19-48.

Fine, S. 2017 From Synagogue Furnishing to Media Event: the Magdala Ashlar. *Ars Judaica* 13: 27-38.

Finkelberg, M. 2021 Roman Reception of the Trojan War. In J.J. Price, M. Finkelberg and Y. Shahar (eds) 2021: 87-99.

Finn, J. 2020 The Ship of Aeneas. *Ancient History Bulletin* 34(1-2): 1-24.

Fitzgerald, W. 2016 *Variety: the Life of a Roman Concept*. Chicago: University of Chicago Press.

Fitzgerald, W. and E. Spentzou (eds) 2018 *The Production of Space in Latin Literature.* Oxford: Oxford University Press.

Fletcher, K.F.B. 2014 *Finding Italy: Travel, Nation and Colonization in Vergil's Aeneid.* Ann Arbor: University of Michigan Press.

Fodorean, F. 2011 Mapping the *Orbis Terrarum*; the Peutinger Map, the Antonine Itinerary and the Cartographic Tradition of the Fourth and Fifth Century A.D.. *Ephemeris Napocensis* 21: 51-62.

Fodorean, F. 2012 Communicating in Antiquity. Itineraries, Geographical Space, Travel and Infrastructure in Roman Dacia. *Ephemeris Dacoromana*, Academia di Romania in Roma Serie Nuova 14: 81-132.

Foss, P.W. 2022 *Pliny and the Eruption of Vesuvius.* London: Routledge.

Fowden, E.K., S. Çağaptay, E. Zychowicz-Coghill, and L. Blanke (eds) 2022 *Cities As Palimpsests? Responses to Antiquity in Eastern Mediterranean Urbanism.* Oxford: Oxbow Books.

Fox, A. 2019 Trajanic Trees: the Dacian Forest on Trajan's Column. *Papers of the British School at Rome* 87: 47-69.

Fox, A. 2023 *Trees in Ancient Rome: Growing an Empire in the Late Republic and Early Principate.* London: Bloomsbury.

French, R. 1994 *Ancient Natural History: Histories of Nature.* London: Routledge.

French, R. and F. Greenaway (eds) 1986 *Science in the Early Roman Empire: Pliny the Elder, His Sources and Influence.* London: Croom Helm.

Fröhlich, T. 2023 *Locus Amoenus* or Elysium? The Landscape in the Tomb. In J. Powers (ed) 2023: 79-90.

Gais, R.M. 1978 Some Problems of River-God Iconography. *American Journal of Archaeology* 82: 355-370.

Gallina, M.A. 2021 *Dall'Immagine Cartografica alla Riconstruzione Storica.* Milan: LED Edizioni.

Gates-Foster, J. 2021 Out of Egypt: Provenance, Racial Representation, and Miniature Images of Nubians in the Menil Collection. In J.N. Hopkins, S.K. Costello, and P.R. Davis (eds) 2021 *Object Biographies: Collaborative Approaches to Ancient Mediterranean Art.* Houston: the Menil Collection: 107-125.

Gazda, E.K. (ed) 1991 *Roman Art in the Private Sphere: New Perspectives on the Architecture and Decor of the Domus, Villa, and Insula.* Ann Arbor: University of Michigan Press.

Gee, E. 2020 *Mapping the Afterlife: From Homer to Dante.* Oxford: Oxford University Press.

Gensheimer, M.B. 2018 *Decoration and Display in Rome's Imperial Thermae: Messages of Power and Their Popular Reception at the Baths of Caracalla.* Oxford: Oxford University Press.

George, M. 2003 Images of Black Slaves in the Roman Empire. *Syllecta Classica* 14: 161-185.

Geus, K. and M. Thiering (eds) 2014 *Features of Common Sense Geography: Implicit Knowledge Structures in Ancient Geographical Texts.* Antike Kultur und Geschichte Bd 16. Berlin: LIT Verlag.

Gianfrotta, P. 2012 La Topografia sulle Bottiglie di Baia. *Rivista di Archeologia* 35: 13-39.

Gilhuly, K. and N. Worman (eds) 2014 *Space, Place and Landscape in Ancient Greek Literature and Culture.* Cambridge: Cambridge University Press.

Giusti, E. 2018 *Carthage in Virgil's Aeneid. Staging the Enemy Under Augustus.* Cambridge: Cambridge University Press.

Gleason, K. 1994 *Porticus Pompeiana*: a New Perspective on the First Public Park of Ancient Rome. *Journal of Garden History* 14.1: 13-27.

Goodfellow, M.S. 1981 North Italian Rivers and Lakes in the Georgics. *Vergilius* 27: 12-22.

Goodman, P.J. 2018 Defining the City: the Boundaries of Rome. In C. Holleran and A. Claridge (eds) 2018: 71-91.
Goodman, P.J. 2020 *In Omnibus Regionibus?* The Fourteen Regions and the City of Rome. *Papers of the British School at Rome* 88: 1-32.
Gordon, A. and B. Klein 2001 *Literature, Mapping, and the Politics of Space in Early Modern Britain*. Cambridge: Cambridge University Press.
Gordon, J.M. 2018 Insularity and Identity in Roman Cyprus: Connectivity, Complexity, and Cultural Change. In A. Kouremenos (ed) 2018 *Insularity and Identity in the Roman Mediterranean*. Oxford: Oxbow Books: 4-40.
Gowers, E. 1993 *The Loaded Table: Representations of Food in Roman Literature*. Oxford: Clarendon Press.
Gowers, E. 1995 The Anatomy of Rome from Capitol to Cloaca. *Journal of Roman Studies* 85: 23-32.
Gowers, E. 2011 Trees and Family Trees in the *Aeneid*. *Classical Antiquity* 30 (1): 87-118.
Graham, E.J. 2020 Pilgrimage, Mobile Behaviours and the Creation of Religious Place in Early Roman Latium. In J. Kuuliala and J. Rantala (eds) 2020: 15-36.
Gray, C. 1974 Leaving the 20th Century. The Incomplete Works of the Situationist International. 1998 Edition/Reprint London: Rebel Press.
Greco, C., M. Osanna, and P. Giulierini (eds) 2016 *Il Nilo a Pompei. Visioni d'Egitto nel Mondo Romano*. Modena: Franco Cosimo Panini Editore.
Green, P. 1997 *The Argonautika: the Story of Jason and the Quest for the Golden Fleece*. Berkeley: University of California Press.
Greene, S. 2023 Recharting Landscapes in the Exhibition *Roma Negata: Postcolonial Routes of the City* (2014) and the Digital Project *Postcolonial Italy; Mapping Colonial Heritage*. In S. Hecker and R. Bedarida (eds) 2023: 211-226.
Grey, C. 2016 Hadrian the Traveller: Motifs and Expressions of Roman Imperial Power in the *Vita Hadriani*. *História* 35, online.
Grey, M.J. and M.D. Ellison 2022 Imagery in Jewish and Christian Ritual Settings. In L.K. Cline and Elkins, N. (eds) 2022: 534-562.
Grig, L. and G. Kelly (eds) 2012 *Two Romes: Rome and Constantinople*. Oxford: Oxford University Press.
Grigorieva, A. 2022 Migrating Mosaics: Transforming Images of *Oceanus* and Marine Environments from the Imperial Period to Late Antiquity. In A. Lampinen and E.M. Ferrándiz (eds) 2022: 26-48.
Gruen, E. 2010 *Rethinking the Other in Antiquity*. Princeton: Princeton University Press.
Gruen, E.S. (ed) 2011 *Cultural Identity in the Ancient Mediterranean*. Los Angeles: Getty Research Institute.
Gullini, G. 1956 *I Mosaici di Palestrina*, Rome: Archeologia Classica Supplementary Volume 1.
Hachlili, R. 1998 Iconographic Elements of Nilotic Scenes on Byzantine Mosaic Pavements in Israel. *Palestine Exploration Quarterly* 130.2: 106-120.
Hales, S. 2003 *The Roman House and Social Identity*. Cambridge: Cambridge University Press.
Halfmann, H. 1986 *Itinera Principum: Geschichte und Typologie der Kaiserreisen im Römischen Reich*. Stuttgart: Franz Steiner.
Hallett, C.H. 2021 The Wood Comes to the City: Ancient Trees, Sacred Groves, and the 'Greening' of Early Augustan Rome. *Religion in the Roman Empire* 7: 221-274.
Hallett, J.P. 1970 'Over Troubled Water': the Meaning of the Title *Pontifex*. *Transactions of the American Philological Association* 101: 219-227.

Hălmagi, D. 2015 Notes on the Dura Europos Map. *Revista CICSA (Centrului de Istorie Comparată a Societăților Antice)* New Series 1: 41-51.
Handley, M. 2011 *Dying on Foreign Shores: Travel and Mobility in the Late-Antique West*. Journal of Roman Archaeology Supplementary Series 86. Portsmouth, RI.
Hannah, R. 2011 The *Horologium* of Augustus as a Sundial. *Journal of Roman Archaeology* 24.1: 87-95.
Hardie, P.R. 1986 *Virgil's Aeneid: Cosmos and Imperium*. Oxford: Clarendon Press.
Hardy, G. and L.M.V. Totelin 2015 *Ancient Botany*. London: Routledge.
Harl, K.W. 1987 *Civic Coins and Civic Politics in the Roman East, A.D. 180-275*. Berkeley: University of California Press.
Harley, J.B. and D. Woodward (eds) 1987 *The History of Cartography. Volume One, Cartography in Prehistoric, Ancient and Medieval Europe and the Mediterranean*. Chicago: University of Chicago Press.
Harrison, S.J. 1999 The Survival and Supremacy of Rome: the Unity of the Shield of Aeneas. *Journal of Roman Studies* 87: 70-76.
Hart, M. 2020 *Extraterritorial: a Political Geography of Contemporary Fiction*. New York: Columbia University Press.
Hartog, F. 1988 *The Mirror of Herodotus: the Representation of the Other in the Writing of History*. Berkeley: University of California Press.
Hartswick, K.J. 2004 *The Gardens of Sallust: a Changing Landscape*. Austin: University of Texas Press.
Hartswick, K.J. 2018 Sculpture in Ancient Roman Gardens. In W.F. Jashemski, K.L. Gleason, K.J. Hartswick, and A.-A. Malek (eds) 2018: 341-365.
Harvey, P.D.A. 1993 *Maps in Tudor England*. Chicago: University of Chicago Press.
Harvey, P.D.A. 2010 *Mappa Mundi. The Hereford World Map: Introduction*. Hereford: Hereford Cathedral.
Haselberger, L. 1997 Architectural Likenesses-Models and Plans of Architecture in Classical Antiquity. *Journal of Roman Archaeology* 10: 77-94.
Haselberger, L., D.G. Romano, E.A. Dumser, and D. Borbonus (eds) 2002 *Mapping Augustan Rome*. Journal of Roman Archaeology Supplementary Series 50. Portsmouth, RI.
Haselberger, L. and J. Humphrey (eds) 2006 *Imaging Ancient Rome: Documentation, Visualization, Imagination*. Portsmouth: RI.
Hawes, G. 2017 *Myths on the Map: the Storied Landscapes of Ancient Greece*. Oxford: Oxford University Press.
Haynes, I. 2013 *Blood of the Provinces: the Roman Auxilia and the Making of Roman Provincial Society from Augustus to the Severans*. Oxford: Oxford University Press.
Hecker, S. and R. Bedarida (eds) 2023 *Curating Fascism: Exhibitions and Memory from the Fall of Mussolini to Today*. London: Bloomsbury.
Helms, K. 2021 Pompeii's Safaitic Graffiti. *Journal of Roman Studies* 111: 203-214.
Hemelrijk, E. 2020 *Women and Society in the Roman World: a Sourcebook of Inscriptions from the Roman West*. Cambridge: Cambridge University Press.
Henig, M. and J. Lundock (eds) 2022 *Water in the Roman World: Engineering, Trade, Religion and Daily Life*. Oxford: Archaeopress.
Herrmann, P. 2007 *Itinéraires des Voies Romaines: de l'Antiquité au Moyen Age*. Paris: Éditions Errance.
Hesecamp, I. 2021 *Das Bild von 'Africa' in der Augusteischen Dichtung. Poetische Konstruktionen Eines Geographischen Raumes*. Berlin: De Gruyter.

Hill, P.V. 1989 *The Monuments of Ancient Rome as Coin Types*. London: Seaby.

Hillner, J. 2015 *Prison, Punishment and Penance in Late Antiquity*. Cambridge: Cambridge University Press.

Hinds, S.E. 2002 Landscape with Figures: Aesthetics of Place in the Metamorphoses and its Tradition. In P. Hardie (ed) 2002 *The Cambridge Companion to Ovid*. Cambridge: Cambridge University Press: 122-149.

Hingley, R. 2005 *Globalizing Roman Culture: Unity, Diversity and Empire*. London: Routledge.

Hinterhöller, M. 2007a Typologie und Stilistische Entwicklung der Sakral-Idyllischen Landschaftsmalerei in Rom und Kampanien Während des Zweiten und Dritten Pompjanischen Stils. *Römische Hitorische Mitteilungen* 49: 17-69.

Hinterhöller, M. 2007b Die Gesegnete Landschaft. Zur Bedeutung Religions und Naturphilosophischer Konzepte für die Sakral-Idyllischen Landschaftsmalerei von Spätrepublikanischer bis Augusteischer zeit. *Jahreshefte des Österreichischen Archäologischen Instituts* 76: 129-169.

Hinterhöller-Klein, M. 2015 *Varietates Topiorum. Perspektive und Raumerfassung in Landscafts und Panoramabildern der Römischen Wandmalerei vom I. Jh. v. Chr. Bis zum Ende der Pompejanischen Stile*. Vienna: Phoibos.

Holleran, C. and A. Claridge (eds) 2018 *A Companion to the City of Rome*. Oxford: Wiley/Blackwell.

Holliday, P. 1997 Roman Triumphal Painting: Its Function, Development, and Reception. *Art Bulletin* 79: 130-147.

Hollis, D. and J.P. König (eds) 2021 *Mountain Dialogues from Antiquity to Modernity*. London: Bloomsbury.

Holmes, B. 2010 Marked Bodies: Gender, Race, Class, Age, Disability, and Disease. In D.H. Garrison (ed) 2010 *A Cultural History of the Body in Antiquity*. New York: Berg: 159-184.

Hopkins, J.N. 2007 The *Cloaca Maxima* and the Monumental Manipulation of Water in Archaic Rome. *The Waters of Rome* 4: 1-15.

Hopkins, J.N. 2012 The "Sacred Sewer". Tradition and Religion in the *Cloaca Maxima*. In M. Bradley and K. Stow (eds) 2012 *Rome, Pollution and Propriety: Dirt, Disease and Hygiene in the Eternal City from Antiquity to Modernity*. Rome: the British School at Rome: 81-102.

Horsfall, N. 1985 Illusion and Reality in Latin Topographical Writing. *Greece and Rome* 32(2): 197-208.

Horster, M. and N. Hächler (eds) 2021 *The Impact of the Roman Empire on Landscapes: Proceedings of the Fourteenth Workshop of the International Network Impact of Empire, Mainz June 12-15, 2019*. Leiden: Brill.

Houghtalin, L. 1996 *The Personifications of the Roman Provinces*. Ann Arbor: University of Michigan Press.

Hudson, J. 2021 *The Rhetoric of Roman Transportation: Vehicles in Latin Literature*. Cambridge: Cambridge University Press.

Hughes, J. 2009 Personification and the Ancient Viewer: the Case of the *Hadrianeum* 'Nations'. *Art History* 32: 1-20.

Hughes, J.D. 2013 Warfare and Environment in the Ancient World. In B. Campbell and L.A. Tritle (eds) 2013 *The Oxford Handbook of Warfare in the Classical World*. Oxford: Oxford University Press: 128-139.

Hughes, J.D. 2014 *Environmental Problems of the Greeks and Romans: Ecology in the Ancient Mediterranean*. Baltimore: Johns Hopkins University Press.

Hunt, A. 2012 Keeping the Memory Alive: the Physical Continuity of the *Ficus Ruminalis*. In M. Bommas, J. Harrison and P. Roy (eds) 2012 *Memory and Urban Religion in the Ancient World*. London: Bloomsbury: 111-128.

Hunt, A. 2016 *Reviving Roman Religion: Sacred Trees in the Roman World*. Cambridge: Cambridge University Press.

Hunt, E.D. 1982 *Holy Land Pilgrimage in the Later Roman Empire AD 312-460*. Oxford: Oxford University Press.

Huskinson, J. 2005 Rivers of Roman Antioch. In E. Stafford and J. Herrin (eds) 2005 *Personification in the Greek World: from Antiquity to Byzantium*. London: Routledge: 247-264.

Hyde, W.W. 1915 The Ancient Appreciation of Mountain Scenery. *The Classical Journal* 11(2): 70-84.

Ingold, T. 1993 The Temporality of Landscape. *World Archaeology* 25(2): 152-174.

Ingold, T. 2004 Culture on the Ground: the World Perceived Through the Feet. *Journal of Material Culture* 9(3): 315-340.

Irby, G.L. 2012 Mapping the World: Greek Initiatives from Homer to Eratosthenes. In R.J.A. Talbert (ed) 2012: 81-108.

Irby, G.L. 2019 The Politics of Cartography: Foundlings, Founders, Swashbucklers, and Epic Shields. In D.W. Roller (ed) 2019: 80-102.

Isaac, B. 2004 *The Invention of Racism in Classical Antiquity*. Princeton University Press: Princeton.

Isaac, B. 2011 Attitudes Toward Provincial Intellectuals in the Roman Empire. In E.S. Gruen (ed) 2011: 491-518.

Isaac, B. 2021 From Rome to Constantinople. In J.J. Price, M. Finkelberg and Y. Shahar (eds) 2021: 17-28.

Iversen, E. 1968-1972 *Obelisks in Exile. The Obelisks of Rome*. Copenhagen: G.E.C. Gad.

Isayev, E. 2009 Unintentionally Being Lucanian: Dynamics Beyond Hybridity. In S. Hales and T. Hodos (eds) 2009 *Material Culture and Social Identities in the Ancient World*. Cambridge: Cambridge University Press: 201-226.

Isayev, E. 2017 *Migration, Mobility and Place in Ancient Italy*. Cambridge: Cambridge University Press.

Isayev, E. 2020 Elusive Migrants of Ancient Italy. In J. Clarkson, P. James, K. McDonald, L. Tagliapietra, and N. Zair *et al.* (eds) 2020: 53-74.

Jaccottet, A.-F. 2020 Les Chemins de Dionysos: la Construction Culturelle du Transfert Cultuel, Entre Textes, Images et Réalité Rituelle. In B. Amiri (ed) 2020 *Migrations et Mobilité Religieuse: Espaces, Contacts, Dynamiques et Intérferences*. Besançon: Presses Universitaires de Franche-Comté: 55-73.

Jacobs, A.S. 2002 'The Most Beautiful Jewesses in the Land': Imperial Travel in the Early Christian Holy Land. *Religion* 32(3): 205-225.

Janni, P. 1984 *La Mappa e il Periplo: Cartografia Antica e Spazio Odologico*. Rome: L'Erma di Bretschneider.

Janrović, M.A. 2014 Violent Ethnicities: Gladiatorial Spectacles and Display of Power. In M.A. Janrović, V.D. Mihajlović, and S. Babić (eds) 2014 *The Edges of the Roman World*. Newcastle Upon Tyne: Cambridge Scholars: 48-60.

Jashemski, W.F., K.L. Gleason, K.J. Hartswick, and A.-A. Malek (eds) 2018 *Gardens of the Roman Empire*. Cambridge: Cambridge University Press.

Jenkyns, R. 2013 *God, Space, and City in the Roman Imagination*. Oxford: Oxford University Press.

Jeskins, P. 1998 *The Environment and the Classical World*. London: Bristol Classical Press.

Johnson, W.R. 1984 Vergil's Bees: the Ancient Romans' Views of Rome. In A. Patterson (ed) 1984 *Roman Images. Selected Papers from the English Institute*, New Series No. 8. Baltimore: Johns Hopkins University Press: 1-22.

Johnston, A.C. 2019 Odyssean Wanderings and Greek Responses to Roman Empire. In T. Biggs and J. Blum (eds) 2019: 211-240.

Jones, F.M.A. 2011 *Virgil's Garden: the Nature of Bucolic Space.* London: Bristol Classical Press.

Jones, F.M.A. 2013a Drama, Boundaries, Imagination, and Columns in the Garden Room at Prima Porta. *Latomus* 72: 997-1021.

Jones, F.M.A. 2013b The Caged Bird in Roman Life and Poetry: Metaphor, Cognition, and Value. *Syllecta Classica* 24: 105-123.

Jones, F.M.A. 2016 *The Boundaries of Art and Social Space in Rome: the Caged Bird and Other Art Forms.* London: Bloomsbury Academic.

Jones, P.J. 2005 *Reading Rivers in Roman Literature and Culture*. Oxford: Lexington Books.

Jones-Lewis, M.A. 2019 *Mutuo Metu aut Montibus*: Mapping Environmental Determinism in the *Germania* of Tacitus. In D.W. Roller (ed) 2019: 135-160.

Joska, S. 2022 Roman Imperial Family on the Road: Power and Interaction in the Roman East During the Antonine Era. In J. Kuuliala and J. Rantala (eds) 2020: 100-121.

Joyce, L. 2014/2015 Roma and the Virtuous Breast. *Memoirs of the American Academy at Rome* 59/60: 1-49.

Juhász, L. 2016 *The Iconography of the Roman Province Personifications and Their Role in the Imperial Propaganda*. Unpublished PhD Thesis, Eötvös Loránd University, Budapest.

Kaldellis, A. 2020 How Was a 'New Rome' Even Thinkable? Premonitions of Constantinople and the Portability of Rome. In Y.R. Kim and A.E.T. McClaughlin (eds) 2020 *Leadership and Community in Late Antiquity: Essays in Honour of Raymond Van Dam.* Turnhout: Brepols: 221-247.

Kammerer, R.C. 1972-1973 The Travel Coins of Hadrian, I. *Journal of the Society of Ancient Numismatists* 4: 69-70.

Katzenellenbogen, A. 1947 The Sarcophagus in S. Ambroglio and St. Ambrose. *The Art Bulletin* 29(4): 249-259.

Keegan, P. 2016 Graffiti as *Monumenta* and *Verba*: Marking Territories, Creating Discourses in Roman Pompeii. In R. Benefiel and P. Keegan (eds) 2016 *Inscriptions in the Private Sphere in the Greco-Roman World.* Leiden: Brill: 248-264.

Keith, A.M. 2014 Imperial Geographies in Tibullan Elegy. *Classical World* 107(4): 477-492.

Keith, A. 2019 Women's Travels in the *Aeneid*. In T. Biggs and J. Blum (eds) 2019: 130-144.

Keith, A. Forthcoming. Women's Travels in Latin Elegy. In E.Z. Damer and M. Myers (eds) Forthcoming *Travel and Geography in Latin Literature*.

Kennedy, R.F. 2015 Airs, Waters, Metals, Earth: People and Land in Archaic and Classical Greek Thought. In R.F. Kennedy and M. Jones-Lewis (eds) 2015: 9-28.

Kennedy, R.F. and M. Jones-Lewis (eds) 2015 *The Routledge Handbook of Identity and Environment in the Classical and Medieval Worlds.* London: Routledge.

Khellaf, K. 2021 Migration and Mobile Memory in the Roman Historical Digression. In A.D. Poulsen and A. Jönsson (eds) 2021 *Usages of the Past In Roman Historiography.* Leiden: Brill: 262-297.

Kienast, H.J. 2014 Der Turm der Wind in Athen. *Archäologische Forschungen* Bd 30. Wiesbaden: Reichert Verlag.

Kiernan, P. 2012 Pagan Pilgrimage in Rome's Western Provinces. *Herom* 1: 79-106.

Kivuila-Kiaku, J.M. and Nkoko, J.-B.N. 2017 *L'Afrique Vue par les Romains: les Écrits de Salluste et de Lucain*. Paris: Éditions L'Harmattan.

Kleiner, D.E.E. 1993 *Roman Sculpture*. New Haven: Yale University Press.

Kleiner, F.S. 1991 The Trophy on the Bridge and the Roman Triumph Over Nature. *L'Antiquité Classique* 60: 182-192

Klotz, A. 1931 Die Geographischen Commentarii des Agrippa und ihre Uberreste. *Klio* 24: 38-58 and 386-466.

Koehler, J. 2013 More Water for Rome: Nothing New in the Eternal City? Water-Related Monuments as Part of the Severan Building Program. In E. De Sena (ed) 2013 *The Roman Empire During the Severan Dynasty: Case Studies in History, Art, Architecture, Economy and Literature*. American Journal of Ancient History 6-8 2013: 117-150.

Koeppel, G.M. 1980 A Military *Itinerarium* on the Column of Trajan. *Mitteilungen des Deutschen Archäologischen Instituts, Römisch Abteilungen* : 301-306.

Koeppel, G.M. 1982 The Grand Pictorial Tradition of Roman Historical Representation. *Aufstieg und Niedergang das Romischer Welt* 12(1): 507-535.

Kolb, A. 2001 Transport and Communication in the Roman State: the *Cursus Publicus*. In J. Adams and R. Laurence (eds) 2001: 95-105.

Kondratieff, E.J. 2014 Future City in the Heroic Past: Rome, Romans, and Roman Landscapes in *Aeneid* 6-8. In A.M. Kemezis (ed) 2014 *Urban Dreams and Realities in Antiquity: Remains and Representations of the Ancient City*. Leiden: Brill: 165-228.

König, J.P. 2016 Strabo's Mountains. In J. McInery and I. Sluiter (eds) 2016: 46-69.

König, J.P. 2018 Fire and Fury: Observing Volcanoes in the Ancient Landscape. *Viewpoint: Magazine of the British Society for the History of Science* 115: 11-13.

König, J.P. 2022 *The Folds of Olympus: Mountains in Ancient Greek and Roman Culture*. Princeton: Princeton University Press.

Kosmin, P.J. 2014 *The Land of the Elephant Kings: Space, Territory, and Ideology in the Seleucid Empire*. Cambridge: Cambridge University Press.

Kosso, C. and A. Scott (eds) 2009 *The Nature and Function of Water, Baths, Bathing and Hygiene from Antiquity Through the Renaissance*. Leiden: Brill.

Kotsidu, H. 1998 *Landschaft im Bild. Naturprojektionen in der Antiken Dekorationkunst*. Worms: Werner.

Kovács, P. 2017 Deities in Trajan's and Marcus Aurelius' Column. *Acta Archaeologica Academiae Scientiarum Hungaricae* 68(1): 47-58.

Kowalski, J.-M. 2012 *Navigation et Géographie dans L'Antiquité Gréco-Romaine: la Terre Vue de la Mer. Antiquité-Synthèse*. Paris: Éditions Picard.

Krebs, C.B. 2006 Imaginary Geography in Caesar's *Bellum Gallicum*. *American Journal of Philology* 127(1): 111-136.

Krebs, C.B. 2010 Borealism: Caesar, Seneca, Tacitus, and the Roman Discourse About the Germanic North. In E.S. Gruen (ed) 2010: 202-221.

Krebs, C.B. 2018 The World's Measure: Caesar's Geographies of *Gallia* and *Britannia* in Their Contexts and as Evidence of His World Map. *American Journal of Philology* 139: 93-122.

Künzl, E. and G. Koeppel 2002 *Souvenirs und Devotionalien: Zeugnisse des Geschäftlichen, Religiösen und Kulturellen Tourismus im Antiken Römerreich*. Mainz: Philipp von Zabern.

Kuttner, A.L. 1999 Culture and History at Pompey's Museum. *Transactions of the American Philological Association* 129: 343-373.

Kuttner, A.L. 1999 Looking Outside Inside: Ancient Roman Garden Rooms. In J.D. Hunt (ed) 1999 *The Immediate Garden and the Larger Landscape.* Special Issue of Studies in the History of Gardens and Designed Landscapes 1: 7-35.

Kuuliala, J. and J. Rantala (eds) 2020 *Travel, Pilgrimage and Social Interaction from Antiquity to the Middle Ages.* London: Routledge.

Laehn, T.R. 2013 *Pliny's Defense of Empire.* London: Routledge.

Laird, M. 2015 *Civic Monuments and the Augustales in Roman Italy.* Cambridge: Cambridge University Press.

Lampinen, A. 2022 Mediterranean as a Contested Environment in Late Antiquity. In A. Lampinen and E.M. Ferrándiz (eds) 2022: 49-68.

Lampinen, A. and E.M. Ferrándiz (eds) 2022 *Seafaring and Mobility in the Late Antique Mediterranean.* London: Bloomsbury.

Lanciani, R. 1901 *Forma Urbis Romae.* Milan: Quasar Reprint.

Landels, J.G. 2000 *Engineering in the Ancient World.* London: Constable and Robinson.

Langdon, M.K. 1999 Classifying the Hills of Rome. *Eranos* 97: 98-107.

Larmour, D.H.J. 2018 Juvenal in the Specular City. In W. Fitzgerald and E. Spentzou (eds) 2018: 95-118.

Larmour, D.H.J. and D. Spencer (eds) 2007 *The Sites of Rome: Time, Space, Memory.* Oxford: Oxford University Press.

La Rocca, E. 2000 L'Affresco con Veduta di Città dal Colle Oppio. In E. Fentress (ed) 2000 *Romanization and the City: Creation, Dynamics and Failures.* Journal of Roman Archaeology Supplement 38: 57-71.

La Rocca, E. 2001 The Newly Discovered City Fresco from Trajan's Baths, Rome. *Imago Mundi* 53: 121-124.

La Rocca, E. 2008 *Lo Spazio Negato: la Pittura di Paesaggio nella Cultura Artistica Greca e Romana.* Florence: Electa.

La Rocca, E., S. Ensoli, S. Tortorella, and M. Papini 2009 *Roma; La Pittura di un Impero.* Milan: Skira.

Latham, J.A. 2016 *Performance, Memory, and Processions in Ancient Rome: the Pompa Circensis from the Late Republic to Late Antiquity.* Cambridge: Cambridge University Press.

Laurence, R. 1999 *The Roads of Roman Italy: Mobility and Cultural Change.* London: Routledge.

Laurence, R. 2004 Milestones, Communications and Political Stability. In L. Ellis and F.L. Kidner (eds) 2004 *Travel, Communication and Geography in Late Antiquity: Sacred and Profane.* Aldershot: Ashgate: 41-60.

Laurence, R. 2015 Towards a History of Mobility in Ancient Rome (300 BC to 100 CE). In I. Östenberg, S. Malmberg, and J. Bjørnebye (eds) 2015 *The Moving City: Processions, Passages and Promenades in Ancient Rome.* London: Bloomsbury: 175-186 and 302-307.

Laurence, R. 2020 The Meaning of Roads: a Reinterpretation of the Roman Empire. In J. Kuuliala and J. Rantala (eds) 2020: 37-63.

Lavan, M. 2020 Devastation. The Destruction of Populations and Human Landscapes and the Roman Imperial Project. In K. Berthelot (ed) 2020 *Reconsidering Roman Power: Christian Perceptions and Reactions.* L'Ecole Francaise de Rome: 179-205.

Layton, R. and Ucko, P. (eds) 1999 *The Archaeology and Anthropology of Landscape: Shaping Your Landscape.* London: Routledge.

Lazzaro, C. 2011 River Gods: Personifying Nature in Sixteenth-Century Italy. *Renaissance Studies* 25(1): 70-94.

Leach, E.W. 1988 *The Rhetoric of Space: Literary and Artistic Representations of Landscape in Republican and Augustan Rome*. Princeton: Princeton University Press.

Lee, M. 2006 Acheloös Pelophoros: a Lost Statuette of a River God in Feminine Dress. *Hesperia* 75: 317-325.

Leech, R. and P. Leech (eds) 2021 *The Colonial Landscape of the British Caribbean*. Martlesham: Boydell Press.

Leemreize, M. 2014 The Egyptian Past in the Roman Present. In J. Ker and C. Pieper (eds) 2014 *Valuing the Past in the Greco-Roman World. Proceedings from the Penn-Leiden Colloquia on Ancient Values 7*. Leiden: Brill: 52-82.

Lefebvre, H. 1991 *The Production of Space*. Oxford: Blackwell.

Le Gall, J. 1953 *Le Tibre, Fleuve de Rome dans l'Antiquité*. Paris: Presses Universitaires de France.

Lehoux, D. 2006 Rethinking Parapegmata: the Puteoli Fragment. *Zeitscrift für Papyrologie und Epigraphik* Bd. 157: 95-104.

Leiris, M. 1934 *L'Afrique Fantôme*. Paris: Éditions Gallimard.

Lemcke, L. 2016 *Imperial Transportation and Communication from the Third to the Late Fourth Century: the Golden Age of the Cursus Publicus*. Brussels: Latomus.

Lemke, R.S. 2002 *Maps and Memory in Early Modern England: a Sense of Place*. New York: Palgrave.

Leone, A., D. Palombi, and S. Walker (eds) 2007 *Res Bene Gestae: Ricerche di Storia Urbana su Roma Antica in Onore di Eva Margareta Steinby*. Lexicon Topographicum Urbis Roma, Supplementum IV. Rome: Quasar.

Lepper, F. and S.S. Frere 1988 *Trajan's Column*. Stroud: Sutton Publishing.

Leuenberger, C. and I. Schnell 2020 *The Politics of Maps: Cartographic Constructions of Israel/Palestine*. Oxford: Oxford University Press.

Levene, D.S. 2019 Monumental Insignificance: the Rhetoric of Roman Topography from Livy's Rome. In M. Loar, S. Murray and S. Rebeggiani (eds) 2019 *The Cultural History of Augustan Rome: Texts, Monuments and Topography*. Cambridge: Cambridge University Press: 10-26.

Levin-Richardson, S. 2021 Sex and Slavery in Pompeian Households: a Survey. In D. Kamen and C.W. Marshall (eds) 2021 *Slavery and Sexuality in Classical Antiquity*. Madison: University of Wisconsin Press: 188-210.

Lewis, M.J.T. 2001. *Surveying Instruments of Greece and Rome*. Cambridge: Cambridge University Press.

Lillie, C. 2017 *The Rape of Eve: the Transformation of Roman Ideology in Three Early Christian Retellings of Genesis*. Minneapolis: Fortress Press.

Lindheim, S. 2021 *Latin Elegy and the Space of Empire*. Oxford: Oxford University Press.

Ling, R. 1977 Studius and the Beginnings of Roman Landscape Painting. *Journal of Roman Studies* 67: 1-16.

Lively, G. 2002 Cleopatra's Nose, Naso and the Science of Chaos. *Greece and Rome* 49(1): 27-43.

Loar, M., C. MacDonald and D-el. Padilla Peralta (eds) 2017 *Rome, Empire of Plunder*. Cambridge: Cambridge University Press.

Lo Cascio, E. 2016 The Impact of Migration on the Demographic Profile of the City of Rome: a Reassessment. In E. Lo Cascio and L.E. Tacoma (eds) 2016 *The Impact of Mobility and Migration in the Roman Empire*. Leiden: Brill: 23-32.

Longfellow, B. 2009 The Legacy of Hadrian: Roman Monumental Civic Fountains in Greece. In C. Kosso and A. Scott (eds) 2009: 211-232.

Lott, J.B. 2004 *The Neighbourhoods of Augustan Rome*. Cambridge: Cambridge University Press.

Lovejoy, A.O. and G. Boas 1935 *Primitivism and Related Ideas in Antiquity*. Baltimore: Johns Hopkins University Press.

Lowenstam, S. 1995 The Sources of the Odyssey Landscapes. *Echos du Monde Classique: Classical Views* Volume 39 New Series 14(2): 193-226.

Lozovsky, N. 2000 *The Earth Is Our Book. Geographical Knowledge in the Latin West ca. 400-1000*. Ann Arbor: University of Michigan Press.

Lozovsky, N. 2008 Maps and Panegyrics: Roman Geo-Ethnographical Rhetoric in Late Antiquity and the Middle Ages. In R.J.A. Talbert and R. Unger (eds) 2008: 169-188.

Luke, T.S. 2010 A Healing Touch for Empire: Vespasian's Wonders in Domitianic Rome. *Greece and Rome* 57(1): 77-106.

Lupher, D.A. 2003 *Romans in a New World*. Ann Arbor: University of Michigan Press.

Lusnia, S.S. 2004 Urban Planning and Sculptural Display in Severan Rome: Reconstructing the *Septizodium* and Its Role in Dynastic Politics. *American Journal of Archaeology* 108: 517-544.

Macaulay-Lewis, E. 2009 Political Museums: Porticos, Gardens and the Public Display of Art in Ancient Rome. In S. Bracken, A.M. Gáldy and A. Turpin (eds) 2009 *Collecting and Dynastic Ambition*. Newcastle: Cambridge Scholars: 1-22.

Maguire, H. 2016 *Nectar and Illusion: Nature and Art In Byzantine Art and Literature*. Oxford: Oxford University Press.

Malek, A.-A. 2018 Mosaics and Nature in the Roman *Domus*. In W.F. Jashemski, K.L. Gleason, K.J. Hartswick, and A.-A. Malek (eds) 2018: 317-340.

Mallan, K. 2021 Taskscapes, Landscapes and the Politics of Agricultural Production in Roman Mosaics. *Theoretical Roman Archaeology Journal* 4(1):8: 1-26. DOI: https://doi.org/10.16995/traj.4340.

Malmberg, S. 2009 Finding Your Way in the Subura. In M. Driessen, S. Heeren, J. Hendriks, F. Kemmers and R. Visser (eds) 2009 *TRAC 2008. Proceedings of the Eighteenth Annual Theoretical Roman Archaeology Conference, Amsterdam 2008*. Oxford: Oxbow Books: 39-51.

Mari, Z. 2008 The 'Egyptian Places' of Hadrian's Villa. In A. Lo Sardo (ed) 2008 *The She-Wolf and the Sphinx: Rome and Egypt from History to Myth*. Rome: Electa: 122-131.

Marion, L. 1965 The Velletri Sarcophagus. *American Journal of Archaeology* 69(3): 207-222.

Markus, R.A. 1994 How On Earth Could Places Become Holy? Origins of the Christian Idea of Holy Places. *Journal of Early Christian Studies* 2: 268-270.

Marretta, A. 2013 The Abstract Mind: Valcamonica. Complex Geometric Compositions in the Light of New Discoveries. *Papers XXV Valcamonica Symposium 2013*: 343-356.

Marzano, A. 2022 *Plants, Politics and Empire in Ancient Rome*. Cambridge: Cambridge University Press.

Matheson, S.B. 1994 The Goddess Tyche. In S.B. Matheson (ed) 1994 *An Obsession with Fortune: Tyche in Greek and Roman Art*. New Haven: Yale University Art Gallery: 18-33.

Mathisen, R.W. and D. Shanzer (eds) 2011 *Romans, Barbarians, and the Transformation of the Roman World: Cultural Interaction and the Creation of Identity in Late Antiquity*. Aldershot: Ashgate.

Maticic, D.A. 2021 Alluvium and Interlude: the Dynamics of Relationality in Ausonius's Mosella. *Arethusa* 54.3: 399-423.

Maticic, D.A. 2021 Hercules, Cacus, and the Poetics of Drains in *Aeneid* 8 and *Propertius* 4.9. In M. Horster and N. Hächler (eds) 2021: 339-352.

Mayer, R. 1986 Geography and the Roman Poets. *Greece and Rome* 33: 47-54.

Mazurek, L.A. 2020 Looking at the Nile from Afar: New Ways of Seeing Imperialism in Roman History and Art. *Journal of Roman Archaeology* 33.2: 679-687.

Mazurek, L.A. 2022 *Isis in a Global Empire: Greek Identity Through Egyptian Religion in Roman Greece*. Cambridge: Cambridge University Press.

McAlpine, L.J. 2023 Roman Villas and Landscapes of Luxury. In J. Powers (ed) 2023: 55-66.

McClaughlin, R. and H.J. Kim 2021 Roman Envoys and Trade Ambassadors in Han China. In H.J. Kim, S.N.C. Lieu and R. McClaughlin (eds) 2021 *Rome and China: Points of Contact*. London: Routledge: 4-30.

McGregor, J.H.S. 2015 *Back to the Garden: Nature and the Mediterranean World from Prehistory to the Present.* New Haven: Yale University Press.

McInerny, J. and I. Sluiter (eds) 2016 *Valuing Landscape in Classical Antiquity: Natural Environment and Cultural Imagination.* Leiden: Brill.

McKittrick, K. 2006 *Demonic Grounds; Black Women and the Cartographies of Struggle.* Minneapolis: University of Minnesota Press.

McLaughlin, R. 2010 *Rome and the Distant East: Trade Routes to the Ancient Lands of Arabia, India and China.* London: Continuum.

McMillan, K. 2019 *Contemporary Art and Unforgetting in Colonial Landscapes.* London: Palgrave Macmillan.

Meiggs, R. 1982 *Trees and Timber in the Ancient Mediterranean World.* Oxford: Clarendon Press.

Meijer, E. 2021 Justifying Civil War: Interactions Between Caesar and the Italian Landscape in Lucan's Rubicon Passage. In B. Reitz-Joose, M.W. Makins, and C.J. Mackie (eds) 2021: 157-176.

Meinecke, K. 2020 Circulating Images: Late Antiquity's Cross-Cultural Visual *Koiné*. In F. Guidetti and K. Meinecke (eds) 2020 A Globalised Visual Culture? Towards a Geography of Late Antique Art. Oxford: Oxbow Books: 321-339.

Melville-Jones, J.R. 2014 Constantinople as "New Rome". *Byzantina Symmeikta* 24: 247-262.

Meneghini, R. 2007 La Cartografia Antica e Il Catasto di Roma Imperiale. In A. Leone, D. Palombi, and S. Walker (eds) 2007: 205-218.

Meneghini, R. and R.S. Valenzani (eds) 2006 *Formae Urbis Romae. Nuovi Frammenti di Piante Marmoree dallo Scavo dei Foro Imperiale.* Rome: L'Erma di Bretschneider.

Menut, N. 2010 *L'Homme Blanc: Les Représentations de l'Occidental dans les Arts Non Européens.* Paris: Éditions du Chêne.

Merrills, A.H. 2005 *History and Geography in Late Antiquity.* Cambridge: Cambridge University Press.

Merrills, A.H. 2017 *Roman Geographies of the Nile: from the Late Republic to the Early Empire.* Cambridge: Cambridge University Press.

Méthy, N. 1992 La Représentation des Provinces dans le Monnayage Romain de l'Époque Impériale (70-235 après J.C.). *Numismatica e Antichita Classiche* 21: 267-289.

Meyboom, P.G.P. 1995 *The Nile Mosaic of Palestrina: Early Evidence of Egyptian Religion in Italy.* Leiden: Brill.

Meyer, M. 2021 Visualizing the Passing of Time: Personifications of the Seasons in Greek and Roman Imagery. In A. Lichtenberger and R. Raja (eds) 2021 *The Archaeology of Seasonality.* Turnhout: Brepols: 323-348.

Meyers, G.E. 2009 The Divine River: Ancient Roman Identity and the Image of *Tiberinus*. In C. Kosso and A. Scott (eds) 2009: 233-247.

Mielsch, H. 2001 *Römische Wandmalerei.* Darmstadt: Wissenschaftliche Buchgesellschaft.

Millar, F. 2005 Last Year in Jerusalem: Monuments of the Jewish War in Rome. In J. Edmondson, S. Mason and J.B. Rives (eds) 2005 *Flavius Josephus and Flavian Rome.* Oxford: Oxford University Press: 101-128.

Miller, P.A. 2007 'I Get Around': Sadism, Desire, and Metonymy on the Streets of Rome With Horace, Ovid, and Juvenal. In D.H.J. Larmour and D. Spencer (eds) 2007: 138-167.

Minchin, E. 2012 Commemoration and Pilgrimage in the Ancient World: Troy and the Stratigraphy of Cultural Memory. *Greece and Rome* 59: 76-89.

Minchin, E. 2021 Homer's Landscapes of War: Spatial Mental Model and Cognitive Collage. In B. Reitz-Joosse, M. W. Makins and C.J. Mackie (eds) 2021: 25-37.

Minchin, E. 2016 Heritage in the Landscape: the Heroic "Tumuli" in the Troad Region. In J. McInerney and I. Sluiter (eds) 2016: 255-275.

Minor, H.H. 1999 Mapping Mussolini: Ritual and Cartography in Public Art During the Second Roman Empire. *Imago Mundi* 51: 147-162.

Mitchell, W.J.T. 2002 Imperial Landscape. In W.J.T. Mitchell (ed) 2002 *Landscape and Power*. Second Edition. Chicago: University of Chicago Press: 1-34.

Molholt, R.M. 2011 Roman Labyrinth Mosaics and the Experience of Motion. *Art Bulletin* 93(3): 287-303.

Montrose, L. 1991 The Work of Gender in the Discourse of Discovery. *Representation* 33: 1-41.

Moorman, E. 2022 Some Observations on the *Templum Pacis*: a *Summa* of Flavian Politics. In M. Heerink and E. Meijer (eds) 2022 *Flavian Responses to Nero's Rome*. Amsterdam: Amsterdam University Press: 127-160.

Morris, K. 2019 *Shifting Ground: Landscape in Contemporary Native American Art*. Seattle: University of Washington Press.

Morstein-Marx, R. 2012 Political Graffiti in the Late Roman Republic: "Hidden Transcripts" and "Common Knowledge". In C. Kuhn (ed) 2012 *Politische Kommunikation und Öffentliche Meinung in der Antike Welt*. Stuttgart: Franz Steiner Verlag: 191-217.

Moynihan, R. 1985 Geographical Mythology and Roman Imperial Ideology. In R. Winkes (ed) 1985 *The Age of Augustus*. Providence: Rhode Island: 149-162.

Mukherjee, N. 2020 *Spatial Imaginings in the Age of Colonial Cartographic Reason: Maps, Landscapes, Travelogues in Britain and India*. London: Routledge.

Munn, M. 2009 Earth and Water: the Foundations of Sovereignty in Ancient Thought. In C. Kosso and A. Scott (eds) 2009: 191-210.

Murphy, T. 2004 *Pliny the Elder's Natural History: the Empire in the Encyclopaedia*. Oxford: Oxford University Press.

Muzzioli, M.P. 2007 Sui Portici Raffigurati nella Lastra di Via Anicia. In A. Leone, D. Palombi, and S. Walker (eds) 2007: 219-237.

Myers, K.S. 2018 Representations of Gardens in Roman Literature. In W.F. Jashemski, K.L. Gleason, K.J. Hartswick, and A.-A. Malek (eds) 2018: 258-277.

Myers, M.Y. 2011 Lucan's Poetic Geographies: Center and Periphery in Civil War Epic. In P. Asso (ed) 2011 *Brill's Companion to Lucan*. Leiden: Brill: 399-415.

Myers, M.Y. and E.Z. Damer 2021 *Travel, Geography, and Empire in Latin Poetry*. London: Routledge.

Najbjerg, T. and J. Trimble 2004 Ancient Maps and Mapping In and Around Rome-A Review of E. Rodríguez-Almeida 2002 Formae Urbis Antiquae. Le Mappe Marmoree di Roma tra la Repubblica e Settimio Severo. *Journal of Roman Archaeology* 17: 577-583.

Najbjerg, T. and J. Trimble 2006 The Severan Marble Plan Since 1960. In R. Meneghini and R. Valenzani (eds) 2006 *Forma Urbis Romae: Nuovi Frammenti di Piante Marmoree dallo Scavo dei Fori Imperiali*. Rome: L'Erma di Bretscheider: 75-102.

Nasrallah, L.S. 2005 Mapping the World: Justin, Tatian, Lucian and the Second Sophistic. *Harvard Theological Review* 98 (3): 283-314.

Nasrallah, L.S. 2010 *Christian Responses to Roman Art and Architecture: the Second-Century Church Amid the Spaces of Empire*. Cambridge: Cambridge University Press.

Newby, Z. 2007 Landscape and Local Identity in the Mosaics of Antioch. In C.E.P. Adams and J. Roy (eds) 2007 *Travel, Geography and Culture in Ancient Greece, Egypt and the Near East*. Oxford: Oxbow Books: 184-205.

Nicolet, C. 1991 *Space, Geography, and Politics in the Early Roman Empire*. Ann Arbor: University of Michigan Press.

Nieto-Hernández, P. 2000 Back in the Cave of the Cyclops. *American Journal of Philology* 121: 345-366.

Nippel, W. 2007 Ethnic Images in Classical Antiquity. In M. Beller and J. Leerssen (eds) 2007 *Imagology: the Cultural Construction and Literary Representation of National Characters. A Critical Survey*. Amsterdam: Editions Rodopi: 33-44.

Noreña, C.F. 2003 Medium and Message in Vespasian's *Templum Pacis*. *Memoirs of the American Academy at Rome* 48: 25-43.

Noy, D. 2000 *Foreigners at Rome: Citizens and Strangers*. London: Duckworth.

Noy, D. 2010 Foreign Families in Roman Italy. In B. Rawson (ed) 2010 *A Companion to Families in the Greek and Roman Worlds*. Oxford: Wiley-Blackwell: 145-160.

O'Connor, C. 1993 Roman Bridges. Cambridge: Cambridge University Press.

O'Gorman, E. 1995 Shifting Ground; Lucan, Tacitus and the Landscape of Civil War. *Hermathena* 158: 117-131.

Olszewski, M. 2015 Les Cadrans Solaires dans les Mosaïques Romaines et Byzantines (1er Siècle Ap. J.C.-IXer Siècle ap. J.C.). In A. Tommaso (ed) 2015 *Ad Fines Imperii Romani*. Krakow: Archeobooks: 449-468.

Opper, T. (ed) 2013 *Hadrian: Art, Politics and Economy*. London: British Museum Press.

Orlin, E.M. 2010 *Foreign Cults in Rome: Creating a Roman Empire*. Oxford: Oxford University Press.

Osborn, R. 2012 How the Gauls Broke the Frame: the Political and Theological Impact of Taking Battle Scenes Off Greek Temples. In V. Platt and M. Squire (eds) 2012: 425-456.

Östenberg, I. 1999 Demonstrating the Conquest of the World: the Procession of Peoples and Rivers on the Shield of Aeneas and the Triple Triumph of Octavian in 29 B.C.. *Opuscula Romana, Annual of the Swedish Institute in Rome* 24: 155-162.

Östenberg, I. 2009a *Staging the World: Spoils, Captives, and Representations in the Roman Triumphal Procession*. Oxford: Oxford University Press.

Östenberg, I. 2009b *Titulis Oppida Capta Leget*: the Role of Written Placards in the Roman Triumphal Procession. *Mélanges de l'École Française de Rome* 121: 463-472.

Östenberg, I. 2014 Animals and Triumphs. In G.L. Campbell (ed) 2014: 491-506.

Östenberg, I. 2015 Augustan Literary Tours: Walking and Reading the City. In I. Östenberg, S. Malmberg and J. Bjørnebye (eds) 2015: 111-122.

Östenberg, I. 2017 Defeated by the Forest, the Pass, the Wind: Nature as an Enemy of Rome. In J. Clark and B. Turner (eds) 2017 *Brill's Companion to Military Defeat in Ancient Mediterranean Society*. Leiden: Brill: 240-261.

Östenberg, I. 2021 The Arch of Titus: Triumph, Funeral, and Apotheosis in Ancient Rome. In S. Fine (ed) 2021 *The Arch of Titus: from Jerusalem to Rome-and Back*. Leiden: Brill: 33-42.

Östenberg, I., S. Malmberg and J. Bjørnebye (eds) 2015 *The Moving City: Passages, Processions and Promenades in Ancient Rome*. London: Bloomsbury Academic.

Ostrow, S. 1979 The Topography of Puteoli and Baiae on the Eight Glass Flasks. *Puteoli, Studi di Storia Antica* 3: 77-140.

Ostrowski, J.A. 1990a *Les Personifications des Provinces dans l'Art Romain*. Warsaw/Krakow: Archeobooks.

Ostrowski, J.A. 1990b Personifications of Rivers as an Element of Roman Political Propaganda. *Études et Travaux* 15: 309-316.

Ostrowski, J.A. 1991 *Personifications of Rivers in Greek and Roman Art*. Warsaw: Nakadem Uniwersytetu Jagielloskiego.

Ostrowski, J.A. 1996 Personifications of Countries and Cities as a Symbol of Victory in Greek and Roman Art. In E.G. Schmidt (ed) 1996 *Griechenland und Rom: Vergleichende Unterschungen zu Entwicklungstendenzen und -höhepunkten der Antiken Geschichte, Kunst und Literatur*. Erlangen: Palm and Enka: 264-272.

O'Sullivan, T.M. 2007 Walking with Odysseus: the Portico Frame of the Odyssey Landscapes. *American Journal of Philology* 128 (4): 497-532.

O'Sullivan, T.M. 2011 *Walking in Roman Culture*. Cambridge: Cambridge University Press.

O'Sullivan, T.M. 2015 Augustan Literary Tours: Walking and Reading the City. In I. Östenberg, S. Malmberg and J. Bjørnebye (eds) 2015: 111-122.

O'Sullivan, T.M. 2019 Epic Journeys on an Urban Scale: Movement and Travel in Vergil's Aeneid. In T. Biggs and J. Blum (eds) 2019: 151-169.

O'Sullivan, T.M. 2023 Mythological Landscapes in Roman Painting. In J. Powers (ed) 2023: 67-78.

Padilla Peralta, D.-el. 2020 Epistemicide: the Roman Case. *Classica: Revista Brasiliera de Estudos Clássicos* 33: 151-186.

Painter, B.W. 2005 *Mussolini's Rome: Rebuilding the Eternal City*. New York: Palgrave MacMillan.

Painter, K. 1975 Roman Flasks with Scenes of Baiae and Puteoli. *Journal of Glass Studies* 17: 54-67.

Panciera, S. 1970 II 7-Regiones, Vici e Iuventus. Tra Epigrafia e Topografia 1.3. *Archeologia Classica* 22: 151-163.

Pandey, N.B. 2018 *The Poetics of Power in Augustan Rome: Latin Poetic Responses to Early Imperial Iconography*. Cambridge: Cambridge University Press.

Pandey, N. 2021 How Foreign Women Have Been Tokenized Since Ancient Roman Times. *Hyperallergic* April 6 2021.

Parker, G. 2002 *Ex Oriente Luxuria*: Indian Commodities and Roman Experience. *Journal of the Economic and Social History of the Orient* 45(1): 40-95.

Parker, H. 1992 Love's Body Anatomized. In A. Richlin (ed) 1992 *Pornography and Representation in Greece and Rome*. Oxford: Oxford University Press: 90-111.

Parry, H. 1964 Ovid's *Metamorphoses*: Violence in a Pastoral Landscape. *Transactions of the American Philological Association* 95: 268-282.

Peakman, J. 2019 *Licentious Worlds: Sex and Exploitation in Global Empires*. London: Reaktion.

Pearson, S. 2021 *The Triumph and Trade of Egyptian Objects in Rome: Collecting Art in the Ancient Mediterranean*. Berlin: De Gruyter.

Peltonen, J. 2020 When Kings and Gods Meet: Agency and Experience in Sacred Travel from Alexander the Great to Caracalla. In J. Kuuliala and J. Rantala (eds) 2020: 78-99.

Pensabene, P. 1980 *Terracotte Votive dal Tevere*. Rome: L'Erma di Bretschneider.

Peters, W. 1963 *Landscape in Romano-Campanian Mural Painting*. Assen: Van Gorcum and Prakke.

Petrain, D. 2014 *Homer in Stone. The Tabulae Illiacae in Their Roman Context*. Cambridge: Cambridge University Press.

Petsalis-Diomidis, A. 2007 Landscape, Transformation, and Divine Epiphany. In S. Swain, S. Harrison, and J. Elsner (eds) 2007: 250-289.

Picard, C. 1959 Pouzzoles et le Paysage Portuaire. *Latomus* 18 (1): 25-51.

Piccirillo, M. 1989 *Madaba: le Chiese ei Mosaici*. Milan: Edizioni Paoline.

Pilhuj, K. 2019 *Women and Geography on the Early Modern English Stage*. Amsterdam: Amsterdam University Press.

Platt, V. 2009 Where the Wild Things Are: Locating the Marvellous in Augustan Wall-Painting. In P. Hardie (ed) 2009 *Paradox and the Marvellous in Augustan Literature and Culture*. Oxford: Oxford University Press: 41-74.

Platt, V. 2012 Framing the Dead on Roman Sarcophagi. In V. Platt and M. Squire (eds) 2012: 353-383.

Platt, V. 2023 Art, Nature and the Material Divine in Roman Landscape Painting. In J. Powers (ed) 2023: 43-54.

Platt, V. and M. Squire (eds) 2012 *The Frame in Classical Art: a Cultural History*. Cambridge: Cambridge University Press

Pollard, E.A. 2009 Pliny's Natural History and the Flavian *Templum Pacis*: Botanical Imperialism in First Century CE Rome. *Journal of World History* 20: 320-324.

Pollini, J. 1993 The *Gemma Augustea*: Ideology, Historical Imagery, and the Creation of a Dynastic Narrative. In P.J. Holliday (ed) 1993 *Narrative and Event in Ancient Art*. Cambridge: Cambridge University Press: 258-298.

Pollini, J. 2003 Slave-Boys for Sexual and Religious Service: Images of Pleasure and Devotion. In A.J. Boyle and W.J. Dominik (eds) 2003: 149-166.

Poole, F. (ed) 2016 *Il Nilo a Pompei: Visioni d'Egitto nel Mondo Romano*. Modena: Franco Cosimo Panini.

Popkin, M.L. 2017 Souvenirs and Memory Manipulation in the Roman Empire. The Glass Flasks of Ancient Pozzuoli. In L. Munteán, L. Plate and A. Smelik (eds) 2017 *Materializing Memory in Art and Popular Culture*. Routledge: London: 45-61.

Popkin, M.L. 2018 Urban Images in Glass from the Late Roman Empire: the Souvenir Flasks of Puteoli and Baiae. *American Journal of Archaeology* 122 (3): 427-462.

Popkin, M.L. 2022a *Souvenirs and the Experience of Empire in Ancient Rome*. Cambridge: Cambridge University Press.

Popkin, M.L. 2022b The Vicarello Milestone Beakers and Future-Oriented Mental Time Travel in the Roman Empire. In M.L. Popkin and D.Y. Ng (eds) 2022 *Future Thinking in Roman Culture: New Approaches to History, Memory, and Cognition*. London: Routledge: 113-132.

Powers, J. 2023 Introduction: Landscapes of the Roman Imagination. In J. Powers (ed) 2023: 17-28.

Powers, J. (ed) 2023 *Roman Landscapes: Visions of Nature and Myth from Rome and Pompeii*. San Antonio: San Antonio Museum of Art.

Price, J.J., M. Finkelberg, and Y. Shahar (eds) 2021 *Rome: an Empire of Many Nations. New Perspectives on Ethnic Diversity and Cultural Identity*. Cambridge: Cambridge University Press.

Purcell, N. 1990 The Creation of Provincial Landscape: the Roman Impact on Cisalpine Gaul. In T. Blagg and M. Millett (eds) 1990 *The Early Roman Empire in the West*. Oxford: Oxbow Books: 7-29.

Purcell, N. 1996 Rome and the Management of Water: Environment, Culture, and Power. In J. Salmon and G. Shipley (eds) 1996 *Human Landscapes in Classical Antiquity: Environment and Culture*. London: Routledge: 18-212.

Purcell, N. 2005 Romans in the Roman World. In K. Galinsky (ed) 2005 *The Cambridge Companion to the Age of Augustus*. Cambridge: Cambridge University Press: 85-105.

Purcell, N. 2012 Rivers and the Geography of Power. *Pallas* 90: 373-387.

Purcell, N. 2015 A Second Nature? The Riverine Landscapes of the Romans. In T.V. Franconi (ed) 2015 *Fluvial Landscapes in the Roman World*. Journal of Roman Archaeology Supplementary Series 104. Journal of Roman Archaeology, Portsmouth, RI: 159-164.

Quarcoopome, N.O. (ed) 2010 *Through African Eyes: the European in African Art, 1500 to the Present*. Detroit: Detroit Institute of Arts.

Rankov, B. 2005 Do Rivers Make Good Frontiers? In Z. Visy (ed) 2005 *Limes XIX. Proceedings of the XIXth International Congress of Roman Frontier Studies held in Pécs, Hungary (September 2003)*. Pécs: University of Pécs: 175-181.

Rapson, J. 2015 *Topographies of Suffering: Bechenwald, Babi Yar, Lidice*. New York: Berghahn Press.

Rapoport, Y. 2021 Islamic *Maps*. Oxford: Bodleian Library.

Rathmann, M. 2016 The *Tabula Peutingeriana* and Antique Cartography. In S. Bianchetti, M. Cataudella, A. Greco, and H.-J. Gehrke (eds) 2015: 337-362.

Rathmann, M. 2022 *Tabula Peutingeriana: die Bedeutendste Weltkarte aus der Antike*. Darmstadt: WBG.

Rea, J.A. 2007 *Legendary Rome: Myth, Monuments and Memory on the Capitoline and Palatine*. London: Bloomsbury.

Readman, P. 2014 "The Cliffs Are Not Cliffs": The Cliffs of Dover and National Identities in Britain c. 1750-c. 1950. *History: the Journal of the Historical Association* 99 (335): 241-269.

Rehak, P. 2006 *Imperium and Cosmos: Augustus and the Northern Campus Martius*. Madison: University of Wisconsin Press.

Reitz-Joosse, B. 2021 Writing a Landscape of Defeat: the Romans in Parthia. In B. Reitz-Joosse, M.W. Makins, and C.J. Mackie (eds) 2021: 177-192.

Reitz-Joosse, B., M.W. Makins, and C.J. Mackie (eds) 2021 *Landscapes of War in Greek and Roman Literature*. London: Bloomsbury Academic.

Reynolds, D. 1996 *Forma Urbis Romae: the Severan Marble Plan and the Urban Form of Ancient Rome*. PhD dissertation, University of Michigan.

Richardson, J. 2008 *The Language of Empire: Rome and the Idea of Empire from the Third Century BC to the Second Century AD*. Cambridge: Cambridge University Press.

Richlin, A. 2017 *Slave Theater in the Roman Republic; Plautus and Popular Comedy*. Cambridge: Cambridge University Press.

Ridgway, B.S. 1997 The Sperlonga Sculptures: the Current State of Research. In N.T. De Grummond and B.S. Ridgway (eds) 1997: 78-91.

Rienjang, W. and P. Stewart (eds) 2020 *The Global Connections of Gandhāran Art: Proceedings of the Third International Workshop of the Gandhāra Connections Project, University of Oxford, 18th-19th March 2019*. Oxford: Archaeopress.

Riess, W. and G.G. Fagan (eds) 2016 *The Topography of Violence in the Greco-Roman World*. Ann Arbor: University of Michigan Press.

Riggsby, A. 2017 The Politics of Geography. In L. Grillo and C.B. Krebs (eds) 2017 *The Cambridge Companion to the Writings of Julius Caesar*. Cambridge: Cambridge University Press: 68-80.

Riggsby, A. 2019 *Mosaics of Knowledge: Representing Information in the Roman World*. Oxford: Oxford University Press.

Rimell, V. 2015 *The Closure of Space in Roman Poetics: Empire's Inward Turn*. Cambridge: Cambridge University Press.

Rimell, V. and M. Asper (eds) 2017 *Imagining Empire: Political Space in Hellenistic and Roman Literature*. Heidelberg: Universitätsverlag Winter.

Rinne, K.W. 2010 *The Waters of Rome: Aqueducts, Fountains, and the Birth of the Baroque City*. New Haven: Yale University Press.

Rizzo, F.P. 1994 Dai Comentarii di Agrippa alla 'Carta di Augusto'. *Seia* 11: 9-45.
Robinson, H. 1974 A Monument of Roma at Corinth. *Hesperia* 43: 470-484.
Rocca, S. 2017 *The Jewish War and Rome's Urban Renewal*. New York: Centro Primo Levi.
Rodríguez-Almeida, E. 1977 Forma Urbis Marmorea: Nuovi Elementi di Analisi e Nuove Ipotesi di Lavoro. *Mélanges de l'Ecole Française de Rome Antiquité* 89(1): 219-256.
Rodríguez-Almeida, E. 1981 *Forma Urbis Marmorea: Aggiornamento Generale 1980*. Rome: Quasar.
Rodríguez-Almeida, E. 2002 *Formae Urbis Antiquae. Le Mappe Marmoree di Roma tra la Repubblica e Settimio Severo*. Rome: École Française de Rome.
Rodriguez-Mayorgas, A. 2010 Romulus, Aeneas and the Cultural Memory of the Roman Republic. *Athenaeum* 98(1): 89-109.
Rogers, D.K. 2020 Aquatic Pasts and the Watery Present: Water and Memory in the Fora of Rome. In N. Chiarenza, A. Haug, and U. Müller (eds) 2020 *The Power of Urban Water: Studies in Premodern Urbanism*. Berlin: De Gruyter: 105-122.
Roller, D.W. 2006 *Through the Pillars of Herakles: Greco-Roman Exploration of the Atlantic*. London: Routledge.
Roller, D.W. 2010 Demolished House, Monumentality, and Memory in Roman Culture. *Classical Antiquity* 29(1): 117-180.
Roller, D.W. 2015 *Ancient Geography: the Discovery of the World in Classical Greece and Rome*. London: I.B. Tauris.
Roller, D.W. (ed) 2019 *New Directions in the Study of Ancient Geography*. Pennsylvania: Pennsylvania State University Press.
Roller, D.W. 2022 *A Guide to the Geography of Pliny the Elder*. Cambridge: Cambridge University Press.
Romer, F.E. 1998 *Pomponius Mela's Description of the World*. Ann Arbor: University of Michigan Press.
Romm, J. 1992 *The Edges of the Earth in Ancient Thought: Geography, Exploration, and Fiction*. Princeton: Princeton University Press.
Rose, C.B. 2005 The Parthians in Augustan Rome. *American Journal of Archaeology* 109: 21-75.
Rose, C.B. 2013 *The Archaeology of Greek and Roman Troy*. Cambridge: Cambridge University Press.
Ross Taylor, L. 1954 The Four Urban Tribes and the Four Regions of Ancient Rome. *Rendiconti della Pontificia Accademia di Archeologia* 27: 225-238.
Rossi, A. 2001 Remapping the Past: Caesar's Tale of Troy (Lucan "BC" 9.964-999). *Phoenix* 53(3/4): 313-326.
Roth, R. 2007 Varro's 'picta Italia' (RR I.ii.1) and the Odology of Roman Italy. *Hermes* 135(3): 286-300.
Rowan, C. 2012 *Under Divine Auspices: Divine Ideology and the Visualisation of Imperial Power in the Severan Period*. Cambridge: Cambridge University Press.
Rowan, E. 2019 Same Taste, Different Place: Looking at the Consciousness of Food Origins in the Roman World. *Theoretical Roman Archaeology Journal* 2(1), p. 5. doi: https://doi.org/10.16995/traj.378.
Russell, A. and M. Hellström (eds) 2020 *The Social Dynamics of Roman Imperial Imagery*. Cambridge: Cambridge University Press.
Russell, B. 2018 Simulacra Gentium (Africanararum). In C. Draycott, R. Raja, K. Welch, and W. Wootton (eds) 2018 *Visual Histories: Visual Remains and Histories of the Classical World*. Turnhout: Brepols: 265-280.

Rutherford, I.C. 2012 Travel and Pilgrimage. In C. Riggs (ed) 2012 *The Oxford Handbook of Roman Egypt*. Oxford: Oxford University Press: 701-716.

Rutledge, S. 2012 *Ancient Rome as a Museum: Power, Identity, and the Culture of Collecting*. Oxford: Oxford University Press.

Sage, M. 2000 Visitors to Ilium in the Roman Imperial and Late Antique Period: the Symbolic Functions of a Landscape. *Studia Troica* 10: 211-232.

Salway, B. 2005 The Nature and Genesis of the Peutinger Map. *Imago Mundi* 57: 119-135.

Salway, B. 2007 The Perception and Description of Space in Roman Itineraries. In M. Rathmann (ed) 2007 *Wahrnehmung und Erfassung Geographischer Räume in der Antike*. Mainz: Philipp von Zabern: 194-201.

Salway, B. 2011 Travel, *Itineraria*, and *Tabellaria*. In C.P. Adams and R. Laurence (eds) 2011: 22-66.

Salway, B. 2012 Putting the World In Order: Mapping in Roman Texts. In R. Talbert 2012: 193-234.

Sandberg, K. 2018 *Monumenta, Documenta, Memoria*: Remembering and Imagining the Past in Late Republican Rome. In K. Sandberg and C. Smith (eds) 2018 *Omnium Annalium Monumenta: Historical Writing and Historical Evidence in Republican Rome*. Leiden: Brill: 351-389.

Sande, S. 2012 The Arch of Constantine-Who Saw What? In S. Birk and B. Poulsen (eds) 2012 *Patrons and Viewers in Late Antiquity*. Aarhus: Aarhus University Press: 277-290.

Sapelli, M. (ed) 1999 *Provinciae Fideles: Il Fregio del Tempio di Adriano in Campo Marzio*. Milan: Electa.

Schefold, K. 1960 Origins of Roman Landscape Painting. *The Art Bulletin* 42(2): 87-96.

Schioler, T. 1994 The Pompeii *Groma* in New Light. *Analecta Romana. Instituti Danici* 22: 45-60.

Schmidt, B. 2002 Inventing Exoticism: the Project of Dutch Geography and the Marketing of the World. In P.H. Smith and P. Findlen (eds) 2002 *Merchants and Marvels: Commerce, Science, and Art in Early Modern Europe*. London: Routledge: 347-369.

Schrijvers, P.H. 2007 A Literary View on the Nile Mosaic at Praeneste. In L. Bricault, M.J. Versluys and P.G.P. Meyboom (eds) 2007: 223-244.

Sciaramenti, G. 2019 *Paesaggi del Dramma: Nelle Metamorfosi di Ovidio e Nella Pittura Romana Coeva*. Rome: Giorgio Bretschneider Editore.

Scodel, R.S. and R.F. Thomas 1984 Virgil and the Euphrates. *American Journal of Philology* 105, p. 339.

Secord, J. 2015 Overcoming Environmental Determinism: Introduced Species, Hybrid Plants and Animals, and Transformed Lands in the Hellenistic and Roman Worlds. In R.F. Kennedy and M. Jones-Lewis (eds) 2015: 210-229.

Seed, P. 1995 *Ceremonies of Possession in Europe's Conquest of the New World 1492-1640*. Cambridge: Cambridge University Press.

Shahar, Y. 2004 *Josephus Geographicus: the Classical Context of Geography in Josephus*. Tübingen: Mohr Siebeck.

Shapiro, H.A. 1993 *Personifications in Greek Art. The Representations of Abstract Concepts 600-400 B.C.*. Zurich: Akanthus.

Sherk, R. 1974 Roman Geographical Exploration and Military Maps. *Aufstieg und Niedergang der Römischen Welt* 2(1): 534-562.

Shipley, G. and J. Salmon (eds) 1996 *Human Landscapes in Classical Antiquity: Environment and Culture*. London: Routledge.

Short, J.R. 2003 *The World Through Maps: a History of Cartography*. Buffalo: Firefly Books.

Silberberg, S.R. 1980 *A Corpus of the Sacral-Idyllic Landscape Paintings in Roman Art*. Los Angeles: University of California Press.

Siwicki, C. 2012 The Restoration of the Hut of Romulus. In P. Emmons, J. Hendrix, and J. Lomholt 2012 *The Cultural Role of Architecture: Contemporary and Historical Perspectives*. London: Routledge: 18-26.

Skempis, M. and I. Ziogas (eds) 2014 *Geography, Topography, Landscape: Configurations of Space in Greek and Roman Epic*. Trends in Classics 22. Berlin: De Gruyter.

Skinner, J.E. 2012 *The Invention of Greek Ethnography: from Homer to Herodotus. Greeks Overseas*. Oxford: Oxford University Press.

Sluyter, A. 2001 Colonialism and Landscape in the Americas: Material/Conceptual Transformations and Continuing Consequences. *Annals of the Association of American Geographers* 91(2): 410-428.

Sluyter, A. 2002 *Colonialism and Landscape: Postcolonial Theory and Applications*. Lanham: Rowman and Littlefield.

Smith, A.C. 1994 Queens and Empresses as Goddesses: the Public Role of the Personal Tyche in the Graeco-Roman World. In S.B. Matheson (ed) 1994 *An Obsession with Fortune: Tyche in Greek and Roman Art*. New Haven: Yale University Art Gallery: 87-105.

Smith, E.C. 2017 *Jewish Glass and Christian Stone: a Materialist Mapping of the 'Parting of the Ways'*. London: Routledge.

Smith, J.Z. 1978 *Map Is Not Territory*: Studies In the History of Religions. Leiden: Brill.

Smith, R.R.R. 1988 *Simulacra Gentium*: the *Ethne* from the *Sebasteion* at Aphrodisias. *Journal of Roman Studies* 78: 50-77.

Smith, R.R.R. 2013 *The Marble Reliefs from the Julio-Claudian Sebasteion*. Aphrodisias 6. Darmstadt: Philipp von Zabern.

Speller, E. 2003 *Following Hadrian: a Second Century Journey Through the Empire*. Oxford: Oxford University Press.

Spence, J.D. 1985 *The Memory Palace of Matteo Ricci*. London: Faber and Faber.

Spencer, D. 2002 *The Roman Alexander: Reading a Cultural Myth*. Exeter: University of Exeter Press.

Spencer, D. 2005 Lucan's Follies: Memory and Ruin in a Civil-War Landscape. *Greece and Rome* 52(1): 46-69.

Spencer, D. 2007 Rome at a Gallop: Livy, On Not Gazing, Jumping, Or Toppling Into the Void. In D.H.J. Larmour and D. Spencer (eds) 2007: 61-101.

Spencer, D. 2011 *Roman Landscape: Culture and Identity*. Cambridge: Cambridge University Press.

Spencer, D. 2015a Vitruvius, Landscape and Heterotopias: How 'Otherspaces' Enrich Roman Identity. In R.F. Kennedy and M. Jones-Lewis (eds) 2015: 171-191.

Spencer, D. 2015b Urban Flux: Varro's Rome-in-Progress. In I. Östenberg, S. Malmberg and J. Bjørnebye (eds) 2015: 99-110.

Spencer, D. 2019 *Language and Authority in De Lingua Latina: Varro's Guide to Being Roman*. Madison: University of Wisconsin Press.

Spencer, D. 2020 Emotional Volume and the Little Things That Make Latin Place. In A. Anguissola, M. Iadanza, and R. Olivitto (eds) 2020 *Paesaggi Domestici. L'Esperienza della Natura nelle Case e nelle Ville Romane-Pompei, Ercolano e L'Area Vesuviana*. Studi e Ricerche del Parco Archeologico di Pompei Vol. 42. Rome: L'Erma di Bretschneider: 4-18.

Spentzou, E. 2018 Propertius' Aberrant Itineraries: Fleeting Moments in the Eternal City. In W. Fitzgerald and E. Spentzou (eds) 2018: 23-44.

Spier, J., T.F. Potts and S.E. Cole (eds) 2018 *Beyond the Nile: Egypt and the Classical World*. Los Angeles: J. Paul Getty Museum.

Squire, M. 2011 *The Iliad in a Nutshell: Visualizing Epic on the Tabulae Iliacae*. Oxford: Oxford University Press.

Squire, M. 2012 Framing the Roman 'Still Life': Campanian Wall-Painting and the Frames of Mural Make-Believe. In V. Platt and M. Squire (eds) 2012: 188-255.

Stafford, E. 2000 *Worshipping Virtues: Personification and the Divine in Ancient Greece*. London: Bloomsbury.

Stafford, E. and J. Herrin (eds) 2005 *Personification in the Greek World from Antiquity to Byzantium*. London: Routledge.

St Clair, A. 1964 The Apotheosis Diptych. *Art Bulletin* 46: 205-211.

Stephenson, J.W. 2012 The Column of Trajan in the Light of Ancient Cartography and Geography. *Journal of Historical Geography* 30: 1-15.

Stewart, A. 2014 *Art in the Hellenistic World: an Introduction*. Cambridge: Cambridge University Press.

Stoiculescu, C.D. 1985 Trajan's Column's Documentary Value from a Forestry Viewpoint. *Dacia* 29: 81-98.

Stoner, J. 2019 *The Cultural Lives of Domestic Objects in Late Antiquity*. Leiden: Brill.

Strother, Z.S. 2016 *Humor and Violence; Seeing Europeans in Central African Art*. Bloomington and Indianapolis: Indiana University Press.

Swain, S., S. Harrison, and J. Elsner (eds) 2007 *Severan Culture*. Cambridge: Cambridge University Press.

Swetnam-Burland, M. 2009 Egypt Embodied: The Vatican Nile. *American Journal of Archaeology* 113(3): 439-457.

Swetnam-Burland, M. 2010 *Aegyptus Redacta*: the Egyptian Obelisk in the Augustan Campus Martius. *The Art Bulletin* 92(3): 135-153.

Swetnam-Burland, M. 2011 "Egyptian" Priests in Roman Italy. In E.S. Gruen (ed) 2011: 336-353.

Swetnam-Burland, M. 2012 Nilotica and the Image of Egypt. In C. Riggs (ed) 2012 *The Oxford Handbook of Roman Egypt*: 684-697.

Swetnam-Burland, M. 2015 *Egypt in Italy. Visions of Egypt in Roman Imperial Culture*. Cambridge: Cambridge University Press.

Syed, Y. 2005 *Vergil's Aeneid and the Roman Self: Subject and Nation in Literary Discourse*. Ann Arbor: University of Michigan Press.

Syme, R. 1988 Journeys of Hadrian. *Zeitschrift für Papyrologie und Epigraphik* 73: 159-170.

Tacoma, L.E. and E. Lo Cascio 2017 Writing Migration. In E. Lo Cascio and L.E. Tacoma (eds) 2017 *The Impact of Mobility and Migration in the Roman Empire. Proceedings of the Twelfth Workshop of the International Network Impact of Empire (Rome, June 17-19, 2015)*. Leiden: Brill: 1-24.

Talbert, R.J.A. 2004 Rome's Provinces as a Framework for World-View. In L. De Ligt, E. Hemelrijk, and H.W. Singor (eds) 2004 *Roman Rule and Civic Life: Local and Regional Perspectives*. Leiden: Brill: 21-37.

Talbert, R.J.A. 2004 Cartography and Taste in Peutinger's Roman Map. In R.J.A. Talbert and K. Brodersen (eds) 2004: 113-141.

Talbert, R.J.A. 2008 Greek and Roman Mapping: Twenty-First Century Perspectives. In R.J.A. Talbert and R. Unger (eds) 2008: 9-28.

Talbert, R.J.A. 2010 *Rome's World: the Peutinger Map Reconsidered*. Cambridge: Cambridge University Press.

Talbert, R.J.A. 2010B The Roman Worldview: Beyond Recovery? In K.A. Raaflaub and R.J.A. Talbert (eds) 2010 *Geography and Ethnography; Perceptions of the World in Pre-Modern Societies*. Oxford: Wiley-Blackwell: 252-272.

Talbert, R.J.A. 2012 *Urbs Roma* to *Orbs Romanus*: Roman Mapping on the Grand Scale. In R.J.A. Talbert (ed) 2012 *Ancient Perspectives: Maps and the Place in Mesopotamia, Egypt, Greece, and Rome*. Chicago: University of Chicago Press: 163-192.

Talbert, R.J.A. 2017 *Roman Portable Sundials: the Empire in Your Hand*. Oxford: Oxford University Press.

Talbert, R.J.A. 2023a Author, Audience, and the Roman Empire in the Antonine Itinerary. In Talbert, R.J.A. (ed) 2023: 100-117.

Talbert, R.J.A. 2023b Communicating Through Maps: the Roman Case. In Talbert, R.J.A. (ed) 2023: 232-258.

Talbert, R.J.A. (ed) 2023 *World and Hour in Roman Minds. Exploratory Essays*. Oxford: Oxford University Press.

Talbert, R.J.A. and K. Brodersen (eds) 2004 *Space in the Roman World: Its Perception and Presentation*. Münster: Lit Verlag.

Talbert, R.J.A. and R. Unger (eds) 2008 *Cartography in Antiquity and the Middle Ages. Fresh Perspectives, New Methods*. Leiden: Brill.

Tan, Z.M. 2014 Subversive Geography in Tacitus' *Germania*. *Journal of Roman Studies* 104: 181-204.

Tarlo, E. 2016 *Entanglement: the Secret Life of Hair*. London: Oneworld Publications.

Taub, L. 2003 *Ancient Meteorology*. London: Routledge.

Taufer, M. 2019 *La Montagna nell'Antichità*. Rombach-Wissenschaften: Reihe Paradeigmata, Band 56. Freiburg: Rombach Verlag.

Taussig, H. 2012 Melancholy, Colonialism, and Complicity: Complicating Counterimperial Readings of Aphrodisias's *Sebasteion*. In A.C. Niang and C. Osiek (eds) 2012 *Text, Image, and Christians in the Graeco-Roman World: a Festschrift in Honour of David Lee Balch*. Eugene, Oregon: Pickwick Publications: 280-295.

Taylor, J.E. and I.N. Gregory 2022 *Deep Mapping the Literary Lake District: a Geographical Text Analysis*. Lewisburg: Bucknell University Press.

Taylor, R.M. 2000 *Public Needs and Private Pleasures: Water Distribution, the Tiber River and the Urban Development of Ancient Rome*. Rome: L'Erma di Bretschneider.

Taylor, R.M. 2009 River Raptures: Containment and Control of Water in Greek and Roman Constructions of Identity. In C. Kosso and A. Scott (eds) 2009: 19-42.

Tekin, O. 2001 River-Gods in Cilicia in the Light of Numismatic Evidence. *Actes de la Table Ronde d'Istanbul 2-5 Novembre 1999*. Publications de l'Institut Française d'Études Anatoliennes 13: 519-551.

Terrenato, N. 2019 *The Early Roman Expansion into Italy. Elite Negotiation and Family Agendas*. Cambridge: Cambridge University Press.

Thagaard Loft, G. 2003 Villa Landscapes in Pompeian Wall Painting: a Different Approach. *Analecta Romana Instituti Danici* 29: 7-28.

Thalmann, W. 2011 *Apollonius of Rhodes and the Spaces of Hellenism*. Oxford: Oxford University Press.

Theweleit, K. 1987 *Male Fantasies Volume 1. Women, Floods, Bodies, History*. Translated by Stephen Conway. Cambridge: Polity Press.

Thill, E.W. 2010 Civilization Under Construction: Depictions of Architecture on the Column of Trajan. *American Journal of Archaeology* 114: 27-43.

Thill, E.W. 2014 The Emperor in Action: Group Scenes in Trajanic Coins and Monumental Reliefs *American Journal of Numismatics* 26: 89-142.

Thill, E.W. 2022 Dispatches from the Home Front: the *Anaglypha* Panels in Rome. *Britannia* 52: 5-30.

Thomas, E. 2007 Metaphor and Identity in Severan Architecture: the *Septizodium*. Between Reality and Fantasy. In S. Swain, S. Harrison, and J. Elsner (eds) 2007: 327-367.

Thomas, J.J. 2021 *Art, Science, and the Natural World in the Ancient Mediterranean, 300 BC to AD 100.* Oxford: Oxford University Press.

Thomas, R.F. 1982 *Lands and Peoples in Roman Poetry: the Ethnographical Tradition.* Cambridge: Cambridge Philological Society Supplementary Series.

Thomas, R.F. 1988 Tree Violation and Ambivalence in Virgil. *Transactions of the American Philological Association* 118: 261-273.

Thommen, H. 2009 *An Environmental History of Ancient Greece and Rome.* Translation of 2012, Cambridge, Cambridge University Press.

Thompson, L. 1989 *Romans and Blacks.* London: Routledge.

Thulin, C. 1971 *Corpus Agrimensorum Romanorum.* Reprint. Stuttgart: B.G. Teubneri.

Tierney, J.J. 1963 The Map of Agrippa. *Proceedings of the Royal Irish Academy Section C-Archaeology Celtic Studies History Linguistics Literature* 63: 151-166.

Totelin, L.M.V. 2012 Botanizing Rulers and Their Herbal Subjects: Plants and Political Power in Greek and Roman Literature. *Phoenix* 66. 1/2: 122-144.

Totelin, L.M.V. 2015 The World in a Pill: Local Specialities and Global Remedies in the Graeco-Roman World. In R.F. Kennedy and M. Jones-Lewis (eds) 2015: 151-170.

Totten, D.M. and K.L. Samuels (eds) 2012 *Making Roman Places, Past and Present.* Journal of Roman Archaeology Supplementary Volume 89. Portsmouth, RI.

Touati, A.-M.L. 2015 Monuments and Images of the Moving City. In I. Östenberg, S. Malmberg and J. Bjørnebye (eds) 2015: 203-224.

Toynbee, J.M.C. 1934 *The Hadrianic School. A Chapter in the History of Greek Art.* Cambridge: Cambridge University Press.

Trentin, L. 2022 Images and Interpretation of "the Other" in Roman Social Practice. In L.K. Cline and Elkins, N. (eds) 2022: 405-424.

Trimble, J. 2006 Rome as Souvenir: the *Septizodium* and the Severan Marble Plan. In C. Mattusch, A.A. Donohue, and A. Brauer (eds) 2006 *Common Ground: Archaeology, Art, Science and Humanities. Proceedings of the XVIth International Congress of Classical Archaeology, Boston August 23-26 2003.* Oxford: Oxbow Books: 196-209.

Trimble, J. 2007 Visibility and Viewing on the Severan Marble Plan. In S. Swain, S. Harrison, and J. Elsner (eds) 2007: 368-384.

Trimble, J. 2008 Process and Transformation on the Severan Marble Plan of Rome. In R.J.A. Talbert and R. Unger (eds) 2008: 67-98.

Trimble, J. 2017 Appropriating Egypt for the *Ara Pacis Augustae*. In M. Loar, C. MacDonald, and D.-el. Padilla Peralta (eds) 2017 *Rome: Empire of Plunder. The Dynamics of Cultural Appropriation.* Cambridge: Cambridge University Press: 109-136.

Trimble, J. 2018 Figure and Ornament, Death and Transformation in the Tomb of the Haterii. In N. Dietrich and M. Squire (eds) 2018 *Ornament and Figure in Graeco-Roman Art: Rethinking Visual Ontologies in Classical Antiquity.* Berlin: de Gruyter: 327-352.

Trousset, P. 1993 La 'Carte d'Agrippa': Nouvelle Proposition de Lecture. *Dialogues d'Histoire Ancienne* 19(2): 137-157.

Tucci, P.L. 2004 Eight Fragments of the Marble Plan of Rome Shedding New Light on the Transtiberim. *Papers of the British School at Rome* 72: 185-202.

Tucci, P.L. 2007 New Fragments of Ancient Plans of Rome. *Journal of Roman Archaeology* 20: 469-480.

Tucci, P.L. 2013 The Marble Plan of the *Via Anicia* and the Temple of Castor and Pollux in *Circo Flaminio*: the State of the Question. *Papers of the British School at Rome* 81: 91-127.

Tucci, P.L. 2018 The Marble Plans. In C. Holleran and A. Claridge (eds) 2018: 13-19.

Turconi, C. 1997 La Mappa di Bedolina nel Quadro dell'Arte Rupestre della Valcamonica. *Notizie Archeoligiche Bergomensi* 5: 85-113.

Tybout, R.A. 2003 Dwarfs in Discourse: the Functions of Nilotic Scenes and Other Roman *Aegyptiaca*. *Journal of Roman Archaeology* 16: 505-515.

Ugolini, F. 2022 Iconography of the Lighthouse in Roman Antiquity: Symbolism, Identity and Power Across the Mediterranean. In M. Henig and J. Lundock (eds) 2022: 6-25.

Van Aerde, M. 2019 *Egypt and the Augustan Cultural Revolution: an Interpretative Archaeological Overview*. Leuven: Peeters.

Van den Hoonaard, W.C. 2014 *Map Worlds: a History of Women in Cartography*. Waterloo: Wilfrid Laurier University Press.

Van der Meer, L. 1987 *The Bronze Liver of Piacenza: Analysis of a Polytheistic Structure*. Amsterdam: J.C. Geiben. Leiden: Brill Reprint 2022.

Van der Vliet, E.C.L. 2006 The Romans and Us: Strabo's Geography and the Construction of Ethnicity. *Mnemosyne* 56: 257-272.

Van Dommelen, P. and N. Terrenato (eds) 2007 *Articulating Local Cultures: Power and Identity Under the Expanding Roman Republic*. Journal of Roman Archaeology Supplementary Series 63, Portsmouth, RI.

Van Haeperen, F. 2020 Roman Places of Collective Worship as Meeting Places. In M.L. Caldelli and C. Ricci (eds) 2020: 229-258.

Van Paasen, C. 1957 *The Classical Tradition of Geography*. Groningen: J.B. Wolters.

Vasaly, A. 1993 *Representations: Images of the World in Ciceronian Oratory*. Berkeley: University of California Press.

Vasta, M. 2007 Flavian Visual Propaganda: Building a Dynasty. *Constructing the Past* 8.1: 107-138.

Vasunia, P. 2001 *The Gift of the Nile: Hellenizing Egypt from Aeschylus to Alexander*. Berkeley: University of California Press.

Vecchi, I. and Vecchi-Gomez, J. 2002 Of Crocodiles and Coins: Roman Egypt Personified. *Minerva* 13(3): 51-53.

Vermeule, C.C. 1968 *Roman Imperial Art in Greece and Asia Minor*. Cambridge: Harvard University Press.

Vermeule, C.C. 1974 *The Goddess Roma in the Art of the Roman Empire*. Revised Edition. London: Spink.

Vermeule, C.C. 1986 The Urban Tyche as an Index of Numismatic Art in the Antonine and Severan Periods. In C.C. Vermeule *Numismatic Art of the Greek Imperial World*. Cambridge: Copyquik: 21-24.

Vermeule, C.C. 1995 Neon Ilion and Ilium Novum: Kings, Soldiers, Citizens, and Tourists at Classical Troy. In J.B. Carter and S.P. Morris (eds) 1995 *The Ages of Homer. A Tribute to Emily Townsend Vermeule*. Austin: University of Texas Press: 467-482.

Versluys, M.J. 2000 *Aegyptiaca Romana: Nilotic Scenes and the Roman Views of Egypt*. Religions in the Graeco-Roman World 144, Leiden: Brill.

Versluys, M.J. 2010 Understanding Egypt in Egypt and Beyond. In L. Bricault and M.J. Versluys (eds) 2010 *Isis on the Nile. Egyptian Gods in Hellenistic and Roman Egypt. Proceedings of the 4th International Conference of Isis Studies*. Leiden: Brill: 7-36.

Versluys, M.J. 2015 Roman Visual Culture as Globalising *Koine*. In M. Pitts and and J.M. Versluys (eds) 2015 *Globalisation and the Roman World: World History, Connectivity and Material Culture*. Cambridge: Cambridge University Press: 141-174.

Volk, K. 2021 *The Roman Republic of Letters: Scholarship, Philosophy, and Politics in the Age of Cicero and Caesar*. Princeton: Princeton University Press.

Volpe, R. 2016 Before and Below the Baths of Trajan (Rome). *Memoirs of the American Academy in Rome* 61: 59-75.

Von Stakelberg, K.T. 2009 *The Roman Garden: Space, Sense, and Society*. London: Routledge.

Vout, C. 2003 Embracing Egypt. In C. Edwards and G. Woolf (eds) 2003 *Rome the Cosmopolis*. Cambridge: Cambridge University Press: 177-202.

Vout, C. 2005 Antinous, Archaeology and History. *Journal of Roman Studies* 95: 80-96.

Vout, C. 2006 *Antinous: the Face of the Antique*. Leeds: Henry Moore Sculpture Trust.

Vout, C. 2007 *Power and Eroticism in Imperial Rome*. Cambridge: Cambridge University Press.

Vout, C. 2012 *The Hills of Rome: Signature of an Eternal City*. Cambridge: Cambridge University Press.

Waite, I. 2012 *Common Land in English Painting 1700-1850*. London: Boydell Press.

Walker, A. 2012 *The Emperor and The World: Exotic Elements and the Imaging of Middle Byzantine Imperial Power, Ninth to Thirteenth Centuries CE*. Cambridge: Cambridge University Press.

Walker, S. 2003 Carry-On at Canopus: the Nilotic Mosaic from Palestrina and Roman Attitudes to Egypt. In R. Matthews and C. Roemer (eds) 2003 *Ancient Perspectives on Egypt*. London: Routledge: 191-202.

Wallace-Hadrill, A. 2003 The Streets of Rome as a Representation of Imperial Power. In L. De Blois, P. Erdkamp, O. Hekster, G. De Kleijn, and S. Mols (eds) 2003 *The Representation and Perception of Roman Imperial Power. Proceedings of the Third Workshop of the International Network Impact of Empire (Roman Empire, c. 200 B.C.-A.D. 476), Rome, March 20-23, 2002*. Leiden: Brill: 189-206.

Wallace-Hadrill, A. 2005 *Mutatas Formas*: the Augustan Transformation of Knowledge. In K. Galinsky (ed) 2005 *The Cambridge Companion to the Age of Augustus*. Cambridge: Cambridge University Press: 55-84.

Warde Fowler, W. 1916 Vergil's Idea of the Tiber. *Classical Review* 30: 219-222.

Warner, V.J. 1917 Epithets of the Tiber in the Roman Poets. *The Classical Weekly* 11(7): 52-54.

Washburn, D.A. 2017 *Banishment in the Later Roman Empire, 284-476 CE*. London: Routledge.

Webb, P.A. 2017 *The Tower of the Winds in Athens: Greeks, Romans, Christians, and Muslims: Two Millennia of Continual Use*. Memoirs of the American Philosophical Society 270. Philadelphia: American Philosophical Society.

Webster, T.B.L. 1954 Personification as a Mode of Greek Thought. *Journal of the Warburg Institute* 17: 10-21.

Weingarten, S. 1999 Was the Pilgrim from Bordeaux a Woman? A Reply to Laurie Douglass. *Journal of Early Christian Studies* 7: 291-297.

Weinstein, L.R. 2021 The Indian Figurine from Pompeii as an Emblem of East-West Trade in the Early Roman Imperial Period. In S. Autiero and M.A. Cobb (eds) 2021 *Globalization and Transculturality from Antiquity to the Pre-Modern World*. London: Routledge: 183-204.

Welch, T.S. 2005 *The Elegiac Cityscape: Propertius and the Meaning of Roman Monuments*. Columbus: Ohio State University Press.

Welch, T.S. 2015 *Tarpeia: Workings of a Roman Myth*. Columbus: Ohio State University Press.
Westall, R. 2015 'Moving Through Town': Foreign Dignitaries in Rome in the Middle and Late Republic. In I. Östenberg, S. Malmberg and J. Bjørnebye (eds) 2015: 23-36.
Whitehouse, H. 1980 *A Catalogue of Nilotic Landscapes in Roman Art*. PhD Thesis, Oxford University.
Whittaker, C.R. 1994 *Frontiers of the Roman Empire: a Social and Economic Study*. Baltimore: Johns Hopkins University Press.
Whittaker, C.R. 2002 Mental Maps: Seeing Like a Roman. In P. McKechnie (ed) 2002 *Thinking Like a Lawyer: Essays on Legal History and General History for John Crook on his Eightieth Birthday*. Leiden: Brill: 81-112.
Wilker, J. 2020 Modelling the Emperor: Representations of Power, Empire, and Dynasty Among Eastern Client Kings. In A. Russell and M. Hellström (eds) 2020: 52-75.
Williams, H. and R. Clare (eds) 2022 *The Ancient Sea: the Utopian and Catastrophic in Classical Narratives and Their Reception*. Liverpool: Liverpool University Press.
Wiseman, T.P. 1974 Legendary Genealogies in Late-Republican Rome. *Greece and Rome* 21: 153-164.
Wiseman, T.P. 1992 Julius Caesar and the *Mappa Mundi*. In T.P. Wiseman (ed) 1992 *Talking to Virgil: a Miscellany*. Liverpool: Liverpool University Press: 22-42.
Witcombe, C.L.C.E. 2018 *Eye and Art in Ancient Greece: a Study in Archaeoaesthetics*. Turnhout: Brepols.
Woolf, G. 2009 Literacy or Literacies in Rome? In W. Johnson and H. Parker (eds) 2009 *Ancient Literacies*. Oxford: Oxford University Press: 46-68.
Woolf, G. 2011 *Tales of the Barbarians: Ethnography and Empire in the Roman West*. Oxford: Wiley-Blackwell.
Woolf, G. 2016 Movers and Stayers. In L. de Ligt and L.E. Tacoma (eds) 2016: 438-461.
Wrigley, R. 2012 *The Roman Campagna Revisited: Art and Environment*. Tate Papers No. 17. London: Tate Gallery.
Yakobson, A. 2021 Rome's Attitude to Jews After the Great Rebellion-Beyond *Raison d'État*? In J.J. Price, M. Finkelberg and Y. Shahar (eds) 2021: 186-202.
Yarden, L. 1991 *The Spoils of Jerusalem on the Arch of Titus: a Re-Investigation*. Stockholm: Svenska Institutet i Rom.
Young, G.K. 2001 *Rome's Eastern Trade: International Commerce and Imperial Policy, 31 BC-AD 305*. London: Routledge.
Zamora, M. 1991 Abreast of Columbus: Gender and Discovery. *Cultural Critique* 17: 127-149.
Zanker, P. 1988 *The Power of Images in the Age of Augustus*. Ann Arbor: University of Michigan Press.
Zarmakoupi, M. 2014 *Designing for Luxury on the Bay of Naples: Villas and Landscapes (c. 100 BCE-79 CE)*. Oxford: Oxford University Press.
Zarmakoupi, M. 2019 Between Conceptual and Perceptual Space: the Representation of Landscape in Roman Wall Paintings. In G. Adornato, E. Falaschi, A. Poggio (eds) 2019 *Pittori, Tecniche, Trattati, Contesti tra Testimonianze e Ricezione Archeologia e Arte Antica*. Milan: LED Edizioni: 173-196.
Zarobell, J. 2009 *Empire of Landscape: Space and Ideology in French Colonial Algeria*. Penn State University Press.
Zerba, M. 2021 *Modern Odysseys: Cavafy, Woolf, Césaire, and a Poetics of Indirection*. Columbus: Ohio University Press.
Zerjadtke, M. (ed) 2020 *Der Ethnographische Topos in der Alten Geschichte: Annäherungen an Ein Omnipräsentes Phänomen*. Stuttgart: Franz Steiner Verlag.
Zwierlein, O. 1986 Lucan's Caesar in Troja. *Hermes* 114(4): 460-478.

Index

Because this book is almost entirely about Rome, the idea of Rome, an idea of Rome, the word does not appear as a stand-alone entry in the Index below, nor does the term 'geography product' used constantly throughout the text.

River Acheloös/Achelous, 70-71
Achilles, 74
Actium, Battle of, 13, 135, 193, 195
Adams, Ansel, xv
Adiatorix, 155
Adulis, 172
Aeneas/*Aeneid*, xxvii, 12, 13, 14, 31, 72, 131, 155-156, 186, 191-192, 201, Figure 18
Aeschylus, 50
Affective spaces, theory of, xxiii
African art, portrayal of Europeans in, 174-175
Agent Orange, 103
Agrimensores, 124-125, 228
Agrimensorum Romanorum, 125
Agrippa, Marcus/Agrippa's world map, 5, 67, 131-134, 148, 193, 220, Figure 76
Alexander, Craig, 4
Alexander the Great/Alexander Mosaic xviii, 17, 161, 179, 189, 192, 195, 232
Alexandria, 62, 195, 212
Altars, 62, 72, 73, Figures 21, 22, and 33
Altar of Zeus, Pergamon, 163
Amelia, 140
River Amenanos, 69-70
Ancona, 213
Antinous, 121, 123, 196, Figure 70
Antioch, 62, Figures 121 and 126
Antonine Itinerary, 144
Antonioni, Michelangelo, xv
Antony, Marc, xxvi, 118-119
Anzio, 70
Apollinaire, Guillaume, 73
Apotheosis, 47-48, Figure 23
Aphrodisias, 157-161, 167-168, 230, Figures 88-89
Appian, 162, 168
Aqueducts, 18, 41, 58, 59, 67, 126, 216
Aquincum, 125
Ara Maxima, 36

Ara Pacis Augustae, xxvi, 8-11, 15, 44, 94, 107, 167, 184, 186, Figures 9-11
Arches: Arch of Claudius, 250; Arch of Constantine, 60, 73, 74, 82, 168-170, 245; Arch of Septimius Severus, 60, 74, 106, 137, 138, 155; Arch of Septimius Severus and Julia Domna, Lepcis Magna, 107, 195; Arch of Tiberius at Orange, 183; Arch of Titus, 56, 60, 74, 81, 140, 181-183, Figure 97; Arch of Titus near Colosseum, 183; Arch of Trajan, Benevento, 74, 103, 195, Figure 41; Arch to Isis, 56; *Arco di Portogallo* 47-48, Figure 23; *Arcus Novus*, 245
Architecture on Roman coins, 56-60, Figures 27-29
Architecture, 54-62, Figures 25-31
Arethusa map, 130-131
Mount Argaeus, 50
Arles Sarcophagus, 214
Armenia, 130, 157-161, Figure 89
Arpinum, 27
Arrian, 194
Artaud, Antonin/Theatre of Cruelty, 17
Artemidorus Papyrus, 131
Artemis Ephesia, 157, Figure 87
Arval brethren, 53
Assertive mimesis, idea of, 175
Asset stripping, 115, 116, 246
Assyrians, xvii
Athenaeus, 184
Atticus, 27
Atrium Libertatus, Rome, 126
Atwood, Margaret, xv
St. Augustine, 50
Augustus, xxvi, xviii, 8-16, 29, 34-35, 41, 47, 54, 62, 67, 106, 134-135, 154, 167, 179, 185, 192, 194, 195, 220, 250, Figures 12 and 13
Autun map, 141

Axum obelisk, 29

Babylon, 195
Bacchus, xxvii, 51-52, 172, 189, 192, 236, Figures 4 and 24
Baiae, 212
Bakhtin, Mikhail, 20
Balbus, L. Cornelius, 68
Baldwin, James, 26
Ballard, J.G., xv
Barbarians, xxii-xxiii, 150-166, 187, 242, Figures 84, 85, and 123
Barberini Diptych, 173, Figure 95
Basanite statues, 121, 140
Base of Tiberius, Figure 34
Basilica Aemilia, 18, 47, 154, 167
Basilca of Maxentius, 147
Baths of Caracalla, 103, 105
Baths of Trajan, 56
Battle paintings, 115
Battles: Actium,13, 135, 193, 195; Carrhae, 106; Milvian Bridge, 73
Baudelaire, Charles, 220
Bay of Naples Bottles, 211-212, Figure 117
Benjamin, Walter, 229
Berenike, 172
Bergson, Henri, 255
Bevan, Lynne, 5
Bianchi, Rino, *29*
Birrus Britannicus, 186
Bishapur, 176
Bloomberg site stylus, 212-213
Boatwright, Mary, 131
Bocca della verita, Figure 114
Bogotá Savanna painting, 117-118
Bordeaux Pilgrim, xxvii-xxviii, 214-216, 222
Borges, Jorge Luis, 230
River Boristhenus, 70
Bourdieu, Pierre, 252
Bovillae, 46
Bradbury, Ray, xviii, 15
Brescia *Capitolium*, plaster plan from, 127
Bridges, 60, 180
Brilliant, Richard, 107
Britannia/Britons, 157-161, 184, Figure 88
Lucius Brutus, 175

Buddha head, 172
Burroughs, William, 156

Cacus, 36
Cadastre of Arausio (Orange), 126
Caligula, 161, 195
Cameos, 163, 217, Figure 118
Campbell, Brian, 67
Campana Plaques, 37, 109, 119, Figure 3
Campus Martius, 47-48, 56, 67, 72, 131
Camuni, 1-6
Canopus of Hadrian, 74, 123, Figure 71
Capitoline Wolf, Figure 16
Capua, 126
Caracalla, viv, 161, 194, 195
Carcer Tullianum, 155
Carey, Sorcha, 116
Carrhae, Battle of 106
Carthage/Cathaginians, xiii, 184, 186, 188
Caryatids, xv-xvi, 155, 166-167, Figure 1
Casagrade-Kim, Roberta, 205, 207
Casa Romuli, 32-34, 37, 205, Figure 17
Cassibry, Kimberly, 210-213
Mount Cassius, 50
Catacomb of Commodilla, 214
Cataloghi Regionari, 33
Cato the Elder, xii, 27
Catullus, 191-192
Cendrars, Blaise, xv
Centuriation, 124, 125, 254
Cerberus, 208, 209, Figures 112-113
Charax, 133
Charon, 209, Figure 111
Cherchel, 180
China, Roman trade with, 251-252
Christians/Christianity/Christian art, 80, 213-216, 244-245, Figures 26 and 56
Chtcheglov, Ivan, 228
Chthonic landscapes, 205-208
Cicero, 20, 23-27, 185, 225, 237-238
Cichorius, Conrad, 97
Circus Maximus, 59-60, 65, 183, 211
Città dipinta, wall painting, 56
Civitas Camunnorum (Cividate Camuno), 6
Cixious, Hélène, 221
Clarke, John, 114-115
Claudian, 43

Index

Claudius, 60, 157-161, 177, 185, Figure 88
Cleopatra, xxvi, 118-119, 135, Figures 66 and 128
Client kingdoms, 177
Cloaca Maxima, 52, 206, Figure 114
Clodius, 27
Coins, 43, 47, 50, 72, 74, 138-140, 177, 181-182, Figures, 20, 27-29, 40, 43-44, 100-102, and 113
Colle Oppio, 140
Colonial landscapes, xiii, 110, 115, 117-118, 123
Colonisation of Italy by Rome, 40
Colosseum, 56, 59-60, 183, 185, 186-187, Figure 27
Column of Antoninus Pius, 48, 62
Column of Marcus Aurelius, 62, 78, 97, 115, 117, 162-166, 235, Figure 47
Commodus, 176
Constable, 88
Constantine, 194, 216
Constantinople, 62, 82, 217-218, 244, Figure 118
Coponius, 16
Corbulo, 130
Corbridge, 62
Cornelia, 17, 232
Corsica, xvii
Crassus, 13-14, 106
Cursus Publicus, 133, 144
Cybele/*Magna Mater*, 157, Figure 86

Dacia/Dacians, 77, 80, 97-105, 115, 117, 168-170, 180
Daly, Herman, 22
River Danube, 69, 74, 80, Figures 42-43
Darabgerd, 176
Darius, 175
Decebalus, 77, 102, 165
Dickens, xxvii
Dio Cassius, 130, 195
Diocletian's Edict on Maximum Prices, 186
Dionysius of Halicarnassus, 33, 67, 162, 168
Domitian, 80, 130, 176, 185, 188
Donne, John, 147
Dougga, 83, 186
Dresken-Weiland, Jutta, 236-239

Dubuffet, Jean, 253
Dueck, Daniela, 236-239
Duffalo, Basil, 191-192
Dura Map, 131
Dying Gaul, Figures 84-85

Eclogues, 125
Egypt, 29, 108-123, 140, 171, 198, 235
Egyptomania, idea of, 108, 171
Elizabeth/Elizabethan period/Elizabethan maps and mapping, ideology of, 147
Elkins, Nathan, 58-59
Elsner, Jaś, 215
Ennius, 72
Erotic Roman art, 118-119, 156-157, Figures 66 and 128
Esquiline Hill Treaty wall painting, xii, 154
Mount Etna, 49, 50
Etruscans, xiii, 6, 7, 190
Eumenius, 141
Euphrates River, 74, 80
Eurydice, 201
Mount Everest, 49
Ewan, Ruth, 222

The Fall, xv
Farnese Atlas, 145, Figure 83
Fasti Triumphales, 148
Faustus, Lucius Aebutius, 125
Fayum, 124
Festus, 206
Ficus Navia/Ficus Ruminalis, 36-38, 53
Fine, Steven, 183
Flaneurs/walkers/walking, xviii-xix, 242
Fluxus, 228
Foreign ambassadors/delegations, 184
Foreign goods, 185-186, 250
Forma Urbis Romae, See Severan Marble Plan
Forum Romanum, 18, 27, 38, 41, 52, 148
Forum of Augustus, 140
Forum Boarium, 129
Forum of Nerva, 140
Foucault, Michel 252
Freedmen/freedwomen art, 15, 250
Frere, Sheppard, 97
Freud, Sigmund, 209

Friedrich, Caspar David, 88
Frontinus, Sextus, 125
Fronto, 72

Gallia/Gaul, 14, 185
Gandhāran art, 81, 252
River Gelas, 69-70
Gemma Augustea, 163
Geographies, ancient, xxvii
Germania, 115, 117, 161
Germanicus, 194-195
Ghosh, Amitav, xv
Gladiators, 6, 183, 210-211
Gladiator ethnotypes, 187
Globes, 145
The Go-Betweens, xv
Goncharova, Natalia, 16
Gordian III, 175-176
Gracchi/Gracchus, Tiberius, 17, 125, 129, 232
Graeffinger, Bob, 250
Graffiti, 175, 185
Grannicus Monument, 17, 232
Great Mosque, Damascus, 110
Greco-mania, 177, 235
Greece/Greeks/Greek art, xiii, 177, 179-180
Groma, 124-125, 254
Gunnar, Kristjana, xv

Hadrian, xxvii, 47-48, 50, 60, 80, 123, 179-180, 194, 196-198, 216,
Hadrianeum, 80, 170-171, Figures 90-92
Hadrian's Villa at Tivoli, 73, 74, 108, 196-198, Figures 71 and 103-105
Hadrian's Wall, 211, Figure 116
Haghia Sophia, 82
Hair/Hair, of barbarian women, 166, 219-220
Halfmann, Helmut, 198
Hall of *Augustales*, Herculaneum, 70
Hapi, 70-71
Haruspex/Haruspes, 6-8
Hauntology, 33-34, 221
Haynes, Ian, 22
Helena, 216
Mount Helikon, 50
Helius, Caius Iulius, shoemaker, 65

Helms, Kyle, 185
Herculaneum, 87-88
Hercules, xxvii, 8, 36, 70, 81, 208, Figures 112-113
Hermia, Aurelius, butcher, 65, Figure 36
Herodotus xv, 194
Hesiod, 50
Hidden transcripts, idea of, 175
High Rochester, 175
Hillner, Julia, 27
River Hipparis, *70*
Hippo Regius, 60
Holliday, Peter, 179
Holy Land/Holy Land landscapes/Holy places, 96-97, 213-216, Figure 56
Homer, xv, 50, 69, 95, 191
Horace, 41, 191, 192
Horologium of Augustus, 48
Horti of Rome, 92
Hostages at Rome, 184
Houses in Pompeii and Herculaneum: House of the Centenary, Pompeii, 51-52, Figure 24; House of the Duke d'Aumale, Pompeii, 109; House of the Ephebe, Pompeii, 109; House of M. Fabius Secundus, Pompeii, 44-45; House of the Faun, Pompeii, 161; House of the Four Styles, Pompeii, 171-172, 252, Figure 93; House of the Mosaic Atrium, Herculaneum, 52; House of Orion, Pompeii, 124; House of the Upper Floor, Pompeii, 96
Hyginus Gromaticus, *125*
Hypogeum of *Viale Manzoni*, 214

Mount Ida, 50, 52
'Illiterate geography', idea of, xix, 226-227, 236-239
Illiad, 12, Figure 99
Illium/Illion/Troy, 194-195, 217
The Illustrated Man, xviii, 15
Imperial journeys, 193-198
India, 93-94, 171-173, 192, Figures 93-94
Indian ivory from Pompeii, 171-172, 252, Figure 93
Indian Ocean trade, 171-173, 251-252
Indian triumph of Bacchus, 172, 189, Figures 4 and 98

Index

Isaac, Benjamin, 179
Isis/Isea, 242-243
Italian/Roman relationships, xii, 18, 30, 154, Figure 15
Itineraries, xx-xxi, xxvii, 143-145, 189, 210, 236, 254
Ivanishvili, Bidzina, 233-234
Ivory panel, 80-81

Jerusalem, xxvii, 130, 213-216
Jesus, 80-81, 96, 109, 214
Jews, 180-184
John the Baptist, 80-81
Jones, Prudence, 67
River Jordan, 80-81
Josephus, 68, 182-183, 232-233
Journeying/Travelling/Migration, xxvii, 189-224, 235
Joy Division, xv
Joyce, James, 82, 198
Judaea, 177, 181-184, 232
Jugurtha, 155
Julian, 176, 194
Julio-Claudian dynasty, 41, 157-161
Julius Caesar, xiii, xxvi, 62, 68, 74, 180, 194-195, 217
Jünger, Ernst, 40
Justinian, 173, Figure 95
Juvenal, 41, 68, 184

Kampen, Natalie Boymel, 180
Keiller, Patrick, xv
Keith, Alison, 191
Kew Gardens, 233
König, Jason, 49
Kötting, Andrew, xv
Kuttner, Ann, 232

Labyrinths, 221
Lacus Curtius and *lacus Iuturnae*, 52
Landscape painting, Roman and Campanian, 68, 87-96, Figures 49-51 and 53-55
Landscapes of defeat, 106-107
Lange, Dorothea, xv
Langlois, Jean-Charles, 117
Laurence, Ray, 216
Sextus Vetulenus Lavicanus, 65

Lazzaro, Claudia, 69
Leander Touati, Anne-Marie, 155
Lee, Chang-Rae, xv
Lentulus, Publius Cornelius, 126
Lepper, Frank, 97
Licinianus, Granius, 126
Liguria, 129
Lillie, Celine, 155-156
Liver of Piacenza, 6-8, Figure 8
Livy, xiii, xv, 35, 129, 130, 192, 194, 238
Locus amoenus/Locus horridus, 222
Loigersfelder, 221
Lorca, Federico Garcia, xxviii
Lucan, 74, 192, 194-195, 217
Lucretius, 50, 191
Luxury, 246
Lysippus, 17

Macaulay-Lewis, Elizabeth, 17, 231
Macedonia, 178-179, 232, Figure 96
Macrobius, 206
Madābā Map, mosaic, 130
Magna Mater, 52, 157, Figure 81
Mainz Celestial Globe, 145
Maionia, 161
Maldoror, Sarah, xv
Male gaze, theory of, 156-158
Map of Bedolina, 1-6, Figures 5-7
Map rooms, 146-147
Maps and Mapping, 124-149
Marcus Claudius Marcellus, 177
Martial, 52, 186-187, 192
Marvell, Andrew, 147
Marzano, Annalisa, 53
Masud, Noreen, xv
Mausoleum of the Plautii, 250
Maxentius, 73
Mediterranean Sea, xiii, xxvii, 227
Pomponius Mela, xiii, xxvi, 50
Menorah, looted from Jerusalem, 140, 181-184, 232-233, Figure 97
Merzbau, 34, 41, 135
Milvian Bridge, Battle of, 73, 180
Mithridates' plant books, 232
Mobility/migration, 213-217, 222-223
Moesia, 180, 242
Montaigne, Michel de, 240

Morrison, Toni, 245-246
Mosaic of the Islands, 130, Figure 75
Mosaics, 60, 70, 124, 130, 221, Figures 46, 59-65, 73-75, 122, and 129
Mount Ventoux, 49
Mountains, 48-52
Mulvey, Laura, 39, 156-157
Munch, Edvard, *88*
Mundus Cereris/Lapis Manalis, 206
Museum, Rome as, 17
Mussolini/Fascism in Italy/Mussolini's maps, 29, 147-148
Myron's cow, 140

Naqsh-i Rustam, 176
Narducci, Roberto, 148
Nasamones, 188
Naumachia of Augustus, 35, 185
Necropoli di Porto, 60, 140
Neighbourhoods of Rome, 39-40
Nero, 17, 46, 60, 70, 157-161, 188, 232, Figure 89
New Romes, idea of, 217-218, 244-245
Nicostratus, Popidius, 125
River Nile, 70-71, 80, 82, Figures 38-39, 40, 46, 59-65, 69 and 103
Nile Mosaic of Praeneste/Palestrina, 74, 96, Figures 59-65 and 110-116
Nilotic scenes, 68, 96, Figures 59-65, 66-67, and 127-129
Northern Britain, landscapes of, 105-106
Notitia Dignitatum, 145

Obelisk of Axum, 29
Obelisks, Egyptian, 29, 48, 116, 119-121
Odysseus/*Odyssey*, 191-192, 198-200, 201, 214, Figures 106-108
Odyssey landscape paintings, 93-94, 95-96
Oil lamps, 212, Figure 128
Mount Olympus, 50, 69
Ondaatje, Michael, xv
Orbe-Boscéaz, 221
River Orontes, 68, 184-185
Orpheus, 201
Ostia, 8, 44, 54, 60, 73
Others viewing Romans, 174-177

Ovid, xxiii, 27, 36, 41-42, 43, 68, 123, 156, 180, 185, 192, 201, 219, 220, 222, 225, 226
Oxyrhyncus, Figure 125

Parabiago plate, Figure 119
Paris-Kugel Celestial Globe, 145
Parthenon, 127
Parthia/Parthians, 13-14, 106-107, 131, 133
Parthian Monument of Lucius Verus, 74, 78-80, 107
Paullus, Lucius Aemilius, 129
Claudia Peloris and Tiberius Claudius Eutychus, 126-127, Figure 72
Peralta, Dan-el Padilla, xv
Perec, Georges, *xv*
Periploi, xiii, 189
Persephone/Proserpina, 164-165, 201-205, Figures 109-110
Personifications, 68-73, 166-171, 178-179, Figures 90-92 and 96
Perugia tomb groundplan, 126-127, Figure 72
Pesaro Wind Rose, 83, Figure 48
Petrarch, 49
Petronius, 185
Peutinger Map, 101, 141-143, Figures 80-81
Pfünz, 124
Philip the Arab, 175-176
Piazzale delle Corporazioni, Ostia, 44, 60, 82, Figure 46
Pictam Italiam, 130
Pilgrimage/Pilgrims, 214-216
Pilhuj, Katja, xvii, 224
Pilkington Bottle, Figure 116
Pindar, 50
Pisco Montano, 47
Plautus, 191, 238
Pliny the Elder, xxiii, xxvi, 16, 35, 36, 38, 41, 51, 53, 73, 89, 116, 130, 131, 133, 140, 184, 231-232, 250
Pliny the Younger, 50-51, 77, 80, 91
Plutarch, xxvi-xxvii, 50, 155, 175, 194, 206
Plutei of Trajan, 36, Figure 19
River Po/*Padus*, 69, 125
Political speeches, modern, xxv
Polybius, 162, 168, 184
Pompeii, xxvii, 44, 51-52, 87-88, 109, 185

Pompey, 16-17
Pompusianus, Mettius, 130
Pons Sublicius, 72
The Pop Group, xv
Porta Fontinalis, Rome, 65
Portable sundials, 8, 85, 145
Portico/Theatre of Pompey, 16-17, 30, 42, 148, 156, 167, 231
Porticus ad Nationes, 136, 148
Porticus Liviae, 42
Porticus Metelli, 17, 231
Porticus Octaviae, 42, 156, 231
Porticus Vipsania, 131, 133
Portus, 54, 60, Figures 30-31 and 122
Prejudice, 184-185
Prima Porta Augustus, statue, xviii, 11-15, 106, Figures 12-13
Procopius of Caesarea, 34-35
Propertius, 130-131, 192
Provinciae fidelis, 80, 170-171, Figures 90-92
Ptolemies, xiii, 117
Ptolemy, xxvi
Punic Wars, xvii
Puteoli, 211-212
Pygmies, 109
Pythagoras, 225

The Raincoats, 191
Rain god, Column of Marcus Aurelius, 83, 235, Figure 47
Ramses Wissa Wassef School of Tapestry, 123
Rape, 18, 155, 162, 163-166
Rape of Sabine women, 18, 154, 193
Rauschenberg, Robert, 28
Renaissance Italy 69, 146-147
Res Gestae of Augustus, 54, 154, 167, 216, 250
River Rhine, 68, 69, 80, Figure 45
River Rhone, 68
Rich, Adrienne, 254
Richman, Jonathan, xxiv
Riggsby, Andrew, 126, 145
Rivers, 67-81
Rivers of Paradise, 82, 109, 214
Rock art in northern Italy, 1-6, Figures 5-7
Rogers, Dylan, 52
Roller, Duane, 124

Roma/Roma Aeterna, 11, 14, 41, 43, 44, 46, 59, 162, 186, Figures 9 and 32-33
Roma Negata exhibition, 29
Roman, to be, 21-42, 43-66, 225-255
Roman army, 22-23
Roman emperors as Pharaohs, 176-177
Roman roads, 41, 216
Romulus and Remus, 32-34, 44, 72, 193, Figures 21-22
Rose of the Winds, Dougga, 83
Rostra/Rostrum, 27, 52
River Rubicon, 74
Rudge Cup, Figure 116
Rufus, Aelius, 125
Russolo, Luigi, 250

Sabina, 47-48
Sacro-idyllic landscapes, Figure 52
Safaitic *graffiti*, 185
S. Ambrogio sarcophagus, 214
St. Jerome, 216
St. Peter, 214, Figure 26
Salgado, Sebastião, xv, 116, 235
Sallust, xiii
Samnites, xii, 190, Figure 15
Sarcophagi, 96, 189, 206-209, 213-214, Figures 26, 56, 98, 111-112, 115, 120, and 130
Sardinia, xiii, 129
Sarmizegetusa, 102
Sassanians/Sassanid reliefs depicting Roman emperors, 173, 175-176, Figure 95
River Scamander, 74
Scego, Igiaba, 29
Schwitters, Kurt, 34
Sebald, W.G., xv
Sebasteion at Aphrodisias, 157-161, 167-168, 230, Figures 88-89
Iulius Secundus, 223
Senate of Rome, 185
Senecas, the 40, 192, 238
Seven hills of Rome, 43-47, 66, Figure 20
Severan Marble Plan/*Forma Urbis Romae*, 5, 65, 73, 134, 136-138, 148, Figures 77-78
Septimius Severus/Severan dynasty, xiv, 106, 137, 138-140, 176-177, 195

Septimontium, 46
Sexual violence against women, xxv-xxvi, 152, 188, 230-231
Shapur I, 175-176
She-Wolf, 44, Figures 21-22
Shield of Achilles, 12
Shield of Aeneas, 12, 72, 109, 193
Ship of Aeneas, 34-35, Figure 18
Sicily, xiii
Silures, 188
Simmel, Georg, xix
Sinclair, Iain, xviii
Situationist International, xviii, 20
Slaves/enslaved, 184-185, 242
Smith, John, xv
Solinus, 33
Sontag, Susan, 28, 230
Souvenirs, 210-213, 235, Figures 116-117
Spencer, Diana, 26
Sperlonga, 198-199, Figures 106-108
Stabiae, 212
Statius, 43, 185
Stewart, Andrew, 117
Strabo, xxvi, 41, 49, 50, 67, 133, 145, 184, 194
Studius, 89
Subterranean Rome, 41, 52, 206, Figure 114
Suetonius, 17, 46, 130
Sullivan, Garrett, 23
Sugambri, 188, 219
Synagogue at Magdala, 183
Synagogue at Hammat Tiberias, 184-185
Syracuse 177
Syria, 178-179, Figure 96

Tacitus, 22-23, 51, 72, 185
Talbert, Richard, 85, 124
Tarpeia/Tarpeian Rock, 18, 47, 154-155
Tarkovsky, Andrei, xv
Tazza Farnese, 74
Tellus, 11, 14, Figures 9 and 32-33
Temples: Temple of Apollo, 42, 156; Temple of Castor and Pollux, 140; Temple of Isis and Serapis, 56, 72; Temple of Jupiter Stator, 56; Temple of *Mater Matuta*, 129; Temple of Peace/*Templum Pacis*, 136-138, 148, 181, 183, 230-231, 232-234, Figures 77-78; Temple of *Tellus*, 15, 130; Temple of Vesta, Figure 28
Terence, 191
Tertullian, xxvii
Theogonus, Q. Fabius, *pigmentarius*, 65
Theophanes, 222
Theweleit, Klaus, 69
Thomas, Joshua, 115
Tiber/*Tiberinus*, 43, 44, 60, 68, 69, 71-73, 80, 82, 108, 119, 137, Figures 37 and 104
Tiberius, 50
Tibullus, 192
River Tigris, 74
Titus, xxiii, 41, 176, 181-183, 186
Tolentino, 213
Tomb of the Haterii, 54-56, Figure 25
Tomb of the Nasonii, 209
Tower of the Winds, Athens, 83
Toynbee, Jocelyn, 206-207
Trajan, 46-47, 54, 60, 62, 74, 80, 176, 195
Trajan's Column, 46-47, 54, 60, 62, 74, 97-105, 107, 115, 117, 162, 165, 180, 195-196, 242, Figures 42 and 57-58
Trajan's Forum, 168-169
Travel coins of Hadrian, 196-198, Figures 100-102
Tree collecting, 232-234
Trees in Rome/Italy, 36-38, 43, 53-54, 232
Trees of Dacia, 102-104
Trier landscape painting, 92, Figure 54
Trimalchio, 185
Trinh T. Minh Ha, 38
Triumphs, xxvi, 68, 115, 134-135, 180-183, 232
Tropaeum Alpium, 6, 154, 250
Trousset, Pol, 133
Troy/Trojans, 31, 72, 74, 190, 194-195, 217
Tugurium Faustuli, 32
Tugurium Romuli, 32
Turconi, Cristina, 3
Tusculum, 27, 126
Tyche/*tychai*, 62, Figures 34-35, 121, and 126
River Tyne, 80

Umbrella Pines of Rome, 53
Underworld, 193, 200-210, 235-236, Figures 109-115

Urbino plan, 126

Valcamonica, 1-6, Figures 5-7
Valerian, 175-176
Varda, Agnès, xv-xvii, 166-167
Varro, 18, 20, 23-27, 43, 129-130
Velletri sarcophagus, 207-208, Figure 115
Vercingetorix, 155
Verlaine, 243
Vespasian, 44, 140, 181-183, 195
Mount Vesuvius, 50, 51-52, 87, Figure 24
Via Anicia marble map fragment, 35, 126, 140, Figure 79
Via Appia, 47
Via Marsala mosaic map, 127, Figures 73-74
Via Salaria, 189
Via Traiana, 59
Vicarello Cups/Beakers, 144, 210, Figure 82
Vicus/vici of Rome, 62-65
Vicus Sandaliarius altar, 62
Vidal, Gore, 225
Vietnam War, 103
Villa del Casale, Sicily, 172-173, Figure 94
Villa of Domitian, 109
Villa of Livia, 11, 41, 93, Figure 55
Villa of Nero, 70
Villa of the Papyri, Herculaneum, 8
Villa of P. Fannius Synistor, Boscoreale, 178-179, Figure 96
Villa/Grotto of Tiberius, 198-200, Figures 106-108
Virgil, xv, 8, 36, 43, 50, 72, 109, 125, 131, 192-193, 201
Vitellius, 22-23
Vitruvius, 33, 167
Voting tribes of Rome, 62
Vout, Caroline, 43

Wall painting, 70, 87-96, 178-179, 212, Figures 15, 24, 49-51, 53-55, 96, and 127
Water/Water management, 52, 68, 81-82
Weather gods, 82, 235, Figure 124
Wels, 172
Weston, Edward, xv
Wire, xv
Wofford, Tobias, 31
Woolf, Virginia, 28

Workers/artisans, 65-66
Workshop of Verus, Pompeii, 124

Xerxes, 175, 194